APPLIED LASER SPECTROSCOPY: TECHNIQUES, INSTRUMENTATION, AND APPLICATIONS

APPLIED LASER SPECTROSCOPY: TECHNIQUES, INSTRUMENTATION, AND APPLICATIONS

David L. Andrews
(Editor)

David L. Andrews
School of Chemical Sciences
University of East Anglia
Norwich NR4 7TJ
ENGLAND

Library of Congress Cataloging-in-Publication Data

Pending

Printed in the United States of America
ISBN 1-56081-023-8
ISBN 3-527-28072-3

Printing History:
10 9 8 7 6 5 4 3 2 1

Published jointly by

VCH Publishers, Inc.	VCH Verlagsgesellschaft mbH	VCH Publishers (UK) Ltd.
220 East 23rd Street	P.O. Box 10 11 61	8 Wellington Court
9th Floor	D-6490 Weinheim	Cambridge CB1 1HZ
New York, New York 10010	Federal Republic of Germany	United Kingdom

Contents

v

9. Laser Mass Spectrometry   **365**
K. W. D. Ledingham and R. P. Singhal

10. Ultrafast Spectroscopic Methods   **401**
P. A. Anfinrud, C. K. Johnson, R. Sension,
and R. M. Hochstrasser

Contributor Affiliation List

D. L. Andrews,
School of Chemical Sciences,
University of East Anglia,
Norwich, NR4 7TJ, UK

M. R. S. McCoustra,
School of Chemical Sciences,
University of East Anglia,
Norwich, NR4 7TJ, UK

W. Demtröder,
Department of Physics,
University of Kaiserslautern,
D-6750, Germany

**B. J. Howard and
J. M. Brown,**
Physical Chemistry Laboratory,
University of Oxford, OX1 3QZ,
UK.

J. Pfab,
Department of Chemistry,
Heriot-Watt University,
Edinburgh, EH14 4AS, UK

M. D. Morris,
Department of Chemistry,
University of Michigan, MI.
Ann Arbor 48109, USA

**H. Berger, B. Lavorel and
G. Millot,**
University of Bourgogne,
Dijon 21100, France.

L. Goodman,
Department of Chemistry,
Rutgers University, NJ.
New Brunswick 08903, USA

J. Philis,
Department of Physics,
University of Ioannina,
GR 45110, Greece.

**K. W. D. Ledingham and
R. P. Singhal,**
Department of Physics and Astronomy,
University of Glasgow, G12 8QQ,
UK

P. A. Anfinrud,
Department of Chemistry,
Harvard University,
Cambridge, MA 02138, USA

C. K. Johnson,
Department of Chemistry,
University of Kansas,
Lawrence, KS 66045, USA

**R. M. Hochstrasser and
R. Sension,**
Department of Chemistry,
University of Pennsylvania, PA.
Philadelphia, 19104, USA.

Preface

Laser spectroscopy has come of age. In many respects the 1980s represented a period in which laser spectroscopy underwent the difficult transition from being a purely academic subject to an analytical discipline with important commercial applications. It is in recognition of the increased significance of the subject in the 1990s that this book has been produced. It provides detailed coverage of current techniques in applied laser spectroscopy and is designed to reflect the full breadth of current applications. Pitched at the advanced pregraduate or graduate level, the treatment generally requires no knowledge of lasers, only an understanding of spectroscopy.

Following a simple introduction to the key features and principles of the subject in Chapter 1 and of the basic instrumentation in Chapter 2, each of the subsequent chapters focuses on one of the major types of technique and covers the principles and more specialized instrumentation and typical applications. The examples that illustrate each method are drawn from a wide range of chemical, physical, and biological studies and serve to highlight the breadth of usage, both academic and commercial. Each chapter is intended to address both the potential and the problems associated with the technique, giving the perspective necessary for the reader to decide which would be most suitable for a particular problem.

It has been a delight to be involved in the compilation of this book. When I initially approached several of its eminently distinguished authors for a chapter, it was without much hope of getting an affirmative response. I was most pleasantly surprised by their ready agreement. I have certainly learned a lot from reading all their manuscripts. It is our joint hope that this book will prove both informative and interesting to its readers and that it will further encourage wider applications of laser spectroscopy.

David L. Andrews
Norwich, January 1992

ix

1 Fundamental Principles of Laser Spectroscopy

D. L. Andrews

1.1. Introduction

The application of spectroscopic techniques based on the interaction of light and matter has for several decades been of central importance in contemporary laboratory practice. Both at the research and the routine analytical level, spectroscopic methods are widely used for the determination of molecular structure, the characterization of unknown compounds, the monitoring of chemical processes, and the quantitative analysis of complex solutions and mixtures. Modern spectroscopy encompasses a breathtaking diversity of techniques, and the distinctive instrumentation and applications of each method are largely determined by the regions of the electromagnetic spectrum involved.

The application of lasers in this context is widely expected to become commonplace before the end of the century. One reason is the increasing sophistication and reliability of laser-based systems. In turn, this is leading to the commercial exploitation of some of the newer laser-based forms of spectroscopic analysis. The well-established uses of lasers as sources for routine Raman studies or as ablation devices for the elemental analysis of heterogeneous solids now represent only part of modern laser spectroscopy.

It is interesting to trace some of the key factors that have led to the increasing utilization of laser-based forms of analysis. A great many spectroscopic methods are ultimately based on the absorption of radiation, where in order to record the spectrum of a given sample a source emitting

1

either broadband or tunable monochromatic output is generally required, and high intensities are mostly not too important. Early fixed-frequency lasers were poorly suited to this kind of job, and they suffered from poorer reliability but greater cost and complexity than traditional sources. Their initial impact on absorption spectroscopy was consequently minimal.

Raman spectroscopy, however, where the spectrum is derived from light inelastically scattered from a necessarily single-frequency source, afforded a unique opportunity for the advantages of laser excitation to become apparent. Although already well established as a spectroscopic tool before the arrival of lasers in the early 1960s, it was immediately evident that Raman scattering was a technique tailor-made for laser instrumentation, with its requirements for an intense monochromatic source. Within a few years Raman spectroscopy thus became essentially identified as a laser technique; indeed, this still represents the application in which most lasers are employed for spectroscopy. Nonetheless, recent developments such as Fourier transform Raman spectroscopy are substantially rejuvenating the technique, as shown in Chapter 6.

Lasers were initially much slower to displace traditional light sources from other kinds of spectroscopy where the needs for high intensities are generally less paramount. However, the development of tunable systems such as the dye laser soon began to open up new possibilities for spectroscopic applications. The attractive combination of high monochromaticity with tunability led to such laser sources gradually making inroads into all the more familiar types of electronic and vibrational spectroscopy. (See Chapters 3, 4, and 5.) Another contributing factor was the increasing use for laser frequency conversion of nonlinear optical processes such as harmonic, sum, and difference frequency generation.

These trends have recently been given a boost by the commercial development of several new tunable solid-state lasers, such as the alexandrite laser.[1] Such lasers represent an increasingly attractive alternative to tunable dye lasers. Their chief advantages are an appreciably more rugged and compact construction and operation without the use of toxic organic chemicals. At the same time the range of available diode laser wavelengths has been encroaching further into the long-wavelength end of the visible region, offering new scope for high-resolution measurements in those kinds of spectroscopy not too dependent on high power and broad tuning range. A graphic illustration of the range of tunable lasers now being marketed is given in Figure 1.1, together with a listing of the wavelengths available from fixed-frequency sources in Table 1.1.

Once the unique attributes of the laser had become fully appreciated a wide range of new spectroscopic techniques rapidly evolved, many characterized by processes in which samples themselves display optical nonlinearity in their response to laser light.[2,3] (See Chapters 7 and 8.) This type of behavior often reflects the fact that photons are not interacting singly

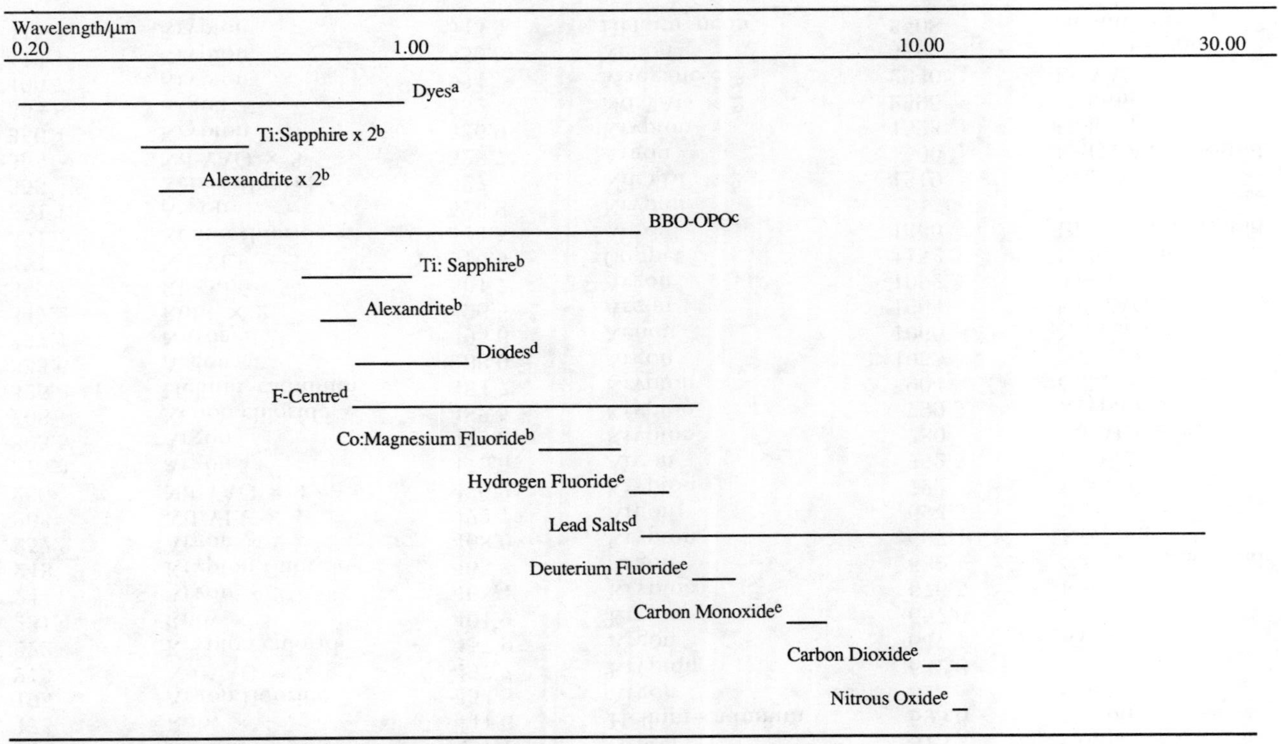

Wavelength/µm
0.20
1.00
10.00
30.00

Dyes[a]

Ti:Sapphire x 2[b]

Alexandrite x 2[b]

BBO-OPO[c]

Ti: Sapphire[b]

Alexandrite[b]

Diodes[d]

F-Centre[d]

Co:Magnesium Fluoride[e]

Hydrogen Fluoride[e]

Lead Salts[d]

Deuterium Fluoride[e]

Carbon Monoxide[e]

Carbon Dioxide[e]

Nitrous Oxide[e]

[a]Continuously tunable over part of range indicated, dependent on dye employed.
[b]Continuously tunable over range indicated.
[c]Continuously tunable over range indicated, dependent on parametric oscillator driver and idler frequencies.
[d]Continuously tunable over part of range indicated, dependent on composition of laser medium.
[e]Line-tunable within range indicated.

Figure 1.1 Tunable laser sources.

Table 1.1 Major Lines Available from Commercial Fixed-Wavelength Laser Sources (in order of increasing wavelength)[a]

λ/nm	Laser	λ/nm	Laser	λ/nm	Laser
157.5	Fluorine	428	Nitrogen	568.2	Krypton
157.6	Fluorine	437.1	Argon	578.2	Copper
173.6	Ruby × 4	441.6	Helium–cadmium	595.6	Xenon
193	Argon fluoride	454.5	Argon	628	Gold
213	Nd:YAG × 5	457.7	Krypton	632.8	Helium–neon
222	Krypton chloride	457.9	Argon	647.1	Krypton
231.4	Ruby × 3	461.9	Krypton	657.0	Krypton
244	Argon × 2	463.4	Krypton	676.5	Krypton
248	Krypton fluoride	465.8	Argon	680	Ga:Al In phosphide
257.2	Argon × 2	468.0	Krypton	687.1	Krypton
263	Nd:YLF × 4	472.7	Argon	694.3	Ruby
266	Nd:YAG × 4	476.2	Krypton	722	Lead
275.4	Argon	476.5	Argon	752.5	Krypton
305.5	Argon	476.6	Krypton	780	Ga:Al:arsenide
308	Xenon chloride	482.5	Krypton	799.3	Krypton
325	Helium–cadmium	484.7	Krypton	904	Gallium arsenide
333.6	Argon	488.0	Argon	1053	Nd:YLF
337.1	Nitrogen	495.6	Xenon	1060	Nd:glass
347.2	Ruby × 2	496.5	Argon	1064	Nd:YAG
350.7	Krypton	501.7	Argon	1092.3	Argon
351	Nd:YLF × 3	510.5	Copper	1152.3	Helium–neon
351	Xenon fluoride	514.5	Argon	1300	In:Ga:As phosphide
351.1	Argon	520.8	Krypton	1315	Iodine
353	Xenon fluoride	527	Nd:YLF × 2	1319	Nd:YAG
355	Nd:YAG × 3	528.7	Argon	1500	In:Ga:As phosphide
356.4	Krypton	530.9	Krypton	1523	Helium–neon
363.8	Argon	532	Nd:YAG × 2	2396	Helium–neon
406.7	Krypton	534	Manganese	2940	Er:YAG
413.1	Krypton	539.5	Xenon	3392	Helium–neon
415.4	Krypton	543.5	Helium–neon	3508	Helium–xenon

[a]Harmonics generated by the more intense pump wavelengths are also included.

with sample molecules as in conventional spectroscopy. Such effects in which photons interact in pairs or even higher-order groupings are only observable under the intense photon flux provided by powerful pulsed lasers. The discovery of these effects and their subsequent implementation for analytical purposes has brought about a major change in emphasis within the field of laser spectroscopy, particularly in connection with laser mass spectrometry.[4] (See Chapter 9.) Two-photon absorption, for example, in which sample atoms or molecules absorb photons in a pairwise fashion, proves useful in spectroscopic studies of a wide range of phenomena such as molecular energy transfer and photodissociation. Here the utilization of either single- or two-color excitation provides an attractive method for accessing regions of the optical spectrum where direct single-photon excitation is difficult.

Despite the wide range of devices now available it is possible to identify a number of characteristics distinctive of the output from laser sources. These are principally their high degree of coherence, collimation, intensity, and monochromaticity. In pulsed systems the shortness of the pulse duration is also a significant parameter. The optics and instrumentation associated with the laser system in any given spectroscopic setup naturally bear on the precise measure of these parameters actually experienced by a sample, and frequently produce significant modifications in pulse and polarization characteristics. In the following section each of these properties is discussed in detail, and their relevance to spectroscopic applications is examined; a more detailed analysis of the instrumentation is given in Chapter 2.

1.2. Laser Characteristics

1.2.1. Coherence

The initially most widely vaunted attribute of the laser, the coherent nature of its emission, is in general only indirectly relevant to spectroscopic applications. The extent of correlation in phase between the photons in a laser beam is characterized by a coherence length l_c, which is the inverse of the linewidth as expressed in wavenumber terms[5]:

$$l_c = \frac{1}{\Delta\bar{\nu}} \tag{1.1}$$

While typical semiconductor lasers have coherence lengths of about 1 mm, l_c values of up to 1 km are achievable in ring dye lasers. The spectroscopic significance of l_c values, however, lies much more in their reflection of the extent of monochromaticity (see Section 1.2.4) rather than coherence. The coherent nature of laser light is also characterized by its photon distribution. The probability of finding N photons in a volume that on a time average contains a mean number M is given by the Poisson distribution[6]:

$$P_N = \frac{M^N e^{-M}}{N!} \qquad (1.2)$$

Theory shows that nonlinear optical processes may occur at different rates when coherent rather than thermally produced light is employed. However, such issues are somewhat academic since lasers are invariably employed for the observation of such interactions.

A number of so-called coherence spectroscopies do exist that are invariably studied with laser light.[7] Some of these, such as self-induced transparency and photon echo, relate to pulse propagation characteristics and are generally not observable in normal chemical media. The coherence of the laser photons here is relevant only in the sense that it enables short (mode-locked) pulses to be created. While such processes have considerable intrinsic interest, they appear to have no real analytical utility. The degree of beam coherence also affects the rate of such nonlinear processes, but mostly to only a small extent.

The term *coherent* is also applied in a different context to a number of optically parametric processes. Many of these are nonlinear analogs of Raman scattering,[8,9] a well-known example being CARS (coherent anti-Stokes Raman scattering) spectroscopy (see Chapter 7), a technique widely used for the study of flames and combustion. The coherence here relates to a relationship between the phases of the photons emitted and those absorbed in individual molecules. The nature of most of these coherent processes is such that, given an intense and well-collimated input, the emission generated also occurs in a well-defined direction and with reasonable intensity, thus displaying laserlike attributes. However, similar observations would be made, in principle, using an incoherent source with the same degree of collimation and intensity.

1.2.2. Collimation

The physically narrow beamwidth and small divergence of a typical laser beam are useful attributes in certain types of application, both in the laboratory and in the field. Most laser beams are well collimated with no more than milliradian divergence, and most of the intensity occurs within a millimeter-scale cross section. The small beam divergence can prove useful by enabling long path lengths through samples to be employed, often by the use of multipass optics. This affords a convenient means of increasing sensitivity in the study of weak spectral features.

The exact distribution of intensity within a laser beam is determined by the mode structure, and in the simplest case consists of an essentially Gaussian distribution.[10] Since there is no hard edge to the beam, its diameter w is usually defined at the transverse distance between points at which the intensity drops to $1/e^2$ (13.5%) of its central peak value. Quantum uncertainty considerations show that any such beam can in general only be focused down to a diffraction limit where its diameter is the same order as the wavelength. In fact, the focused beam waist w_0 is determined by the relation

$$w_0 = \frac{2M^2\lambda}{\pi\Theta} \tag{1.3}$$

where λ is the wavelength and Θ the angle of convergence of the focused beam. The quantity M^2 is generally accepted as a good measure of *beam quality* and has a diffraction-limited value of unity for a fundamental Gaussian-mode beam.[11] In practice, M^2 values below 1.2 are regarded as optimum.

The tight focusing afforded by lasers is especially significant for microsampling applications, where sample volumes of the order of 1 μl are entirely feasible. Indeed in laser-based methods for the analysis of HPLC eluant, sample volumes of a few nanoliters are possible. In the microanalysis of heterogeneous solids such as minerals or corroded metals, the precision of the laser beam as probe is also significant, especially where it is used for the ablation of a small region of surface for analysis.[12] In conjunction with conventional or laser-based spectroscopic detection, this affords a useful technique for characterizing the composition of heterogeneous solids. Scanning a focused laser beam over the surface enables different areas to be analyzed, typically with micron resolution; where high-power pulses are used to ablate successive layers of such materials, the analysis additionally provides depth profile data. The ability to deliver laser beams into optical fibers also facilitates a number of biological applications such as remote fiber fluorimetry.

The other extreme of scale in remote sensing is represented by the various spectroscopic adaptations of lidar (light detection and ranging), in which atmospheric species are targeted by a ground-based laser system and identified telescopically by their optical response.[13] These techniques, which depend crucially on the minute divergence of the beams delivered by most large laser systems, are of particular value in the study of industrial stack gases and atmospheric pollutants and are currently much in vogue in connection with the study of stratospheric ozone depletion.

1.2.3. Intensity

For a wide variety of spectroscopic applications, far more important than collimation is the high intensity often associated with laser output. There are in fact several parameters in terms of which the high photon flux delivered by lasers can be characterized. For a continuous-wave (CW) laser the output power is mostly no more than a few watts or even milliwatts, certainly less than that of a common electric light bulb.

Indeed, only recently has an average power of 1 kW been obtained from a solid-state laser head.[14] Substantially higher powers are obtainable from gas dynamic lasers, but these are of little spectroscopic interest, being more suited to heavy industrial applications.

Higher output powers can be achieved by pulsing techniques, based on the simple fact that the power is increased if a given energy E is delivered within a decreased time. Here the pulse energy is an equally important consideration, and the corresponding power $W = dE/dt$. A good idea of the mean power per pulse is obtained by dividing the pulse energy by the pulse length. Peak power levels here depend very much on the type of pulsing employed, with the record[15] currently standing at 20 TW (2×10^{13} W). For comparison purposes, this is about 500 times the entire power-generating capacity of the United Kingdom.

The best measure of the intensity actually experienced by a sample, however, is the *irradiance*, which is the power delivered per unit cross-sectional area of the beam. It is principally the small dimensions of the beam that give rise to the very high irradiances delivered by laser sources. A good CW argon ion laser, for example, produces an irradiance of about 10^7 W m^{-2}. As mentioned earlier, this intensity can be greatly increased by focusing to the diffraction limit; in this instance irradiances of 10^{13} W m^{-2} can be achieved, representing intensities 10^{10} times higher than that of typical sunlight on the earth's surface. This possibility of focusing to produce intensities greater than the source itself is one of the most useful and distinctive characteristics of laser light.

In terms of the irradiances produced by pulsed lasers, the highest figures are obtained through use of mode-locking and allied techniques, characterized by pulses measured on the picosecond (10^{-12} s) or even femtosecond (10^{-15} s) time scale. Intensities of the order of 10^{16} W m^{-2} are readily attainable from a focused mode-locked argon laser, for example, representing a light flux equal to that to be found in the interior of stars. Spectroscopically, such high intensities prove useful for both the identification of weak spectral features and the study of intrinsically weak processes.

1.2.4. Monochromaticity

It is the high degree of monochromaticity of laser light that carries the most significance for spectroscopic applications, with the continual advances in cavity design reflected in a gradual reduction in achievable laser linewidth. The parameter that most effectively characterizes the degree of monochromaticity is the quality factor Q, defined as the ratio of the laser emission frequency to its linewidth. It is also expressible as the coherence length divided by the emission wavelength:

$$Q = \frac{\nu}{\Delta \nu} = \frac{l_c}{\lambda} \qquad (1.4)$$

In absorption-based laser spectroscopy this parameter represents an upper limit on the achievable resolution, which in practice may be reduced by other features of the instrumentation. Quality factors of 10^7 are not uncommon in the lasers most widely used for spectroscopy.

The impact of the continued improvements in source monochromaticity is well illustrated by the tremendous developments that have taken place in high-resolution electronic spectroscopy based on dye lasers, where resolutions of below 0.1 cm^{-1} are now routinely attainable. With suitable instrumentation frequency-stabilized ring dye lasers can in fact offer a spectral resolution better than 10^{-5} cm^{-1}, corresponding to Q values of about 10^{11}. In the infrared region tunable diode lasers can offer linewidths of less than 10^{-3} cm^{-1}, albeit over a limited spectral range, providing a means for the collection of high-resolution vibrational and rotation–vibrational spectra.

1.2.5. Pulsing

Time-resolved studies constitute an increasingly significant area of application for laser spectroscopy. (See Chapter 10.) Principally, this reflects the progressive and remarkable reduction in achievable pulse duration, particularly with regard to the mode-locked lasers producing what have become known as ultrashort (subnanosecond) pulses. (See Figure 1.2.) These lasers operate by establishing a phase relationship between a large number of longitudinal modes within the laser. With N_L modes in a cavity of optical length L we obtain a pulse repetition frequency.

$$f = \frac{c}{2L} \tag{1.5}$$

typically around 100 MHz, and a pulse duration given by

$$\Delta t = \frac{4\pi L}{(2N_L + 1)c} \tag{1.6}$$

where c is the speed of light. Pulse lengths are typically on the picosecond scale, although by use of pulse compression techniques some off-the-shelf commercial systems now offer reproducible femtosecond pulses. There is an inevitable price to pay for the time resolution thus achieved, however, and that is a loss of spectral resolution in accordance with the inexorable operation of the quantum mechanical uncertainty principle. Nonetheless, such pulses are very significant for the study of ultrafast processes.

It is not a coincidence that as far as genuine chemistry is concerned most such processes directly involve interaction with light. The reason is simply that the range of atomic motion over picosecond and subpicosecond times is necessarily very limited, and only unimolecular processes such as electron transfer or bond fission can occur on such a time scale. Mostly, such processes require a substantial input of energy, and photoabsorption thus provides a convenient initiation stage. In fact, the primary processes of vision and of photosynthesis have both been shown to occur on a picosecond time

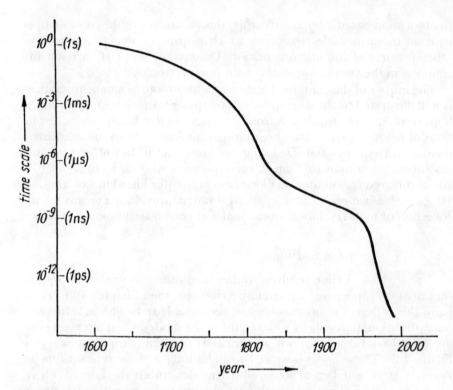

Figure 1.2 Development of time resolution, most recently based on mode-locked laser instrumentation. Adapted from Ref. 16.

scale.[19] At the other end of the scale in chemical complexity, the utility of such methods is nicely illustrated by the work on the photodecomposition of ICN.[20] Time-resolved measurements of laser-induced fluorescence following pulsed excitation with a femtosecond pump has shown that I-CN bond fission occurs on a time scale of ~200 fs. Clearly, such information can only be derived using ultrashort pulsed laser instrumentation.

To capitalize to the fullest extent on the high resolution potential afforded by laser spectroscopy, it is, of course, necessary to minimize the effects of phenomena such as collision broadening and Doppler broadening in the sample. This can be accomplished by a number of means, including molecular beam methods[21] and the utilization of various kinds of nonlinear optical effects such as saturation and two-photon absorption. Such studies are not primarily of analytical importance, but they are unparalleled in the wealth of fine structure they can reveal for the detailed characterization of molecular systems and their chemical reactions.

Having seen how the various characteristics of laser sources may be exploited for spectroscopic applications, it is appropriate to consider the physics of laser interactions in more detail, to understand better the basis for each specialized type of laser spectroscopy. To do so, we shall have to focus on what factors determine the possibility of spectroscopically useful coupling between the photons of the laser radiation and the atoms or molecules of the sample.

1.3. Interactions with Laser Photons

The characteristic features of the various kinds of laser spectroscopy, in terms of both their instrumentation and the distinctive analytical information they provide, are largely determined by two considerations. The first concerns the physical nature of the basic processes by which laser light interacts with samples (e.g., by photoabsorption or light scattering). Generally, such interaction results in the promotion of sample species such as atoms, ions, or molecules to excited quantum levels. The secondary processes that occur subsequent to the sample excitation (e.g., molecular fragmentation or fluorescence) represent the other factor determining the spectroscopic methodology. The enormous variety with which these factors can be combined is reflected in the diversity in the specialized spectroscopies described in later chapters. However, it is possible to identify some of the features these techniques hold in common by simple virtue of their involving interaction with laser radiation.

The issues involved in determining whether a given sample molecule, for example, will undergo a spectroscopically useful transition through its interaction with a given laser beam can be enumerated as follows. First and perhaps most fundamentally, there is the question of the likelihood of the molecule being intercepted by the necessary number of photons. For most types of spectroscopy, a single photon induces the molecular transition, but where nonlinear interactions are involved, two or more laser photons may need to traverse the molecule almost simultaneously. Next, the laser photon needs to have an appropriate energy to accommodate the desired transition. There are also selection rules to be satisfied, in terms of both molecular symmetry and photon polarization. Finally, in the case where sample molecules are not in free rotation, the orientation of the sample also has a bearing on whether the required transition can occur. Each of these factors is now examined in more detail.

1.3.1. Spatial Considerations

First, we have to consider the probability of finding a single photon in the volume of space V_1 occupied by an atom, molecule, or chromophore (i.e., the species that is of spectroscopic interest). For laser light whose intensity fluctuations approximate to a Poisson distribution, the result for the mean photon number $\langle N \rangle$ is given by[22]

$$M = \frac{IV_1\lambda}{hc^2} \tag{1.7}$$

where I is the beam irradiance and λ its wavelength. At first encounter it is surprising to find how small $\langle N \rangle$ generally is. The likelihood of finding

a laser photon traversing a sample molecule under representative laboratory conditions can be illustrated by considering a sample of water (mean
molecular volume $V_1 = 3 \times 10^{-29}$ m^3) irradiated by a 10^{15} W m^{-2} pulse
from a mode-locked argon laser operating at a wavelength of 488 nm. Us-
ing equation (1.7) we find that $\langle N \rangle = 2.5 \times 10^{-4}$. In other words, one
molecule in 4000 has a photon passing through the volume it occupies at
any given instant in time.

The average time each photon takes to pass through any given molecule
is given by $t = V_1^{1/3}/c$; the mean interval between photons, $\tau = t/\langle N \rangle$, is
therefore

$$\tau = \frac{hc}{IV_1^{2/3}\lambda} \tag{1.8}$$

In the example given earlier, $t = 10^{-18}$ s (10^{-3} fs) and $\tau = 4$ fs. To put this
into perspective and give reference to more common intensity levels, we
can apply equations (1.7) and (1.8) to sunlight on the surface of the ocean.
Using $I = 1$ kW m^{-2} and averaging over wavelengths in the visible spectrum, we find that the mean number of photons passing through a mole-
cule of water is 2×10^{-16} (i.e., at any instant only one molecule in 5×10^{15} is experiencing the transit of a single photon of sunlight, and the mean
interval between photons is 5 ms).

If we question the probability of finding *two* photons, we need to evaluate $\langle N(N-1) \rangle$, as the detailed theory shows.[23] The $(N-1)$ factor arises
because after detecting one photon there are only $(N-1)$ left. Here the
answer is $\langle N(N-1) \rangle = M^2$, which incidentally reveals the quadratic dependence of two-photon absorption on light intensity. So in the case of the
mode-locked argon laser considered previously, only one water molecule
in 16 million experiences the simultaneous transit of two photons. This
goes some way to explaining why nonlinear effects that require two or more
photons to be present give such weak signals; it also explains why pulsed
laser sources are almost invariably necessary for observations of optical
nonlinearity. To complete the comparison, we can note that two photons of
sunlight will be found together in only one molecule in every million *liters*
of water on the surface of the ocean.

1.3.2. Energetics

Given a sample molecule with the right number of pho-
tons present, all other issues concern whether or not a spectroscopic transition can occur. For processes involving only the absorption of a single
laser photon, the laser photon energy $h\nu$ must obviously match the difference between two quantum states of the molecule, $E_f - E_i$, as in Figure
1.3a. The extension of this energy conservation principle to single-beam n-
photon absorption processes, as in Figure 1.3b,

$$n h\nu = E_f - E_i \tag{1.9}$$

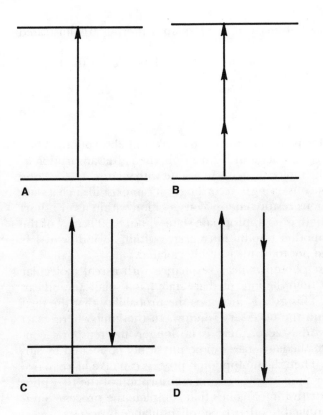

Figure 1.3 Energy level diagrams for (a) single-photon absorption, (b) nonresonant three-photon absorption, (c) nonresonant (Stokes) Raman scattering, (d) nonresonant four-wave mixing.

follows. It is important to note that multiphoton absorption of this nonresonant kind (see later) is a *concerted* rather than a multistep process, as there is no physically identifiable intermediate stage between the absorption of successive photons. For other processes, where the emission of light also contributes to the overall transition, as in the Raman effect (Figure 1.3c), or where two or more laser beams interact in sample molecules, as in four-wave mixing processes (Figure 1.3d), laser wavelength again plays an important role, but one that is often more subtle.

It is not generally necessary in processes involving photons of two or more different frequencies for the energy of any lesser number of photons from a given source to satisfy equation (1.9). Overall energy conservation simply demands that the difference between the sum of the absorbed and emitted photon energies equals a molecular transition energy. However, when equation (1.9) is *also* satisfied, it generally transpires that the rate of laser-induced transitions increases markedly, often by many orders of magnitude. This situation is known as *resonance enhancement,* and examples of resonance in multiphoton absorption and Raman scattering are illustrated in Figures 1.4a and b.

The reason for resonance behavior can be understood on the basis of the time–energy uncertainty principle:

$$\Delta E \, \Delta t \geq \frac{h}{4} \tag{1.10}$$

For simplicity we can focus on the case of multiphoton absorption. When resonance conditions are absent, as in Figure 1.3b, then after absorption of the first laser photon the molecule exists in a state with a large ΔE (i.e., one that is badly energy nonconserving). Accordingly, it can exist in such a state only if further absorptions restore energy conservation within a very short time. For example, if there is no appropriate state within 8,000 cm^{-1} of the photon wavenumber, another photon must arrive within a window of 1 fs for energy conservation not to be measurably violated.

However, if the first photon is near resonance with a real molecular state, as in Figure 1.4a, then the intermediate state has a small ΔE and can exist for much longer. This greatly increases the probability that the next photon will arrive within the necessary window. In the limit where exact resonance occurs and ΔE is zero, there is no longer any temporal constraint, because the intermediate state can be physically populated by single-photon absorption. Then the multiphoton process can take place in two stages, with an arbitrary time interval between absorption of the first photon and subsequent radiative interactions that complete the process. Similar arguments can be applied to other types of resonance processes.

Figure 1.4 Energy levels for (a) three-photon absorption with single photon resonance, (b) resonance (Stokes) Raman scattering.

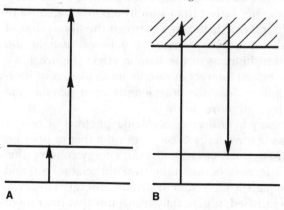

A   B

1.3.3. Transition Strength and Quantum Selection Rules

If a photon or photons of the right energy to match the gap between two molecular states are present at a given sample molecule, there is still only a finite and usually small probability that transition will occur and contribute to the generation of a spectral feature. Normally, each radiative interaction occurs by electric dipole coupling with the photon electric field, and the quantum mechanical transition moment dictates the strength of the coupling. Symmetry restrictions may entirely rule out any such coupling between the radiation and the molecule, though often a much weaker coupling mediated by magnetic dipole or electric quadrupole interactions permits such "forbidden" interactions to occur with a probability reduced by a factor of 10^4 or more.

In the case of single-photon absorption processes, the detailed theory reveals the relationship between the transition moment and the molar absorption coefficient, ε.[24] Laser excitation here offers no distinctive features at the molecular level, other than the increased flux effect examined in Section 1.3.1. This clearly has no effect on the measurements of *relative* absorption, $\Delta I/I$, which form the basis for the calculation of ε through $\Delta I/I = l - 10^{-\varepsilon C}$, l being the path length through the sample and C the sample concentration. As mentioned earlier, lasers do present the possibility of a multipass beam configuration to increase l in studying weak transitions (i.e., those with unusually low absorption coefficients). The high photon flux provided by a laser source is directly useful where *absolute* (zero background) rates are measured, as, for example, in laser-induced fluorescence (LIF). (See Chapter 4.)

It should be noted that the relationship between $\Delta I/I$ and ε given earlier, based on Beer's law, does not hold under certain conditions, as, for example, when laser intracavity enhancement or other saturation conditions apply. This is because Beer's law applies only when there is a direct proportionality between the probability of absorption by individual molecules and the local intensity they experience. This condition totally fails to be satisfied in processes involving the absorption of two or more photons from a laser source, where there is nonlinear intensity dependence. The strength of transitions can be calculated quantum mechanically, but results cannot be given in terms of conventional absorption coefficients.

The quantum mechanical selection rules for conventional photoabsorption are well known and apply with equal force to laser-induced single-photon absorption. Probably most familiar is the Laporte selection rule for centrosymmetric molecules, which allows transitions only between states of opposite parity, gerade (g) $\leftrightarrow$ ungerade (u). Each successive stage in a multiphoton process is also subject to the same rules, given that each radiative interaction almost invariably occurs by the same electric dipole coupling. However, the product selection rules depart substantially from the normal results; for example, if an even number of photons is involved, as in Raman scattering, two-photon absorption, and four-wave mixing, then the selection rules $g \leftrightarrow g$ and $u \leftrightarrow u$ apply. In fact, for molecules of reasonably high symmetry, far more states can be accessed by multiphoton processes than those that show up in conventional single-photon spectra, as Table 1.2 illustrates.

1.3.4. Polarization Effects

Photon polarization is also involved in the selection rules for processes involving more than one photon. This contrasts strongly with the case of single-photon absorption, where, at least for isotropic media, there is essentially no dependence on beam polarization. Such dependence as does exist arises only for optically active (chiral) compounds and is associated with the interference of magnetic dipole and electric quadrupole interactions. These weak effects produce the small polarization dependence that is manifest in the phenomena of circular dichroism and optical rotation.

For two-photon and higher-order processes, the spectra of reasonably symmetrical molecules display a marked dependence on polarization. This is well known in connection with Raman scattering, where depolarization ratio measurements, which reflect a dependence on the relative polarizations of the laser input and the scattered beam, also yield further information on excited-state symmetries.[26] In fact, the polarization behavior itself attests to the concerted nature of the photon absorption and emission involved in the Raman process. However, polarization dependence also arises in connection with processes such as two-photon absorption, where it has similar analytical applications.[27]

Laser beam polarization plays an obvious role in the orientational effects displayed by anisotropic media such as crystals, liquid crystals, and surfaces. However, it should also be noted that strong laser pumping can induce anisotropy in an isotropic liquid sample. This results from the fact that the probability of molecular photoabsorption depends on $\cos^2 \Theta$, where Θ is the angle between the molecular transition moment and the

Table 1.2 Symmetries of States in Benzene (D_{6h}) Directly Accessible by n-Photon Processes[a]

n	A_{1g}	A_{2g}	B_{1g}	B_{2g}	E_{1g}	E_{2g}	A_{1u}	A_{2u}	B_{1u}	B_{2u}	E_{1u}	E_{2u}
1								√				√
2	√	√*			√	√						
3							√*	√	√	√	√	
4	√	√*	√	√	√	√						

Adapted from Ref. 25.

[a]To access the states denoted with an asterisk, at least one photon must differ from the other $(n - 1)$.

photon electric field vector. Where high intensity pulsed laser radiation is employed, it is thus possible to create a preferentially oriented population of excited molecules, a process termed *photoselection*.[28] In many cases rotational relaxation destroys this anisotropy within a matter of picoseconds after the inducing laser pulse, but by use of subpicosecond mode-locked pulses the kinetics of this process can be discerned.

1.4. Conclusion

In this chapter the essential characteristics of laser spectroscopy have been delineated by reference to the unique properties of laser light and their effect on interactions with sample molecules. Subsequent chapters lucidly illustrate the enormous range of techniques now available to exploit these characteristic features. The amount of information each type of laser-based spectroscopy provides often far outstrips anything that can be offered by conventional spectroscopy. Speed of analysis, high sensitivity, microsampling capability, and unsurpassed spectral resolution are just a few of the attractions offered by laser spectroscopy, and applications are demonstrated in a broad range of disciplines in science, medicine, engineering, and industry. It is also shown how the application of such methods has permitted enormous advances to be made in the understanding of a number of fundamentally important processes. With the move toward solid-state technology holding the promise of less expensive laser instrumentation, the future place of these techniques in laboratory practice seems assured.

REFERENCES

1. J. C. Walling, in *Tunable Lasers*, L. F. Mollenauer and J. C. White, eds. (Springer, Berlin, 1987), p. 331.
2. V. S. Letokhov and V. P. Chebotayev, *Nonlinear Laser Spectroscopy* (Springer, Berlin, 1977).
3. M. D. Levenson and S. S. Kano, *Introduction to Nonlinear Laser Spectroscopy*, 2nd ed. (Academic, New York, 1988).
4. V. S. Antonov and V. S. Letokhov, in *Laser Analytical Spectrochemistry*, V. S. Letokhov, ed. (Adam Hilger, Bristol, 1986).
5. W. Demtröder, *Laser Spectroscopy* (Springer, Berlin, 1982), p. 66.
6. R. Loudon, *The Quantum Theory of Light* (Oxford University Press, Oxford, 1973).
7. S. J. Smith and P. L. Knight (eds.), *Multiphoton Processes* (Cambridge University Press, Cambridge, 1988).
8. A. B. Harvey (ed.), *Chemical Applications of Nonlinear Raman Spectroscopy* (Academic, New York, 1981).

9. R. J. H. Clark and R. E. Hester (eds.), *Advances in Nonlinear Spectroscopy* (Wiley, Chichester, 1988).

10. A. E. Siegman, *Lasers* (Oxford University Press, Oxford, 1986), p. 626.

11. T. F. Johnston, *Laser Focus World* 26(5), 173 (1990).

12. L. Moenke-Blankenburg, *Laser Microanalysis* (Wiley, New York, 1989).

13. B. L. Sharp, *Chem. Br.* 18, 342 (1982).

14. *Photonics Spectra*, 24(4) 34 (1990).

15. C. Sauteret, G. Mainfray, and G. Morou, *Laser Focus World* 26(10), 85 (1990).

16. J. Herrmann and B. Wilhelmi, *Lasers for Ultrashort Light Pulses* (North-Holland, Amsterdam, 1987).

17. G. R. Fleming, *Chemical Applications of Ultrafast Spectroscopy* (Clarendon, Oxford, 1986).

18. V. Brückner, K.-H. Feller, and U.-W. Grummt, *Applications of Time-Resolved Optical Spectroscopy* (Elsevier, Amsterdam, 1990).

19. R. R. Alfano (ed.), *Biological Events Probed by Ultrafast Laser Spectroscopy* (Academic, New York, 1982).

20. M. K. Rosker, M. Dantus, and A. H. Zewail, *Science* 241, 1200 (1988).

21. R. B. Bernstein, *Chemical Dynamics via Molecular Beam and Laser Techniques* (Clarendon, Oxford, 1982).

22. D. L. Andrews, *Lasers in Chemistry*, 2nd ed. (Springer, Berlin, 1990), p. 8.

23. D. P. Craig and T. Thirunamachandran, *Molecular Quantum Electrodynamics* (Academic, New York, 1984).

24. J. M. Hollas, *High Resolution Spectroscopy* (Butterworths, London, 1982), p. 42.

25. D. L. Andrews, *Spectrochimica Acta*, 46, 871 (1990).

26. D. A. Long, *Raman Spectroscopy* (McGraw-Hill, New York, 1977).

27. W. M. McClain and R. A. Harris, in *Excited States,* Vol. 3, E. C. Lim, ed. (Academic, New York, 1977), p. 1.

28. D. S. Kliger, J. W. Lewis, and C. E. Randall, *Polarized Light in Optics and Spectroscopy* (Academic, New York, 1990), p. 201.

2 General Aspects of Laser Instrumentation

M. R. S. McCoustra

2.1. Introduction

In any laser-based spectroscopic experiment, the laser itself obviously plays a significant role, and considerable thought must be put into the correct choice of system for the particular experiment. Although lasers can, in general, be regarded as a historically mature technology, even today significant developments in the field are still occurring; for example, it is likely that in the next 2 or 3 years broadly tunable, completely solid-state visible laser systems will become generally available. As a consequence, much of the laser instrumentation discussed in the classics of laser spectroscopy are somewhat dated, although Demtröder's presentation of the fundamental concepts of laser physics and of laser spectroscopy remains unsurpassed.[1] In this chapter we review the current state of the art in laser technology, highlighting some significant recent developments, and consider the general instrumentation required in typical laser-based spectroscopic experiments in an effort to provide an updated source of information for both current laser spectroscopists and budding users.

2.2. Tunable Coherent Radiation Sources

Some idea of the vast number of coherent light sources available is given in Table 2.1 and Figure 2.1. We do not propose to discuss each of these laser systems, but rather to highlight some of those of more

20

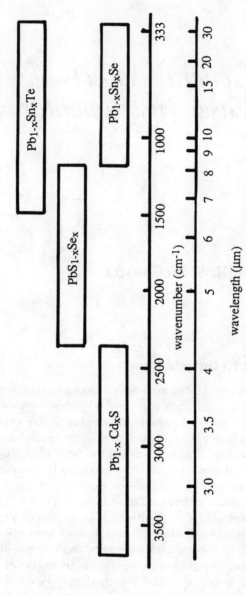

Figure 2.1 Variation of emission characteristics with semiconductor compound. (Reproduced from Ref. 25 with permission.)

Table 2.1 Some FIR Laser Transitions

Molecule	Formula	Wavelength (μm)	Pump Line[a]	Power[b]
D$_1$-Methanol	CH$_3$OD	4.1	10R18	M
Methanol	CH$_3$OH	67.5	9R18	M
Methanol	CH$_3$OH	70.5	9P34	M
^{15}N Ammonia	^{15}NH$_3$	78.5	10P24(^{13}C)	W
D$_1$-Methanol	CH$_3$OD	86.7	9R28(^{13}C)	M
Methanol	CH$_3$OH	96.5	9R10	S
^{15}N Ammonia	^{15}NH$_3$	112.3	10P14(^{13}C)	M
Methanol	CH$_3$OH	118.8	9P36	S
D$_3$-Methanol	CD$_3$OH	143.4	9P18(^{13}C)	W
^{15}N Ammonia	^{15}NH$_3$	152.7	10R18(^{13}C)	S
Methanol	CH$_3$OH	163.0	10R38	S
Difluoromethane	CH$_2$F$_2$	184.3	9R32	S
Difluoromethane	CH$_2$F$_2$	193.9	9R22	M
Difluoromethane	CH$_2$F$_2$	214.6	9R34	S
D$_3$-Methanol	CD$_3$OH	221.0	9P36(^{13}C)	W
D$_1$-Methanol	CH$_3$OD	255.0	10R36	M
Chloromethane	CH$_3$Cl	333.9	9P42	W
^{15}N Ammonia	^{15}NH$_3$	375.0	10R42	M
D$_3$-Chloromethane	CD$_3$Cl	383.3	9R34	M
Ammonia	NH$_3$	388.5	10R10(^{13}C)	M
Formic acid	HCOOH	393.6	9R18	M
Formic acid	HCOOH	432.6	9R20	M
D$_3$-Chloromethane	CD$_3$Cl	443.3	9P10	M
Iodomethane	CH$_3$I	447.1	10P18	M
Fluoromethane	CH$_3$F	496.1	9P20	M
Formic Acid	HCOOH	513.0	9R28	M
D$_3$-Methanol	CD$_3$OH	530.4	9P20(^{13}C)	W
Methanol	CH$_3$OH	570.6	9P16	M
1,1-Difluoroethene	C$_2$H$_2$F$_2$	890.1	10P22	W
D$_1$-Methanol	CH$_3$OD	917.9	9R24(^{13}C)	W
Chloromethane	CH$_3$Cl	944.0	9R12	W
1,1-Difluoroethene	C$_2$H$_2$F$_2$	1020.0	10P14	W
^{13}C Fluoromethane	^{13}CH$_3$F	1221.8	9P32	M

Source: Adapted with permission of Edinburgh Instruments Ltd., Edinburgh.

[a] CO$_2$ laser transition employed as the pump radiation. 10 indicates the 10.6-μm CO$_2$ band, and 9 indicates the 9.6-μm band. *P* and *R* have their usual meanings and the *J* quantum number is after the branch indicator.

[b] W = weak; M = medium; S = strong.

direct relevance to the spectroscopist. Detailed descriptions of the more specialized instrumentation relevant to the discussion of particular spectroscopic techniques may be found in the following chapters.

2.2.1. Coherent Radiation Sources in the Region Beyond 20 μm

The region beyond 20 μm (500 cm^{-1}), known as the *far infrared* (FIR), is perhaps the most poorly served in terms of tunable coherent radiation sources. There are, however, two commonly used approaches to the generation of tunable coherent radiation in the FIR. One involves an optically pumped molecular gas laser, which is probably the most widely used type of source in this region and provides line tunable output from a large number of rovibrational transitions of a variety of simple polyatomic gases,[2-4] as illustrated by Table 2.1. The other involves sum- and difference-frequency lasers.[5] Let us briefly consider each of these in turn.

The optically pumped molecular gas laser can in a certain sense be regarded as an FIR analogue of the tunable dye laser. Radiation from an intense pump laser, typically a pulsed or CW CO or CO_2 laser operating in the mid-IR at around $900-1,100$ cm^{-1}, is used to create a population inversion in an appropriate molecular vibration in the lasing gas. Methanol is commonly the chosen laser medium, providing output on numerous rovibrational transitions in the range from ~25 μm to $\sim1,670$ μm using a variety of isotopically substituted methanols (CH_3OH, CH_3OD, CD_3OH, and CD_3OD).[6-9] However, various mono- and di-halomethanes have shown promise as laser media and are consequently now in widespread use,[10-15] and some even more complex molecules, such as propargyl fluoride, vinyl fluoride, and vinyl chloride,[16] have exhibited FIR laser action. Output from such lasers can be either pulsed, with multimillijoule pulse energies, or CW, with possible output powers in excess of 1 W.

Although the output of any such laser is restricted to those frequencies where molecular rovibrational transitions occur (i.e., the laser is only *line-tunable*), this has not prevented spectroscopic application. If the laser cannot be tuned into resonance with a molecular energy level, then the molecular energy levels must be tuned into resonance with the laser. This can be achieved in two manners, shifting molecular energy levels using either an electric field (*Stark tuning* or *Stark spectroscopy*) or a magnetic field (*Zeeman tuning* or *Zeeman spectroscopy*). The latter technique, commonly known as *laser magnetic resonance* (LMR) *spectroscopy*,[1] has found great favor in the study of the spectroscopy of free radicals in both the FIR and mid-IR, where the line-tunable CO and CO_2 lasers are normally employed.[17,18]

For continuously tunable FIR radiation, the spectroscopist must turn to sum- and difference-frequency lasers. As outlined by Evenson et al.,[5] there are essentially three varieties of coherent radiation source that can be placed within this group; harmonic generation of tunable, coherent microwave sources,[19] frequency mixing of a line-tunable FIR laser with continuously tunable microwave radiation,[20] and difference-frequency generation using CO_2 lasers with and without additional microwave frequency addition in either GaAs diodes[21] or metal–insulator–metal diodes.[22,23] In the latter group of techniques, two approaches are commonly used,[5] radiation in the appropriate region being generated using either a second-order method (mixing a fixed-frequency CO_2 laser with a tunable waveguide CO_2 laser in a tungsten–nickel diode) or a third-order method (summing the difference of two fixed-frequency CO_2 lasers with tunable microwave radiation in a tungsten–cobalt diode). Using any of the methods described briefly here, it is possible to generate FIR radiation that is continuously tunable over regions of the order of 100–150 MHz wide, at powers from a few microwatts to over 100 mW.

2.2.2. Coherent Radiation Sources in the 1–20-μm Region

In assigning the wavelength limits of this section, most of the mid-IR region, which is normally assumed to stretch from ~500 cm^{-1} to ~4,000 cm^{-1} (i.e., the region normally covered by simple benchtop dispersive or Fourier transform IR spectrophotometers), is considered for ease of discussion together with the region beyond 4,000–10,000 cm^{-1} (1 μm), which is normally regarded as part of the near-IR. This region is considerably better served by tunable coherent radiation sources than the FIR region, discussed previously. However, few of the sources span the complete range; rather, they give excellent coverage in limited regions.

As in the FIR, line-tunable molecular gas lasers play an important role in the spectroscopic instrumentation of this region. These are dominated by the CO_2 laser and the CO laser. Commercial CO_2 lasers can provide immense quantities of IR radiation as either discrete pulses of up to several joules, at repetition rates of a few hertz, or alternatively CW powers in the multikilowatt range. Consequently, such lasers have found consider application as industrial cutting and welding tools. At lower powers, both pulsed and CW CO_2 lasers provide ideal sources of line-tunable radiation in either the 9.6-μm or 10.6-μm regions. The use of isotopically substituted CO_2 can naturally be used to extend these regions. Moreover, if the laser medium has a high-pressure discharge, then pressure broadening of the rovibrational lines can lead to the gain profile of the laser appearing continuous rather than discretely line-tunable. Details of the excitation mechanism and molecular states relevant to CO_2 laser action can be found in Demtröder.[1]

Probably the second most common line-tunable laser in this region is the CO laser. In contrast to the preceding example, CO lasers are operated normally in CW mode. Line-tunable outputs of several watts are possible from several hundreds of rovibrational lines in the 4.8–8.4-μm region of the spectrum.[24,25] In contrast to the CO_2 laser, however, the CO laser system cannot be pressure-broadened to provide a pseudocontinuous source of radiation. This has not limited its application to spectroscopy, given the success of the LMR technique briefly referred to earlier.[1,17,18]

In contrast to the FIR, continuously tunable coherent sources in this region are quite numerous. Furthermore, such sources tend to be much simpler to operate than those in the FIR, directly providing continuously tunable radiation within the range of wavelengths under discussion. Probably the most widely available of these sources is the semiconductor diode laser.[1,25,26] Diodes constructed from lead salts exhibit laser action in the mid-IR as illustrated by Figure 2.1. The exact range of emission wavelengths is determined by the exact composition of the semiconductor material (illustrated by Figure 2.2) and by the operating temperature of the diode (Figure 2.3). Extension of the region of application of semiconductor lasers to the near-IR and the long-wavelength visible is possible, using devices based on compounds of group III elements and group V elements (III–V semiconductors) and has been driven by the desire for compact laser sources for telecommunications and consumer goods. Outputs of several milliwatts at discrete wavelengths in the 630–680-nm region are possible from the current generation of devices, and consequently such semiconductor devices represent a threat to the dominance of the He–Ne laser in some applications.[27] For spectroscopic applications, however, the lead salt diode remains the only commonly used semiconductor laser.

Figure 2.2 Dependence of the emission frequency on composition for Pb-salt lasers of various alloy compositions at a temperature of 10 K. (Reproduced from Ref. 1 with permission. Original source material, Spectra-Physics, Inc.)

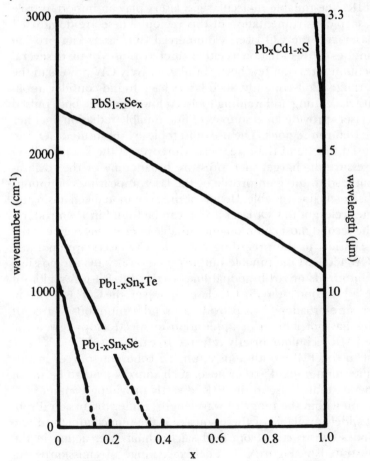

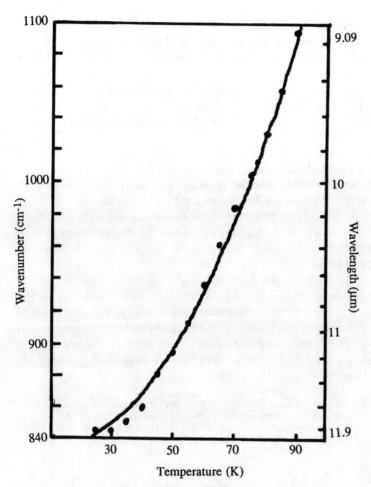

Figure 2.3 Temperature tuning of a Pb–Sn–Se laser. (Reproduced from Ref. 1 with permission. Original source material, Spectra-Physics, Inc.)

The basic physics of the lead salt diode laser is that of a *p–n* junction diode and is akin to that of simple light-emitting diodes (LED). As is illustrated by Figures 2.1 and 2.2, the wavelength range of interest can be covered by constructing diodes of differing compositions. Each individual diode is capable of covering a wavelength range of up to 100 cm^{-1} about a region defined by its composition, at a bandwidth of typically 20 MHz (about 0.0005 cm^{-1}), which is ideal for Doppler-limited spectroscopy in the mid-IR. Consequently, a large library of diodes is required to cover the whole of the mid-IR region. Each diode can then be more precisely tuned within its particular wavelength range either by changing the operating temperature of the diode (see Figure 2.3) and/or by changing the driving current of the diode. Although many of the early diode lasers were of limited usable lifetime and tuning range and could only be operated using a liquid helium cryostat, molecular beam epitaxial growth methods have recently been applied to the construction of the diodes and this has led to significant improvements in lifetime and tuning range. More important, this has produced diodes that can be operated using liquid nitrogen cryostats. These improvements, as well as increase in output power to a few milliwatts, should result in increased use of these devices in mid-IR spectroscopy.

Moving toward the shorter wavelengths region, *color-center* or *F-center lasers* become a viable alternative to the semiconductor diode laser.[1,25] Although alkali metal halide single crystals are normally transparent throughout much of the ultraviolet, visible, and mid-IR regions of the electromagnetic spectrum, damage induced by radiation, for example, hard UV, X-rays, γ-rays, or electron beams, can lead to the formation of defects that absorb radiation in the visible and near-IR regions. These are known as color or F centers.

The simplest F center results from the replacement of an anion in the lattice with an electron. Physically, this system is readily described by the simple particle-in-a-box model from quantum mechanics. The dynamics of the simple F center, however, is such that rapid nonradiative relaxation of the excited state prevents its application as a laser system. However, some of the more complicated F centers, as illustrated in Figures 2.4a and 2.4b,

Figure 2.4 (a) Color centers in a cubic alkali halide structure. Squares designate anion vacancies; circles denote dopants. (Reproduced with the permission of Burleigh Instruments, Inc.) (b) Schematic representation of type (II) laser behavior showing both ground and excited states. Key as in (a). (Reproduced with the permission of Burleigh Instruments, Inc.)

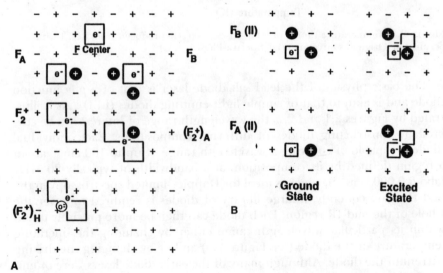

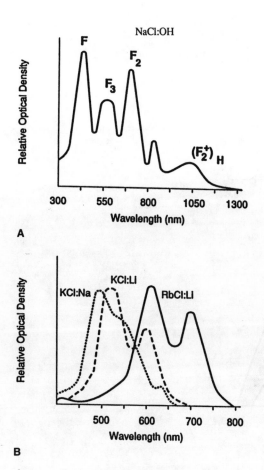

Figure 2.5 Absorption spectra of some color-center laser materials. (Reproduced with the permission of Burleigh Instruments, Inc.)

are ideal as laser systems and are readily produced by doping small quantities of alkali metals into an alkali halide. This is particularly true of the $F_A(II)$ and $F_B(II)$ centers, as represented by Li:KCl (Li:RbCl) and Na:KCl, respectively, which possess the high fluorescence efficiency and low nonradiative relaxation rates that make them ideal laser media. As the absorption spectra of these materials fall across much of the visible and near-IR region of the spectrum, Figures 2.5a and 2.5b, and the media exhibit sufficient gain to sustain CW laser action, excitation by argon ion, krypton ion, and CW Nd–YAG or Nd–YLF is favored, although UV excitation is also feasible.[28] These lasers exhibit gain in the regions illustrated by Figures 2.6a and 2.6b, providing both reasonable output powers (up to 150 mW) and narrow bandwidth (1 MHz is possible) in a region of considerable spectroscopic interest, covering as it does the region of X–H stretching vibrations and the overtone region.

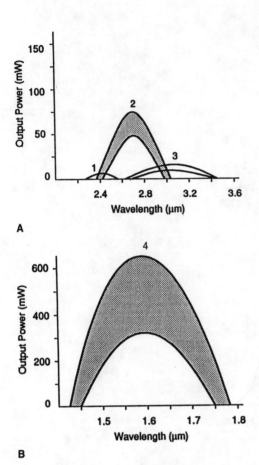

Figure 2.6 Tuning curves for some color-center laser materials: (1) KCl:Na; (2) KCl:Li; (3) RbCl:Li; (4) NaCl:OH. (Reproduced with the permission of Burleigh Instruments, Inc.)

At the short-wavelength limit of this region and extending into the region below 1 μm, considerable interest and experimental effort is being placed into a group of laser systems known as *transition metal vibronic solid-state lasers*.[29-33] These contain transition metal ions doped into an essentially inert matrix and are typified by the systems listed in Table 2.2. While the original ruby laser system of Maiman[34,35] may be classed as a transition metal ion laser, the ground state of the Cr^{3+} ion does not couple to the vibrations of the support matrix as the orbital angular momentum of the ion is quenched by the crystal field. The lower lasing level is therefore a single, narrow energy state, and laser action occurs at a single wavelength. In a vibronic solid-state laser, the dopant ion orbital angular momentum is not quenched and strong coupling occurs between the ground electronic state of the ion and lattice vibrations. Consequently, instead of a single en-

Table 2.2 Vibrational Solid-State Lasers

Dopant Ion	Matrix	Tuning Range^a/μm
Co^{2+}	MgF_2	1.5–2.5
Ni^{2+}	MgO	1.0–2.0
	MgF_2	1.6–1.7
Cr^{3+}	$BeAl_2O_4$ (Alexandrite)	0.7–0.8
Ti^{3+}	Al_2O_3 (Sapphire)	0.67–1.1

^{a}Approximate limits. Actual tuning range is dependent on mode of operation, pump source, and so on.

ergy level in the ground lasing state, a broad band composed of many closely lying states is formed. Laser action in these systems then resembles that of organic molecular dye lasers in the visible in that they exhibit broad emission profiles.

In the region under current discussion, two transition metal vibronic state lasers show particular promise. The first of these is the Ni^{2+} system doped either into MgO, in which laser action occurs between 1 and 2 μm[36] or into MgF_2, where laser emission occurs over the narrower 1.6–1.73-μm range.[37] The second system of interest operates on Co^{2+} doped into MgF_2, which lases in the range from 1.5 to 2.5 μm.[39,38] In pulsed operation, pumped by a fixed-frequency near-IR source such as a Nd–YAG laser, output energies in excess of 70 mJ over this region have been achieved from commercial Co:MgF_2 laser systems. Further developments in this region are inevitable.

2.2.3. Coherent Radiation Sources in the 0.35–1-μm Region

In the region between 0.35 and 1 μm, the dominant laser medium is at present the *organic molecular dye laser*.[40] These systems are vibronic lasers, (cf. the preceding discussion) in that emission from the upper laser state is into a broad range of ground state vibrational levels determined by the Franck–Condon envelope.[41] An individual dye typically covers a wavelength region around 20–30 nm wide, and thus to cover the entire region numerous individual dyes are required, as illustrated in Plate 1. Typically, dye lasers are capable of converting between 5% and 25% of the input pump radiation into tunable visible radiation in either a CW or pulsed operating mode and are, therefore, probably the most efficient type of laser system available to the spectroscopist.

In CW operation, argon or krypton ion lasers are the traditional pump sources. However, recent developments in nonlinear media have seen a growth in the use of frequency-doubled CW Nd–YAG an Nd–YLF pump sources. This is especially true in the case of synchronously pumped mode-locked ultrafast (ps and fs) laser systems, where the traditional model-locked Ar^+ and Kr^+ lasers have been replaced by mode-locked Nd–YAG and Nd–YLF. It seems unlikely, however, given that second harmonic outputs from such systems are only of the order of a few hundred milliwatts to 1 or 2 W, unless Q-switched or mode-locked, that such frequency doubled Nd–YAG and Nd–YLF lasers will completely replace Ar^+ and Kr^+ lasers, which with current technology are achieving powers in excess of 25 W (Ar^+) and 10 W (Kr^+) lasing on all lines in the visible. More important, the current generation of ion lasers is also capable of producing in excess of 7 W (Ar^+) and 2 W (Kr^+) on lines lying in the near UV (333.6–363.8 nm from Ar^+ and 337.5–356.4 nm from Kr^+) and nearly 1 W of output in the deep UV (274.4–305.5 from Ar^+). The latter wavelength groups are ideal for pumping dyes that lase in the violet, blue, and green regions of the visible spectrum.

CW dye lasers can operate in one of two optical configurations, the *standing-wave laser* and the *ring laser,* as illustrated in Figure 2.7. Each of these configurations has its advantages and disadvantages as outlined by Demtröder.[1] Typically, a simple standing wave cavity will produce a broadband laser output with a linewidth of between 30 and 300 GHz (1–10 cm^{-1}) at each set wavelength, while the simple ring cavity produces a naturally narrower line of around 3 GHz (0.1 cm^{-1}). In both configurations, the addition of further line narrowing optics (etalons) to the cavity allows essentially single-frequency operation at a linewidth of around 500 kHz (0.00002 cm^{-1}).

For pulsed operation, the simplest excitation source is the traditional xenon flashlamp. Such *flashlamp–pumped dye lasers* are capable of output energies in excess of 1 J at low repetition rates (typically, 1–10 Hz). Pulsewidths are, of course, determined by the temporal profile of the exciting flashlamp and the photophysics of the laser dye. Such lasers are, however, also capable of producing pulses a few nanoseconds wide using electro-optic Q switches (cf. pulsed Nd–YAG lasers). The longer microsecond pulses find favor with the major use of these systems as medical lasers, as the peak powers of the microsecond pulses cause less secondary tissue damage than the considerably higher peak powers of nanosecond laser systems. As spectroscopic tools, these lasers are capable of producing narrow-bandwidth output, typically around 0.04 cm^{-1}, but their low duty cycle prevents their widespread use as spectroscopic tools, and they are thus restricted to a few specialist spectroscopic applications, such as those in chemical kinetics.[42,43]

By far the most common type of pulsed dye laser is the *pulsed laser–pumped dye laser* operating in the nanosecond regime. A number of pumped sources have been tried successfully but at low repetition rates (up to ~250 Hz), the second and third harmonics of the Nd–YAG (532 and 355 nm, respectively), xenon chloride (308 nm), and xenon fluoride (351 nm) exciplex lasers and the nitrogen laser (337 nm) are the most commonly employed pump sources. At higher repetition rates of up to 20 or 30 kHz, frequency-doubled Q-switched CW Nd–YAG and Nd–YLF lasers and cop-

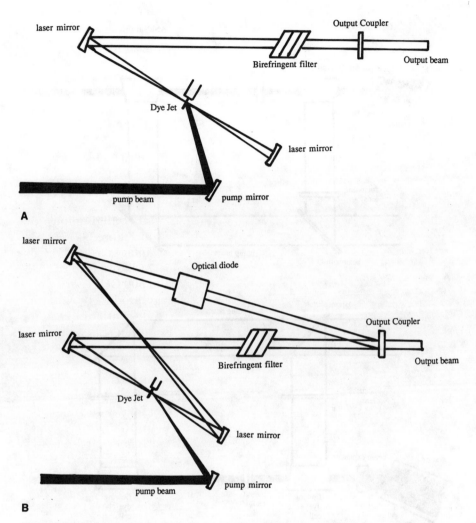

Figure 2.7 Continuous-wave dye laser configurations: (a) standing-wave and (b) ring dye laser configurations.

per vapor lasers are employed. With both of these groups of laser systems, it is possible to cover most of the region beyond 0.35 μm with tunable dye laser output as shown in Plate 1. Output pulse energies can reach many hundreds of millijoules for dye lasers with large pump lasers, but pulse energies of up to about 100 mJ are normally produced.

As in the case of CW dye lasers, commercial pulsed dye laser systems have settled on one of two basic cavity designs, the *Littrow dye laser* and the *grazing incidence dye laser,* as illustrated in Figure 2.8. The former cavity design is perhaps the easier of the two to align but produces a broad linewidth of ~0.15–0.20 cm^{-1}. The addition of a further intracavity etalon can

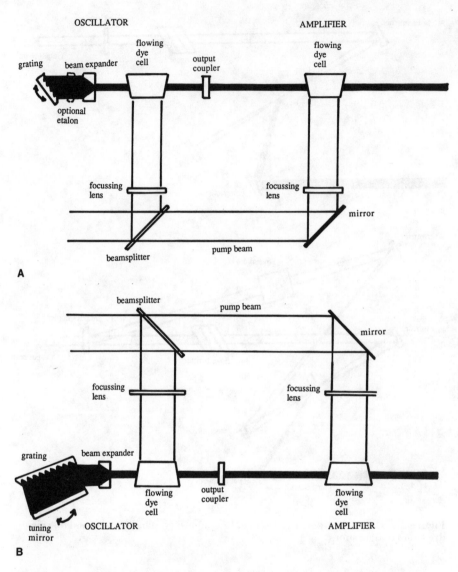

Figure 2.8 Common pulsed dye laser configurations: (a) Littrow-type laser; (b) grazing incidence dye laser.

narrow this to ~0.08 cm^{-1} or narrower, but at the expense of restricting the tuning range to within the free spectral range (FSR) of the etalon. The grazing incidence cavity, however, produces light with a naturally narrow linewidth of around 0.05–0.08 cm^{-1}, without additional intracavity components, over the entire tuning range of a dye. Many dye laser manufacturers now seem to favor this latter cavity design as users are demanding narrower linewidth combined with simplicity of tuning. In contrast to a CW dye laser, there is, however, a natural limit on the linewidth of the light that can be emitted from a pulsed dye laser. This is determined by the Heisenberg uncertainty principle and is the so-called Fourier transform limited linewidth. For a 10-ns laser pulse, this is around 300–500 MHz (0.01–0.015 cm^{-1}). A number of dye laser cavity designs are capable of producing transform limited linewidth from a single laser cavity mode, but usually over a limited tuning range. Recent work, however, seems to promise the possibility of a dye laser capable of producing transform limited linewidth from single-mode output over the entire gain curve of a dye without mode hopping.[44] This is obviously one area where considerable effort will be placed by both researchers and manufacturers.

One particular design of cavity that has proved popular in the laboratory but that has yet to be commercialized is the *distributed-feedback dye laser* (DFDL). This cavity design offers some advantages over the more widely used grating-tuned designs. It offers low amplified spontaneous emission and both higher lasing efficiencies and wider gain curves than the grating-tuned designs.[45] More important, however, it offers the possibility of producing output pulsewidths of 70–250 ps using a simple nanosecond pump source and a surprisingly simple cavity design.[46,47] (See Figure 2.9.) The addition of a second quenching cavity to the basic DFDL design results in the production of transform-limited pulses of the order of 80 ps duration.[48]

Figure 2.9 A simple distributed-feedback dye laser cavity. This laser is tuned by rotating the mirrors and/or altering the refractive index of the dye solution.

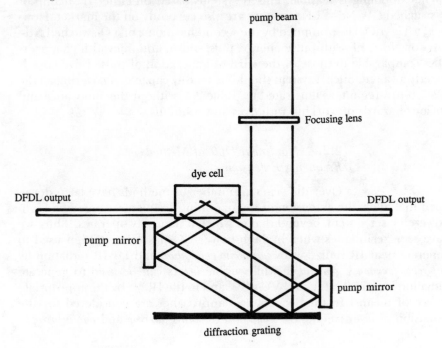

(See Figure 2.10.) When pumped by a high-repetition-rate source such as a copper vapor laser or the second harmonic of a Q-switched CW Nd–YAG or Nd–YLF, such a dye laser will provide moderately narrow pulses at moderate repetition rates, at a fraction of the cost of a synchronously pumped mode-locked dye laser system. It does not, however, represent a viable alternative to these lasers for the production of much shorter pulses unless pumped by a short-pulse laser.

As discussed in the previous section, transition metal ion vibronic state lasers are now making rapid inroads into the upper limits of the visible region of the spectrum.[29–33] Two systems in particular are proving popular, the alexandrite laser (Cr^{3+} doped into $BeAl_2O_4$), with a tuning range of 700–800 nm, and the titanium:sapphire laser (Ti^{3+} doped into Al_2O_3) with the wider range of 670–1,100 nm. Current commercial versions of the former are capable of producing several hundreds of millijoules in the fundamental region at a repetition rate of ~10 Hz with Q switching. Linewidths of ~150 MHz (0.005 cm^{-1}) are possible. These pulse energies mean that efficient frequency doubling, tripling, and Raman shifting (see later) can be used to produce tunable radiation that gives complete coverage from ~0.2 to 2.5 μm, at pulse energies from upward of 2 or 3 mJ. (See Figure 2.11.) In contrast, operation of the $Ti:Al_2O_3$ system is currently largely limited to operation in CW standing-wave or ring lasers. Such lasers are capable of producing two to three times the CW power of dye lasers in the corresponding region from the same pump power. Intracavity frequency doubling is feasible. Pulsed systems based on either flashlamp or Q-switched CW Nd–YLF pumping are also currently on the market. However, a $Ti:Al_2O_3$ laser pumped by the second harmonic of a Q-switched Nd–YAG or Nd–YLF with pulse energy, pulsewidth, and linewidth characteristics comparable to those of the current generation of pulsed dye laser is eagerly awaited. Such a system should be readily capable of covering all the 0.2–1.0-μm region, as illustrated by Table 2.3, without the annoyance and obvious hazards of working with dyes and solutions.

2.2.4. Nonlinear Optical Methods of Frequency Extension

Over the years, a number of methods have been developed that allow the extension of the operating frequency range of a high-powered laser system, beyond that in which it naturally operates. Thus, we have seen techniques whereby high-power CO_2 lasers have been used to generate near-IR radiation by *harmonic generation* and far-IR radiation by *difference-frequency generation,* while visible lasers can be used to generate radiation both in the UV and VUV[49–51] and in the IR[25,52] by the appropriate choice of means. In this section, two approaches are considered for the extension of laser frequency: *stimulated Raman scattering* and *nonlinear optical mixing.*

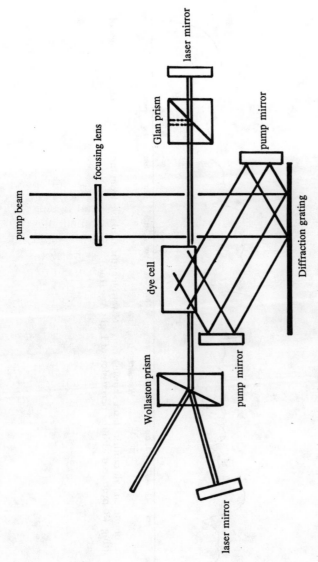

Figure 2.10 A more complex design of distributed feedback dye laser capable of producing pulses a few tens of picoseconds wide from a nanosecond pump source.

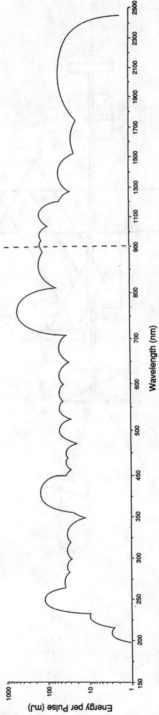

Figure 2.11 Tuning curves for a commercial alexandrite laser showing range of wavelengths available from the fundamental, harmonics of the fundamental, and Raman shifting. (Reproduced with the permission of Light Age, Inc.)

Table 2.3 Wavelength Coverage of a Hypothetical Pulsed Ti^{3+}: Sapphire Laser Pumped by the Second Harmonic of a Nd–YAG Laser

Wavelength Region (nm)[a]	Mechanism
700–1000	Laser fundamental
542–706	Raman-shifted fundamental, first anti-Stokes in H_2
442–546	Raman-shifted fundamental, second anti-Stokes in H_2
416–506	Fundamental mixed with Nd–YAG fundamental (1064 nm)
350–500	Second harmonic
306–414	Raman-shifted second harmonic, first anti-Stokes in H_2
271–353	Raman-shifted second harmonic, second anti-Stokes in H_2
261–336	Second harmonic mixed with Nd–YAG fundamental (1064 nm)
234–334	Third harmonic
213–293	Raman-shifted third harmonic, first anti-Stokes in H_2
196–261	Raman-shifted third harmonic, second anti-Stokes in H_2
190–252	Third harmonic mixed with Nd–YAG fundamental (1064 nm)

[a]Approximate values.

The basic principles of the Raman and stimulated Raman effects are discussed in detail in Chapters 6 and 7. Suffice it to say that with a sufficiently intense laser source, stimulated Raman scattering can occur into, or at a small angle to, the propagation direction of the stimulating laser. If the stimulated beam, which can be either Stokes or anti-Stokes shifted in frequency, is itself of sufficient intensity, then it too may stimulate further scattering. The net effect of this multiple scattering is to produce a series of Stokes frequencies:

$$\nu_{Stokes} = \nu_{pump} - n\nu_{vib} \tag{2.1}$$

and of anti-Stokes frequencies:

$$\nu_{anti\text{-}Stokes} = \nu_{pump} + n\nu_{vib} \tag{2.2}$$

where ν_{pump} is the frequency of the pumping laser, ν_{vib} is the vibrational frequency of the most Raman active of the vibrational modes within the scattering molecule, and the integer n is the *order* of the scattering process. Clearly, stimulated Raman scattering offers the possibility of frequency shifting in integral multiples of a Raman active vibrational frequency to either lower or higher frequency of a pump frequency. Thus, the technique can be used to generate either UV and VUV light by anti-Stokes scattering or IR light by Stokes scattering simply by focusing the pump radiation into the scattering medium. Nonetheless, it has been observed that more efficient Stokes scattering is obtained when the pump radiation is focused into a silica tube within the scattering medium. The tube is thought to behave as an *optical waveguide*, increasing the efficiency of the scattering.

By far the most commonly employed medium for Raman shifting is H_2 (ν_{vib} = 124.6 THz; $\tilde{\nu}_{viv}$ = 4155 cm^{-1}), although D_2, N_2, and CH_4 are also employed. Experimental studies of the process abound and are well illustrated by Figure 2.12, from the work of Schomberg et al.[53] More recent work by Baldwin et al. has shown how the efficiency of this scattering process depends both on the pump wavelength, with the shorter wavelengths appearing more efficient, and on the pressure of the scatterer.[54]

Although Raman scattering from molecular vibrations is the most commonly used means of Raman shifting, scattering from atomic electronic states is possible. The process of *stimulated electronic Raman scattering* has also proved to be of considerable use in frequency shifting. In particular, the $6s$–$5d$ $^2D_{5/2}$ transition in atomic cesium has proved especially useful in the generation of tunable IR radiation from both nanosecond[55] and picosecond[56] sources. In the former case, it is possible to generate radiation covering the entire 1,000–4,000-cm^{-1} (2.5–10-μm) region using a tunable, Nd–YAG pumped dye laser, as illustrated in Figure 2.13. In the 3,000–3,300 cm^{-1} region, quantum conversion efficiencies of around 10% have been achieved (2 mJ output for 120-mJ input), while at longer wavelengths, ~6 μm (1,600 cm^{-1}) efficiencies of 1.2% are possible (0.1 mJ from 80 mJ).[55]

Figure 2.12 Measured energy of first Stokes and up to 13th anti-Stokes emission (138 nm) in H_2 gas (at 2–3 atm pressure) excited by 100-mJ laser radiation at 545 nm. (Reproduced from Ref. 51 with permission.)

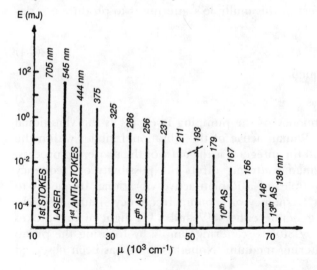

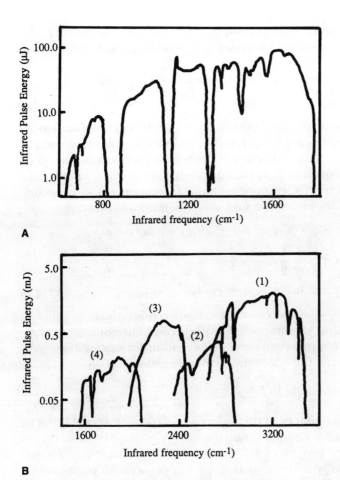

Figure 2.13 (a) Infrared tuning curves obtained by stimulated electronic Raman scattering in cesium vapor with rhodamine dyes in the pump laser. Vertical scale is logarithmic. (1) Rhodamine 590 in methanol (oscillator) and methanol/water (amplifier). (2) Rhodamine 610 in methanol. (3) Kiton Red in methanol/water. (4) Rhodamine 640 in methanol. (Reproduced from Ref. 55 with permission.) (b) Infrared tuning curve obtained by stimulated electronic Raman scattering in cesium vapor with DCM in the pump laser. Vertical scale is logarithmic. (Reproduced from Ref. 47 with permission.)

Let us now consider the second approach to frequency extension, the use of nonlinear optical mixing techniques. The dielectic polarization P of a nonlinear medium with susceptibility χ can, in the presence of an electric field E, be written as a power series expansion:

$$P = \varepsilon_0 \left[\chi^{(1)}E + \chi^{(2)}E^2 + \chi^{(3)}E^3 + ...\right] \tag{2.3}$$

where the $\chi^{(i)}$ are the ith-order susceptibilities. Clearly, by considering the preceding, we can define several orders of nonlinear effects. But for the most part, practical applications of use to the laser spectroscopist in fre-

quency shifting use only second-, third-, and perhaps fourth-order processes.[50,51] The second-order processes will now be considered in some
detail.

Consider an electromagnetic wave composed of two components

$$E = E_1 \cos(\omega_1 t - k_1 z) + E_2 \cos(\omega_2 t - k_2 z) \qquad (2.4)$$

incident on a nonlinear medium, where the ω_i are the circular frequencies
and k_i are the corresponding wavevectors. The induced polarization at $z = 0$ is generated by the combined action of the two components. So the quadratic term $\chi^{(2)} E^2$ in equation (2.3) includes contributions, as demonstrated
in equation (2.5), from a DC component and AC components at the second
harmonics $2\omega_1$ and $2\omega_2$ and at the sum and difference frequencies:

$$E^2(z = 0) = E_1^2 \cos^2 \omega_1 t + E_2^2 \cos^2 \omega_2 t + 2E_1 E_2 \cos \omega_1 t \cos \omega_2 t$$

$$= \frac{1}{2}(E_1^2 + E_2^2) + \frac{1}{2} E_1^2 \cos 2\omega_1 t + \frac{1}{2} E_2^2 \cos 2\omega_2 t$$

$$+ E_1 E_2 [\cos(\omega_1 + \omega_2)t + \cos(\omega_1 - \omega_2)t] \qquad (2.5)$$

Of course, if $\omega_1 = \omega_2 = \omega$ and $E_1 = E_2 = E$, then this reduces to simply a
DC component and an AC component at the second harmonic.

As the electric fields are vectors and the second-order susceptibility $\chi^{(2)}{}_{ijk}$
is a tensor, we can write the quadratic term in equation (2.3) as

$$P_i = \varepsilon_0 \sum_{j,k} \chi^{(2)}_{ijk} E_j E_k \qquad (i, j, k = x, y, z) \qquad (2.6)$$

where P_i is the ith component of the dielectric polarization vector $\mathbf{P} = \{P_x,
P_y, P_z\}$. This may seem complex, but considerable simplification occurs
when the symmetry properties of the nonlinear medium are considered.[57,58] For a centrosymmetric species, $\chi^{(2)}$ has no nonzero components;
therefore, no second-order nonlinear processes are possible, although
higher-order processes may well be. However, second-order processes are
observed in birefringent crystals such as quartz, potassium dihydrogen
phosphate (KDP), lithium iodate, and a number of other species. While $\chi^{(2)}$
may in the limit possess 27 individual components, symmetry arguments
may significantly reduce that number. Thus, in the case of KDP there are
only three nonvanishing components of the second-order susceptibility.

The nonlinear polarization induced in an atom, molecule, or ion within
the nonlinear medium by an incident electromagnetic wave, as given in
equation (2.4), acts as a source of waves at frequencies $2\omega_1$, $2\omega_2$, and $\omega_1 \pm
\omega_2$, which propagate through the medium with *phase velocities* $v_{ph}(\omega_i)$ given
by equation (2.7):

$$v_{ph} = \frac{\omega_i}{k_i}$$

$$= \frac{c}{n(\omega_i)} \qquad (2.7)$$

where $n(\omega_i)$ is the refractive index of the medium at circular frequency ω_i. The microscopic contributions from individual components of the medium can only combine to generate a macroscopic wave with significant intensity if the phase velocities of the incident and polarization waves are matched. This condition is known as *phase-matching* and can be written for second harmonic generation as

$$\mathbf{k}(2\omega) = 2\mathbf{k}(\omega) \tag{2.8}$$

and for sum- and difference-frequency generation as

$$\mathbf{k}(\omega_1 \pm \omega_2) = \mathbf{k}(\omega_1) \pm \mathbf{k}(\omega_2) \tag{2.9}$$

which can be interpreted in terms of momentum conservation for the three photons involved in the process. Detailed classical descriptions of the phase-matching conditions for second harmonic generation and sum- and difference-frequency generation, using both focused and unfocused beams, exist within the literature[59–61] and can be used as a prescription for estimating the efficiency of a nonlinear mixing process. It can, however, be generally stated that the intensity of the output mixed wave depends on the nonlinear coefficients of the active medium and the length of the interaction zone within the medium. It is also proportional to the square of the incident intensity for second harmonic generation, or the product of the two incident intensities, in the case of sum- and difference-frequency generation.

The second-order nonlinear mixing processes described previously are readily applied to both pulsed and CW laser systems. The high peak powers available from pulsed laser systems in particular mean that significant conversions are possible; values of up to 25% are not uncommon. Both second harmonic generation and sum-frequency generation options are widely available from manufacturers of pulsed laser systems as a means of extending the range of the dye laser into the region between 0.4 and 0.25 μm. Pulse powers of many millijoules can be obtained from high-power lasers in this region. Until recently, the region below 0.25 μm proved difficult to reach using traditional nonlinear materials that exhibit significant optical absorption in that region. However, new materials, in particular β-phase barium borate (BBO), have pushed the lower limit to the edge of the vacuum ultraviolet at 190 nm.[62–64]

In frequency shifting into the near- and mid-IR, manufacturers offer difference-frequency generation as an option for the production of radiation from 1 μm to ~4 or 5 μm. Bethune and Luntz have, in fact, demonstrated the feasibility of using multiple difference-frequency generation for the production of radiation covering the entire 1.4–22-μm region.[65,66] However, the output pulse energies are significantly lower than those achievable using stimulated electronic Raman scattering in Cs vapor.[55]

The application of second-order nonlinear processes to CW lasers is less routine than with pulsed systems. The low powers of CW lasers mean that conversion efficiencies are limited. For second harmonic generation, the placement of the nonlinear medium within the laser cavity is favored, as the higher intensity of radiation circulating within the cavity enhances the generation of the frequency-doubled light. Output powers of a few tens to a few hundreds of milliwatts are achievable using this approach using CW dye lasers pumped by large-frame ion lasers. However, such manufacturers' specifications for intracavity doubling as exist are generally for single-frequency, narrow-band operation. Specifications for operation of a broadband ring or standing wave dye laser are not so readily obtained. Both sum-frequency generation[67] and difference-frequency generation[54] have been demonstrated for CW dye lasers. However, these must both be performed outside the dye laser cavity, and hence output powers are typically only a few hundreds of microwatts. Moreover, it is theoretically possible to enhance both sum- and difference-frequency generation by constructing a second cavity around the nonlinear medium,[68] although clearly this presents practical difficulties if tunable sum- or difference-frequency light is required.

In extending the useful range of tunable lasers in the 0.35–1-μm region into the vacuum ultraviolet and beyond, it is generally necessary to employ third- and higher-order nonlinear effects. Symmetry considerations show that even simple rare gas or metal atoms in the gas phase will exhibit third-order nonlinear effects,[50,51,69] and much of the current work in this area is focused on these systems. The tuning ranges of such third-order mixing systems are, unfortunately, limited. However, initial studies using small gas-phase molecules such as ethyne[70] show that their use holds some promise for the future.

Finally, we should mention a device that represents one of the most active areas of research in the generation of coherent radiation from solid-state materials, the *optical parametric oscillator* (OPO).[71,72] When a medium with a large nonlinear susceptibility interacts with a strong pump light source, a single pump photon can be split into two other photons, the *signal* and *idler*, which satisfy the energy conservation requirement

$$\hbar\omega_p = \hbar\omega_s + \hbar\omega_i \tag{2.10}$$

where p is the pump, s is the signal, and i is the idler. In essence, the process is the reverse of sum-frequency generation and is constrained by the same phase-matching requirements:

$$\mathbf{k}_p = \mathbf{k}_i + \mathbf{k}_s \tag{2.11}$$

In construction, the OPO somewhat resembles a tunable dye laser. The nonlinear medium is contained within a cavity that can be resonant to either the signal or idler wavelengths or both. The bandwidth of the output, determined by the dispersion of the medium and by the pump power, is typically in the range 0.1–5 cm^{-1}, although the addition of an etalon can lead to a reduction in bandwidth. Tuning of the OPO is achieved simply by rotation of the nonlinear crystal with respect to the pump beam.

Although low-loss cavities do allow the operation of CW OPOs,[73] most commonly, high-power pulsed systems are employed as pump sources. A variety of nonlinear media have been employed to produce both near-IR and visible radiation. In the near-IR, lithium iodate ($LiIO_3$)[74] and lithium niobate ($LiNbO_3$)[75] have proved especially successful, providing tunable radiation in the 1.3–1.7- and 0.55–4-μm regions, respectively. Visible output has also been achieved from a urea OPO over the range 0.5–0.9 μm using the 355-nm third harmonic of a pulsed Nd–YAG laser as the pump source.[76] The use of newer nonlinear materials, especially BBO, promises the possibility of tunable coherent radiation covering the entire 0.4–2-μm region from UV-pumped OPOs and may thus represent a viable alternative to solid-state vibronic lasers in the future.

2.2.5. The Free-Electron Laser[77,78]

The classical Maxwell equations for an electron under acceleration indicate a loss of energy and deceleration by the electron through the emission of a continuum of electromagnetic radiation. When a bunch of electrons is contained within a storage ring and continually pumped with energy to raise its velocity to relativistic speeds and maintain it there, the radiation that is emitted, *synchrotron radiation*, represents a spectroscopically useful source. Such synchrotron sources have a number of properties tht make them ideal for spectroscopy. They are bright and widely tunable, with output radiation ranging in wavelength from the far-IR to the deep X-ray region. After filtering out the wavelength regions not required, by the use of appropriate blocking materials, wavelength selection is made simply by the use of traditional monochromatization techniques. The tunability in the X-ray region has proved a particular boon to spectroscopists. The radiation emitted from the synchrotron source is also highly collimated, with beam divergencies of around 1 mrad. Moreover, the output radiation is pulsed, with the pulsewidth and repetition rate depending simply on the length of the electron bunch, the dimensions of the storage ring, and the velocity of the electrons. This makes synchrotron sources ideal for some time-resolved measurements. Although in many ways synchrotron radiation sources are comparable to laser sources, there is one notable difference, the scale of the source. While a commercial pulsed Nd–YAG pumped dye laser with nonlinear mixing capabilities may cost on the order of $100,000 and sit on a small optical table, a synchrotron storage ring will cost several millions and fill a moderately sized building. Thus, synchrotron sources are commonly established as multiuser facilities as opposed to a single-user resource.

Given the synchrotron emission from relativistic electrons, it is natural to consider the possibility of constructing a suitable optical cavity around a section of the storage ring, using the relativistic electrons as a gain medium and so constructing a *free-electron laser* (FEL). Detailed descriptions of the physics of FELs and of the problems associated with their operation are given in the excellent review by Leach.[77] Although it may be possible theoretically to construct a FEL that will operate in any region of the electromagnetic spectrum from the millimeter wave to the X-ray region, much of the work on FEL systems appears to have focused on the two extremes with numerous experiments in both the millimeter and FIR regions beyond 10 μm, and in the extended UV region (XUV) below about 100 nm. A few studies, however, have considered the properties of FELs operating in the visible and near-IR. Output powers from FELs operating in the far-IR and XUV regions exceed those currently possible from broadly tunable laser sources in those regions. It is likely that future development of the FEL will continue toward the X-ray region, where coverage by traditional laser sources is poor. Such developments will obviously be assisted by future developments of conventional short-wavelength lasers, since the FEL can be operated in an optical amplifier mode when injection seeded with coherent radiation of the appropriate wavelength.

2.3. Wavelength Calibration and Measurement

With the definition of the base SI unit of time in terms of the cesium atomic clock[79] and the fixing of the velocity of light,[80] the direct measurement of wavelength is no longer of primary metrological importance. Instead, wavelength can now be calculated, using the expression

$$\lambda = \frac{c}{v} \tag{2.12}$$

from measurements of the corresponding optical frequency, v, to a precision limited only by the precision of measurement of the optical frequency and of the defined value of the velocity of light.

As a consequence of this fundamental change, Evenson et al. have proposed a laser frequency synthesis chain, illustrated in Figure 2.14, that would allow optical frequency measurements in a frequency range stretching from that of the cesium atomic clock to that of the 1.52-μm line of the He–Ne laser.[81] Extension of this chain into the visible has been demonstrated by Pollock et al. and is shown in Figure 2.15.[82] The technique employed relies extensively on nonlinear frequency mixing and beat frequency generation in devices such as metal–insulator–metal (MIM) diodes.[83]

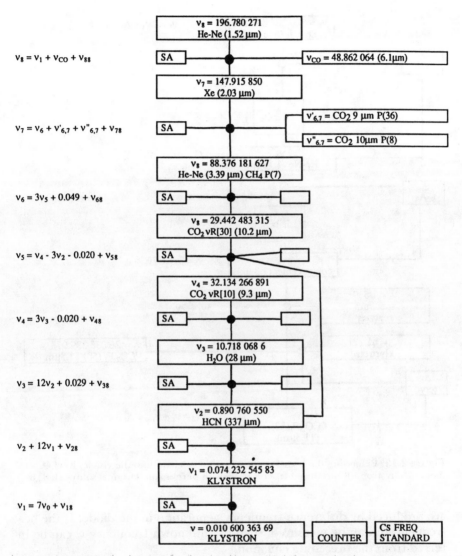

Figure 2.14 Laser frequency synthesis chain from the Cs frequency standard to a He–Ne laser at $\nu = 197$ THz. Key: SA–spectrum analyzer, ●-MIM diode. □-klystron. (Reproduced from Ref. 1 with permission. Original source, Ref. 81.)

Consider two known laser frequencies ν_1 and ν_2 (e.g., from two FIR lasers). If these are allowed to impinge on an MIM diode, the nonlinear response results in generation of harmonics of the known frequencies, $m\nu_1$ and $n\nu_2$. If the unknown frequency ν_x is also allowed to fall on the diode, beat frequencies, ν_b, given by

$$\nu_b = \pm\, n\nu_x \pm m\nu_1 \pm n\nu_2 \tag{2.13}$$

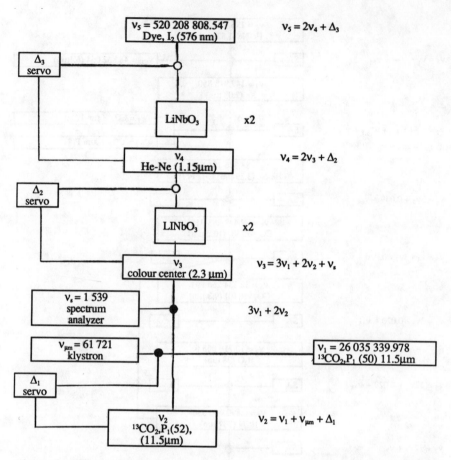

Figure 2.15 Extending the laser frequency synthesis chain into the visible. Key: ●-MIM diode; ○-klystron. (Reproduced from Ref. 95 with permission. Original source, Ref. 82.)

are produced by difference-frequency generation in the diode. If the beat frequency v_b can be measured, then the unknown frequency v_x can be inferred from the preceding equation.

Although this technique is clearly of fundamental importance in the experimental determination of laser frequencies, the techniques employed are difficult and consequently have application restricted to national fundamental metrology laboratories. However, in the majority of laser laboratories, secondary measurement techniques that can ultimately be traced back to national and international standards are the rule of the day. These techniques can be based on the use of either standard spectra or instrumental methods. We now consider each of these in turn.

2.3.1. Spectroscopic Calibration and Measurement Techniques

The recording of standard spectra provides perhaps the simplest approach to the measurement of laser wavelength. In each of the spectroscopic regions, workers have employed measurements of direct absorption and/or of emission spectra of simple, well-characterized systems that have themselves been compared previously with primary standards. The choice of secondary standard clearly depends on the region of the spectrum under study, but in general the species employed are simple one-, two-, or three-atom species that possess relatively simple spectra. Thus, in the far-IR region, the pure rotational spectra of CO, HCl, and HF[84] are frequently employed as a standard. Similarly, in the mid-IR, the readily resolved rovibrational bands of CO, HF, CO_2, and their isotopomers are frequently employed as secondary wavelength standards.

In the visible, the situation is more complicated. In emission studies the dispersed emission of iron or uranium atomic hollow-cathode lamps is frequently employed, while in absorption the rovibronic spectral atlases of iodine[85] and tellurium are widely employed. The spectrum of nitrogen dioxide (NO_2) also provides a useful secondary wavelength standard for such studies.[86] Although these atlases provide detailed Doppler-limited spectra of the aforementioned molecules, many experiments are in fact performed at a resolution greater than that of the Doppler limit. Consequently, it is difficult to compare coarsely structured spectra recorded at degraded resolution with the finely structured, Doppler-limited spectra. An ideal alternative would be to record the widely spaced, narrow features of an atomic spectrum. *Optogalvanic spectroscopy* (OGS) provides such a tool and is now widely employed as a wavelength calibration technique in laser spectroscopy.

In a hollow-cathode discharge of, for example, iron or uranium, a distribution of quantum states lying near the ionization threshold of the relevant atomic species is produced. Photoexcitation of these excited atomic species with radiation of the appropriate wavelength can lead to subsequent photoionization of the excited state. This photoinduced current gives rise to the optogalvanic signal. By tuning the light source a variety of excited states can be ionized, providing a number of narrow spectral lines, as illustrated by Figure 2.16. These are ideal for the wavelength calibration of either CW or pulsed lasers. The apparatus employed in OGS is relatively simple, as illustrated by Figure 2.17. A number of atomic species have been employed as calibrants using OGS, but by far the most commonly employed are neon and argon.[87–92] These provide a variety of lines in the visible, ultraviolet, and infrared and are commonly used as buffer gases in commercial atomic absorption spectrometer lamps. However, such lines are often widely spread, making calibration in many regions of the visible and ultraviolet difficult. Ideally, one would choose to use a heavier atom pos-

48

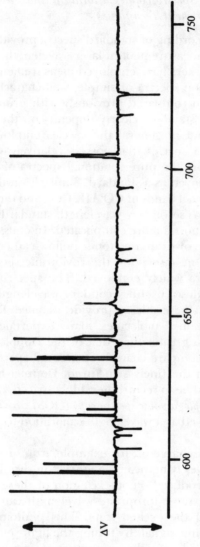

Figure 2.16 Complete first-order optogalvanic effect spectrum of a neon discharge in the visible region. Vertical scale is relative voltage deviation from the quiescent value. Horizontal scale is the laser wavelength in nanonmeters. (Reproduced from Ref. 87 with permission.)

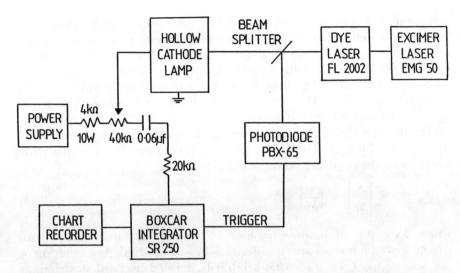

Figure 2.17 Experimental arrangement for the calibration of a commercial pulsed dye laser using the optogalvanic effect. (Reproduced from Ref. 88 with permission.)

sessing many more excited atomic states to increase the number of observable optogalvanic transitions. Uranium is ideal for this purpose and is already widely accepted as a calibrant in emission studies. One problem remains in the use of uranium, however. Commercial uranium hollow-cathode lamps are best operated at low excitation currents. However, at such currents the number of uranium atoms in the short, hollow-cathode discharge region is too low to obtain significant optogalvanic signals. A longer discharge region and much higher discharge currents are required, although these can considerably shorten the lifetime of the lamp. Both these design aims have been achieved in the lamp design of Dovichi et al.,[93] which has been employed successfully in calibrating fundamental and frequency-doubled pulsed dye lasers operating in the visible and near ultraviolet.

2.3.2. *Instrumental Measurement Techniques*[94,95]

The simplest instrument to measure laser wavelengths is the traditional prism or grating monochromator. While widely used in laser laboratories, such devices have a number of drawbacks and are usually limited to applications in which wavelength needs to be determined only to within a few tenths of a nanometer, although determination to within a few hundredths of a nanometer can be achieved using larger instruments. The major drawbacks of monochromators are threefold. Perhaps most important, the high average powers of CW lasers and, more seriously, the high peak powers of pulsed lasers can lead to optical damage within such instruments. Diffraction gratings in particular can be readily destroyed by high-energy pulsed laser light. This problem and a second associated with incomplete illumination of the prism or grating by a narrowly collimated source, leading to wavelength inaccuracy in the device, are simply solved by the use of a ground glass or silica plate in front of the entrance slit of the monochromator. The diffuse, scattered radiation from such a plate can then be imaged onto the entrance slit to illuminate the dispersive optic fully. The intensity of the light reaching the optic is also significantly reduced by this operation.

The third and most obvious problem is that monochromators are limited in resolution for a particular set of physical dimensions. For any dispersing instrument, the resolving power R is given by

$$R = \left|\frac{\lambda}{\Delta\lambda}\right| = \left|\frac{v}{\Delta v}\right| \tag{2.14}$$

where $\Delta\lambda$ is the minimum separation of the central wavelengths of two closely spaced lines that are considered to be just resolved. The criterion of *just resolved* has been somewhat arbitrarily defined by Lord Rayleigh, as illustrated in Figure 2.18. Lines are considered to be resolved provided the central dip between the maxima drops to $8/\pi^2$ of the maximum intensity I_{max}. This *Rayleigh criterion* has become widely accepted as a working definition for resolved spectral features and can be seen to represent a limit on the accuracy with which laser wavelength may be determined using a dispersive instrument.

The actual resolving power of a particular instrument depends on the dispersive element employed. For prism-based devices, it can be shown that

$$R = g\frac{dn}{d\lambda} \tag{2.15}$$

where g is the baselength of the prism and $dn/d\lambda$ is the dispersion of the prism material. Clearly, to obtain higher resolving power in such a situation

Figure 2.18 Rayleigh's criterion for the resolution of two nearly overlapping lines. (Reproduced from Ref. 1 with permission.)

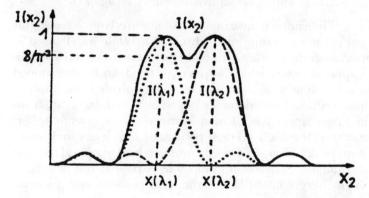

the baselength of the prism can be increased and/or the dispersion of the material increased. The latter occurs near the absorption edge of the material; thus, quartz is ideal in the 180–350-nm region, while glass is favored to the red of this region. Although the dispersion of quartz is greatest in the range 150–180 nm, significant absorption occurs in this region as well, which is further enhanced if large prisms are employed. This absorption can, of course, lead to optical damage when measurements on pulsed lasers are attempted. The use of larger prisms can also mean considerable expense. Consequently, the use of large prism-based instruments is not favored for the measurement of laser wavelength. In contrast for instruments based on diffraction gratings, the resolving power is given simply by

$$R = mN \tag{2.16}$$

where m is the order of the diffraction and N is the total number of grooves illuminated in the grating. Thus, increasing m and/or N will lead to an increase in resolving power. However, there are practical limits to the density of grooves that can be ruled on a grating, and there are also limits on the physical size set by construction costs. Thus, while monochromators based on diffraction gratings may offer significant improvements in resolving power over comparable prism-based instruments, the other factors discussed earlier limit their use in measuring laser wavelengths to localizing the wavelength to within a few hundredths of a nanometer. A more precise determination of the laser wavelength can then be made using interferometric methods.

There are essentially three commonly used interferometric methods of laser wavelength measurement, although a fourth based on a rotating parallele-piped interferometer has also been described in the literature.[96] By far the most widely employed method, both in the UV/visible and in the near- and mid-IR, is based on a *Michelson interferometer,* as illustrated in Figure 2.19. Light from the laser is split by a beamsplitter between two nearly equal optical paths, one with a fixed mirror or corner cube and one with a mirror or corner cube that is driven by a motorized translation stage. When the beams are recombined on a simple photodetector, interference

Figure 2.19 Schematic representation of a scanning Michelson wavemeter. (Reproduced from Ref. 94 with permission.)

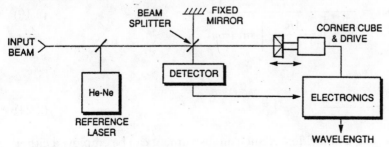

produces a sinusoidally varying signal with a peak for each half-integral wavelength traversed by the moving mirror or corner cube. This signal is described by

$$I(t) = I_0 \left\{ \frac{1 - \sin(2\pi vt/\lambda)}{2} \right\} \tag{2.17}$$

where v is the velocity of the moving mirror. The modulation frequency of the signal is thus

$$f = \frac{2\pi v}{\lambda} \tag{2.18}$$

If the velocity of the moving component is fixed for both a reference laser wavelength and the unknown wavelength, then the unknown is obtained simply from the ratio of the two modulation frequencies. A number of such Michelson-based wavelength measurement instruments (*wavemeters* or *lambdameters*) have been described in the literature for use both in the visible[97-101] and in the infrared,[102,103] and the technique has shown itself to be sufficiently mature that commercial instruments are available. However, since in such instruments the reference laser of choice is the He–Ne laser, wavelength accuracy is typically restricted to around 1 part of 10^6 as a consequence of the gain profile width of standard commercial He–Ne lasers. To achieve any significant improvement in the accuracy of the wavemeter, a frequency-stabilized He–Ne must be employed. Using such an approach, accuracies of around 1 in 10^8 have been achieved in visible wavelength measurements.[104,105]

Although ideal for CW laser wavelength measurement, the Michelson-based wavemeter is impractical for use with pulsed laser systems, and alternative methods must be found. These basically involve using either Fabry–Perot or Fizeau interferometers, as reviewed by Reiser.[106] The basic optics of the *Fabry–Perot interferometer* has been well described by Demtröder.[1] Basically, the device consists of two plane-parallel, loss-free surfaces separated by an air space of width d. If the reflectivity of the plate surfaces is R, then the transmitted intensity is given as a function of the order number m by

$$I(m) = \frac{I_0}{1 + \left[\dfrac{4R}{(1 - R)^2} \right] \sin^2(\pi m)} \tag{2.19}$$

where

$$m = \frac{2d \cos \Theta}{\lambda} \tag{2.24}$$

and Θ is the angle of incidence. Such an instrument can be employed either as a static or scanning device. In the latter, the interferometer is illuminated with a parallel beam of light at fixed angle of incidence and the plate spacing is varied. The variation of the transmitted intensity is coupled to the variation of the in ferometer spacing and can be displayed on a synchronized oscilloscope. Such a scanning Fabry–Perot interferometer is ideal for mode structure determination in CW lasers but is of little practical application for pulsed laser wavelength determination.

The static Fabry–Perot interferometer does, however, provide a useful means of obtaining wavelength information from pulsed sources. In such devices, the plate spacing is held constant and the interferometer is illuminated by either a diverging or a converging light beam. This produces a wide range of angles of incidence and it is found that maxima occur for those angles that give integer m, meeting the condition expressed by equation (2.20). If the resulting fringes are focused onto an array detector such as a charge-coupled device (CCD) camera, then for a lens of focal length f, the pth ring has a diameter D, given by

$$D = 2f \left\{ 2 - \frac{p\lambda}{d} \right\}^{1/2} \tag{2.21}$$

By sampling the fringe pattern with the CCD, the laser wavelength can be estimated from the measured fringe diameter and the known parameters, f and d, of the instrument. Comparison of an unknown laser wavelength against a reference laser wavelength is thus easy, provided that the two wavelengths share the same integral component of m. This is obviously true if the two wavelengths are very close to each other [i.e., within what is known as the *free spectral range* (FSR = λ^2/d) of the interferometer], but it is not obviously so for distinctly different wavelengths. Clearly, the smaller the value of d, the larger the FSR of the interferometer and consequently the wider the range of wavelengths that can be determined, but at reduced precision. Thus, the approach commonly employed is to use more than one Fabry–Perot interferometer, as in Figure 2.20, to provide the necessary accuracy and precision in wavelength measurement. A narrow etalon with

Figure 2.20 Schematic representation of a multiple Fabry–Perot wavemeter. (Reproduced from Ref. 94 with permission.)

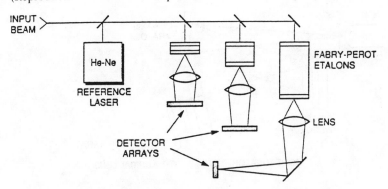

large FSR or a monochromator provide an initial wavelength estimate, and further etalons of decreasing FSR are used to attain the required precision in the measurement. A number of such instruments that have been described for the visible and UV regions provide accuracies for pulsed wavelength determination only slightly poorer than that obtained using Michelson interferometers for CW lasers.[107-109]

The final type of interferometer in common use for laser wavelength measurement, particularly that of pulsed laser, is the *Fizeau interferometer*. The basic principle of applying the Fizeau interferometer to pulsed laser wavelength measurement has been described by Synder.[110,111] The device, as illustrated in Figure 2.21, is based on two uncoated plates with a spacing *d* of about 1 mm and a wedge angle *a* of around 1 mrad. When illuminated by a collimated beam of light, interference within the wedge gives rise to a set of straight, evenly spaced fringes that are ideal for detection by an array photodetector. The intensity of the fringe pattern is found to vary with the distance along the wedge as

$$I(x) = \frac{I_0}{2}\left\{1 + \cos\left[2\pi\left(fx + \frac{m}{2}\right)\right]\right\} \tag{2.22}$$

where *f* is the spatial frequency of the pattern. Snyder has shown that an initial estimate of the laser wavelength can be obtained from this spatial frequency.

$$f = \frac{2\tan\alpha}{\lambda} \tag{2.23}$$

This can then be used as a basis for a calculation of the phase of the fringe pattern from which the accurate wavelength can be determined.[112] A number of Fizeau wedge-based wavemeters have been constructed since the original work of Snyder[113-118] and, like the multiple Fabry–Perot waveme-

Figure 2.21 Schematic representation of a Fizeau wavemeter. (Reproduced from Ref. 94 with permission.)

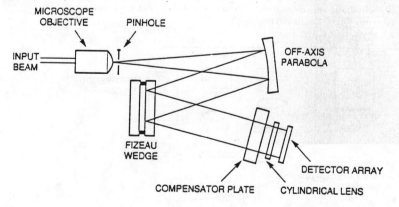

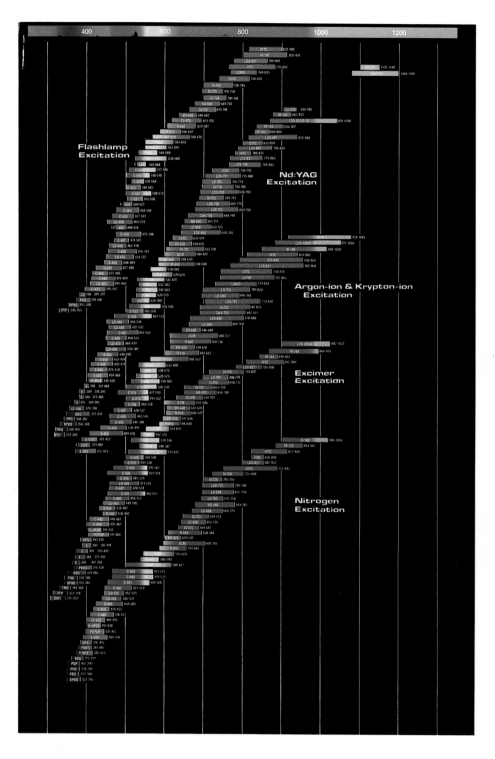

Plate 1 Tuning ranges available from commercial dyes by a variety of pumping schemes. (Reproduced by the permission of Exciton, Inc.)

ter, have reached commercial fruition. Measured wavelength accuracy is comparable to both the Michelson and multiple Fabry–Perot devices at around 1 part in 10^6, although the Fizeau instrument is much simpler in its construction.

2.4. Laser Power, Energy, and Beam Profile Measurement

Although measurement of wavelength is, of course, of primary importance in laser spectroscopy as in traditional spectroscopic studies, for most laser-based methods the measurement of the laser power or pulse energy plays an equally important role. This is especially true where multiphoton absorption plays a significant part in producing the observed spectroscopic signal, be it emission or ionization in nature. A high-order nonlinear dependence on the laser power generally reflects either the multiphoton nature of the process or possibly a feature of the excitation geometry.[119] In this concluding section it is therefore appropriate to consider some of the techniques available for the measurement of laser power and pulse energy, and how that energy is distributed over the profile of the laser beam.

2.4.1. Some Basic Definitions

In considering the output powers of laser systems, we must again consider the basic division of systems into those with pulsed and CW operation. Consider first a CW laser system with a power P measured in watts. Assuming for simplicity a uniform distribution of intensity across the beam, we can define for such a system as *irradiance*, I measured in W m^{-2}, as

$$I = \frac{4P}{\pi d^2} \tag{2.24}$$

where d is the beam diameter. In contrast, for pulsed lasers we must consider two definitions of laser power. The first of these is the *average power*, P_{av}, given by

$$P_{av} = E_{pulse} R \tag{2.25}$$

where E_{pulse} is the pulse energy and R is the repetition rate of the laser. For many pulsed laser systems, we can also define a *peak power*, P_{peak}, as

$$P_{peak} = \frac{E_{pulse}}{\tau_{pulse}} \tag{2.26}$$

where τ_{pulse} is the pulse duration. Note that since laser power is not constant throughout a pulse, the *instantaneous* power may briefly exceed P_{peak} as given in equation (2.26). From the preceding definitions of power, we can define an irradiance as given by equation (2.24). Furthermore, we can define the *photon flux, I_{ph}*, in quanta $m^{-2} s^{-1}$, for both types of laser source as

$$I_{ph} = \frac{I}{h\nu} = \frac{I\lambda}{hc} \tag{2.27}$$

where ν is the frequency and λ the wavelength of the laser radiation.

To end this brief section, it is worth considering some of the figures for irradiance that can be obtained from lasers, contrasting these with the solar irradiance of around 1.5 kW m^{-2}. For a small He–Ne laser with an output of around 5 mW and 1-mm beam diameter, the corresponding irradiance is around 5 kW m^{-2}. Clearly, even small lasers are brighter sources than sunlight and consequently must be treated with respect. The irradiance figures rapidly rise when we consider large frame ion lasers and pulsed lasers to in excess of 10^9 W m^{-2} for unfocused beams. Focusing can, of course, increase these figures enormously. For this reason great care should be taken over the choice and cleanliness of materials for laser optics, since these high irradiances can lead to catastrophic optical damage, even to apparently transparent materials.

2.4.2. Measurement of Laser Power

The techniques available for the measurement of laser power and/or pulse energy can be broadly classified into two group. The first are those methods based on the direct detection of the quanta of electromagnetic radiation, *photon detectors*. The second group comprises those methods based on measurement of the thermal effects of the interaction of laser radiation with matter, *thermal detectors*. We will consider each of these groups briefly in turn.

First, let us consider the photon or quantum detectors. We can divide this group of detectors into three subgroups that specify the mode of operation of the detector. These are *photoemissive, photovoltaic,* and *photoconductive* devices. Photoemissive detectors, as well illustrated by *photomultiplier tube (PMT) detectors,* represent a practical application of the photoelectron effect. When light falls on materials that have low value for their work function, such as alkali metal alloys, then the solid can emit electrons, one electron for each incident photon with sufficient energy ($h\nu$). The electrons so emitted from the *photocathode* of such a detector are then accelerated down the PMT toward the collecting *anode*. En route, the photoelectrons collide with the *dynodes* within the PMT, each collision leading to the production of a cascade of *secondary electrons,*[120] which themselves are accelerated down the tube and can collide with other dynodes. Consequently, at the anode, for a single photon incident on the primary photocathode, a cascade of 10^6–10^8 electrons may be produced, representing a *gain* or *am-*

plification of 10^6–10^8. It is in this high gain that the extreme sensitivity of the PMT lies. Light levels of as low as 10^{-12} W (CW) can readily be detected and PMTs can exhibit linear response to powers of around 10^{-7} W (CW).[121] Above this, however, severe nonlinearity arises and damage to the photocathode is distinctly possible. The design of the standard photomultiplier is also such that measurement of the temporal characteristics of laser pulses is possible on time scales longer than 500 ps. For all their advantages at low power levels, PMTs do have some limitations. The first is the relatively low *quantum efficiency* of most photocathode materials, typically around 20% to 30%. This figure represents the efficiency with which photons are converted to electrons and can be significantly higher in some of the devices to be discussed later. Furthermore, the optimal spectral response of PMTs lies in the region below 600 nm, and extends into the near-UV and ultimately the VUV, again in contrast to semiconductor devices. This clearly restricts the application of PMTs to the detection of radiation from sources emitting in that region.

Photovoltaic and photoconductive detectors represent two different applications of the same basic detector type, the semiconductor photodiode. When light of the appropriate wavelength is incident on a semiconductor photodiode, promotion of an electron to the conduction band of the semiconductor occurs. Direct measurement of this *photocurrent* generated by the incident radiation in the absence of a bias potential is the basis of operation of photovoltaic devices. If a small reverse bias potential is applied across the device, photoexcitation leads to a change in conductivity of the device, providing the basis of photoconductive operation. This latter mode is favored, as it provides for a larger linear sensitivity range of the photodiode. Simple diodes typically show linear response of the power range from 10^{-7} to 10^{-3} W and are clearly less sensitive than photoemissive devices. However, they are considerably more compact. Some photodiode devices, known as *avalanche photodiodes*, do offer somewhat greater sensitivity (10^{-10}–10^{-6} W) but still fall somewhat short of PMTs.[121] The real advantage of the photodiode devices lies in their spectral response. Simple silicon-based devices have peak quantum efficiencies of around 60% to 70% in the range 800–1,000 nm, a region where PMTs are difficult to operate. Furthermore, devices based on other semiconducting materials have extended the range of photoconductive detectors into the near-IR (germanium 0.8–1.6 μm; indium arsenide, 1.0–3.6 μm; indium antimonide, 1.0–5.5 μm) and the mid-IR (mercury cadmium telluride, 2–22 μm; germanium–gold, 2–10 μm; germanium–mercury, 2–15 μm; germanium–copper, 2–30 μm; germanium–zinc, 4.5–42 μm) with comparable sensitivity to that of visible detectors. Further developments in the construction of silicon and other semiconductor-based photodetectors, particularly IR detectors, can be expected with improvements in materials-handling techniques and semiconductor growth technology. The response of silicon photodiodes to UV and shorter-wavelength radiation is, however, poor.

Consequently, photoemissive detectors are favorable for these regions. The simple trick of using a *quantum converter,* a solution of a UV-absorbing dye that fluoresces at longer wavelengths adsorbed on a solid support, can nonetheless extend the useful wavelength range of silicon photodiodes into the near-UV.

Thermal detectors comprise the second general group of detectors used in power measurement. As with the photon detectors, we can further divide this class of detectors into *calorimetric* and *pyroelectric* detectors. The simplest but least sensitive devices are calorimetric. At their very simplest, these consist of an absorber, be it a *volume absorber,* such as a glass filter or metal block, or a *surface absorber,* such as a graphite coating, thermally connected to a thermistor or platinum resistance thermometer that is part of a balanced bridge circuit. Measurement of the temperature change associated with the absorption of radiation can then be used to infer the laser power through knowledge of the optical and thermal properties of the absorber. For measurement of CW powers in the range 10^{-5}–10^2 W, surface absorbers are ideal, as they have flat spectral response over the entire 0.2–35-μm range. However, significant damage can occur to surface absorbers when measurements are made of pulsed lasers with peak irradiances in excess of 10^{10} W m^{-2}. Volume absorbers, in contrast, can sustain peak irradiances of 10^{12}–10^{15} W m^{-2} over a similar range of wavelengths. It should be noted that even volume absorbers become effectively surface absorbers when attempts are made to measure short-wavelength UV radiation. Consequently, their damage thresholds are reduced as the wavelength is reduced.

Calorimeters are ideal devices where the slow response of the detector can average out short-term variations in laser output. Consequently, they find widespread use in routine *tweaking up* of laser systems. For faster measurements, particularly true shot-to-shot energy measurements, faster devices must be employed, either the quantum detectors discussed earlier or the fast pyroelectric devices now available. These devices employ a pyroelectric crystal, typically lithium tantalate ($LiTaO_3$), which can directly convert the heating effect of a laser pulse into a fast electrical signal. As with the calorimeters, the spectral response is essentially flat from 0.2 to beyond 50 μm as shown in Figure 2.22, although the use of window materials obviously limits this. Energies in the range 10^{-9}–10^{-3} J can be readily measured using such devices, although the upper limit is determined simply by the damage threshold of the material and the laser pulse characteristics. In many respects, pyroelectric detectors complement the quantum detectors described earlier by allowing for true shot-to-shot pulse measurement at energies that would saturate a photodiode or destroy a PMT.

Ideally, any laboratory operating laser systems should be well equipped with well-calibrated power and energy meters. In particular with pulsed laser systems, it is suggested that users employ both averaging devices, such as calorimeters or thermopiles, and shot-to-shot devices, such as pyroelec-

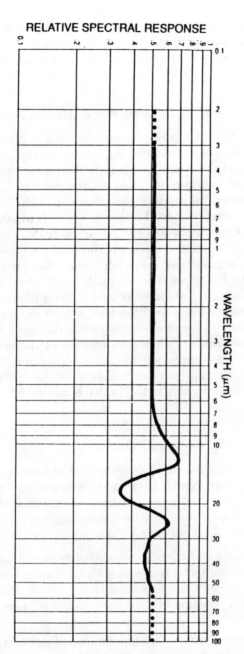

Figure 2.22 Wavelength response of a typical commercial pyroelectric joulemeter. (Reproduced with the permission of Molectron Detector, Inc.)

tric and photodiode-based detectors. The use of PMTs is not generally recommended for measurement of laser power, except in those situations that demand the extreme sensitivity of these devices, for example in XUV and VUV generation.

2.4.3. *Laser Beam Profile Measurements*[122-124]

With the growing use of nonlinear optical techniques for laser frequency extension, a developing interest in the use of fiber-optic delivery systems and the continual trend to obtain the minimal focal spot dimensions in numerous laser applications, the past few years have seen a growing trend toward the routine measurement of the energy or power distribution across a laser beam. The key question is whether a given laser is truly operating single mode in TEM_{00} or whether there are contributions to the energy/power distribution from higher-order modes.

Initially, these measurements were performed using simple devices that employed a knife edge scanned across the beam.[125] However, with the development of array detectors (initially using silicon photodiode array and more recently using pyroelectric element arrays) equipped with frame stores and microcomputer processing systems, the last 2 or 3 years have been the arrival of commercial devices capable of measuring single-shot pulse profiles.[123,125] By employing modeling techniques, estimates of the mode purity of the beam profile can then be performed by microcomputer. Further expansion of this area in the future is inevitable.

2.5. Conclusions

With the rapid development in laser technology and in the support technology that all laser spectroscopists must employ, it is inevitable that some detail of what has been discussed earlier will rapidly become dated. However, it is worthwhile noting a few general points of advice to the prospective laser user.

1. Laser systems cost a considerable amount of money. Always choose carefully. Newly evolving technologies may look exciting, but beware the pitfalls of purchasing a "first-generation" laser system, such as the lack of an established user community and commonly a reliance on a single supplier.

2. Always consider the future development of a laser-based program and purchase your laser system accordingly. Thoroughly research the market. Talk to users of existing systems and do not be afraid to approach manufacturers with queries.

3. The components that go into the construction of a laser do have limited lifetimes. Spares are essential and users should carry a limited stock of

some components. However, some components, such as thyratrons, may have a limited shelf life. A balance between reactive replacement of failed components that may be stock items and planned replacement of aging components that cannot be retained for long periods on the shelf is required.

4. Always budget for the running costs of the laser. Consider discharge lamps a consumable item. Consider the costs of rare gases and halogen gas mixes for exciplex lasers, and of course the necessary plumbing for the safe handling of these toxic gas mixtures. Currently, the costs of xenon, in particular, are high. Consider the possibility of recycling. For dye lasers, the cost of dyes and solvents can be considerable. Again, is there sufficient knowledge within your institution to consider recovery?

5. It is no use having the most modern laser possible if you cannot effectively measure its power/pulse energy or determine the output wavelength. Consideration of these support facilities at the budgeting stage can be just as important as the laser.

REFERENCES

1. W. Demtröder, *Laser Spectroscopy: Basic Concepts and Instrumentation* (Springer, Berlin, 1982).

2. M. Inguscio, G. Moruzzi, K. M. Evenson, and D. A. Jennings, *J. Appl. Phys.* **60,** R162 (1986).

3. C. T. Gross, J. Kiess, A. Mayer, and F. Keilmann, *IEEE J. Quantum Electron.* **QE-23,** 377 (1987).

4. R. Wessel, T. Theiler, and F. Keilmann, *IEEE J. Quantum Electron.* **QE-23,** 385 (1987).

5. K. M. Evenson, D. A. Jennings, and M. D. Vanek, in *Frontiers of Laser Spectroscopy of Gases*, A. C. P. Alves, J. M. Brown, and J. M. Hollas, eds. (Kluwer Academic, Dordrecht, 1988), p. 43.

6. M. Fourrier and A. Kreisler, *Appl. Phys. B* **B41,** 57 (1986).

7. J. O. Henningsen, *Int. J. Infrared Millim. Waves* **7,** 1605 (1986).

8. G. Carelli, N. Ioli, A. Moretti, D. Pereira, and F. Strumia, *Appl. Phys. B* **B44,** 111 (1987).

9. I. Mukhopadhyay, R. M. Lees, W. Lewis-Bevan, J. W. C. Johns, F. Strumia, and G. Moruzzi, *Int. J. Infrared Millim. Waves* **8,** 1483 (1987).

10. J. A. Golby, N. R. Cross, and D. J. E. Knight, *Int. J. Infrared Millim. Waves* **7,** 1309 (1986).

11. J. R. Izatt, *Proc. SPIE* **666,** 15 (1986).

12. J. C. Deorche, E. K. Benichou, G. Guelachvili, and J. Demaison, *Int. J. Infrared Millim. Waves* **7,** 1653 (1986).

13. P. Cheng-Zhi, Q. Yi, and H. Shao-Ping, *Rev. Roum. Phys.* **31,** 893 (1986).

14. H. Ohkuma and S. Yamaguchi, *Jpn. J. Appl. Phys. Part I* **25,** 1956 (1986).

15. C. Gastard, M. Redon, and M. Fourrier, *Int. J. Infrared Millim. Waves* **8,** 1069 (1987).

16. G. Taubmann, H. Jones, and P. B. Davies, *Appl. Phys. B* **B41,** 179 (1986).

17. W. Rohrbeck, A. Hinz, P. Nelle, M. A. Gondal, and W. Urban, *Appl. Phys. B* **B31,** 139 (1983).

18. A. Hinz, D. Zeitz, W. Bohle, and W. Urban, *Appl. Phys. B* **B36,** 1 (1985).

19. P. Helminger, J. K. Messer, and F. C. DeLucia, *Appl. Phys. Lett.* **42,** 309 (1983).

20. D. D. Bicanic, B. F. J. Zuidberg, and A. Dymanus, *Appl. Phys. Lett.* **32,** 367, (1978).

21. H. R. Fetterman, P. E. Tannenwald, B. J. Clifton, W. D. Fitzgerald, and N. R. Erickson, *Appl. Phys. Lett.* **33,** 151 (1978).

22. K. M. Evenson, D. A. Jennings, and F. R. Petersen, *Appl. Phys. Lett.* **44,** 576 (1984).

23. K. M. Evenson, D. A. Jennings, K. R. Leopold, and L. R. Zink, in *Laser Spectroscopy,* Vol. 7 (Springer, Berlin, 1983).

24. J. X. Lin, R. Rohrbeck, and W. Urban, *Appl. Phys. B* **B26,** 73 (1981).

25. W. Urban, in *Frontiers of Laser Spectroscopy of Gases,* A. C. P. Alves, J. M. Brown, and J. M. Hollas, eds. (Kluwer Academic, Dordrecht 1988), p. 9.

26. W. Lo, *Proc. SPIE* **484,** 144 (1984).

27. P. Mortensen, *Laser Focus World* **26**(7), 67 (1990).

28. G. Litfin, G. Heise, and H. Welling, *Opt. Commun.* **59,** 137 (1986).

29. P. F. Moulton, in *Laser Handbook,* Vol. 5, M. Boss and M. L. Stitch, eds. (North-Holland, Amsterdam, 1985), p. 203.

30. P. F. Moulton, *Proc. SPIE* **622,** 120 (1986).

31. J. Hecht, *Lasers Appl.* **3**(9), 77 (1984).

32. L. Holmes, *Laser Focus/Electro-Opt.* **22**(4), 70 (1986).

33. L. G. DeShazer, *Laser Focus/Electro-Opt.* **23**(2), 54 (1987).

34. T. H. Maiman, *Nature* **187,** 493, (1960).

35. T. H. Maiman, *Br. Commun. Electr.* **7,** 674 (1960).

36. T. Benyattou, R. Moncarge, J. M. Breteau, and F. Auzel, *Crys. Latt. Def. Amorph. Mater.* **15,** 157 (1987).

37. J. M. Breteau, D. Meichenin, and F. Auzel, *Rev. Phys. Appl.* **22,** 1419 (1987).

38. A. M. Fox, A. C. Maciel, and J. F. Ryan, *Opt. Commun.* **59,** 142 (1986).

39. D. Welford and P. F. Moulton, *Opt. Lett.* **13,** 975 (1988).

40. F. P. Schaefer, (ed.) *Dye Lasers,* 3rd ed. (Springer, Berlin, 1990).

41. J. M. Hollas, *Modern Spectroscopy* (Wiley, Chichester, 1987), p. 323.

42. M. J. Frost and I. W. M. Smith, *J. Chem. Soc., Faraday Trans. II* **86,** 1751 (1990).

43. M. J. Frost and I. W. M. Smith, *J. Chem. Soc., Faraday Trans. II* **86,** 1757 (1990).

44. P. Ewart and D. R. Meacher, *Opt. Commun.* **71,** 197 (1989).

45. Z. Bor, *Opt. Commun.* **39,** 383 (1981).

46. Z. Bor, *Opt. Commun.* **29,** 103 (1979).

47. Z. Bor and A. Müller, *IEEE J. Quantum Electron.* **QE-22,** 1524 (1986).

48. J. Hebling, *Opt. Commun.* **64,** 539 (1987).

49. V. I. Gladushchak, S. A. Moshkalev, G. T. Razdobarin, and E. Ya. Shreider, *Sov. Phys.–Tech. Phys.* **31**, 855 (1986).

50. R. Wallenstein, in *Frontiers of Laser Spectroscopy of Gases*, A. C. P. Alves, J. M. Brown, and J. M. Hollas, eds. (Kluwer Academic, Dordrecht, 1988), p. 53.

51. B. P. Stoicheff, in *Frontiers of Laser Spectroscopy of Gases*, A. C. P. Alves, J. M. Brown, and J. M. Hollas, eds. (Kluwer Academic, Dordrecht, 1988), p. 63.

52. T. Oka, in *Frontiers of Laser Spectroscopy of Gases*, A. C. P. Alves, J. M. Brown, and J. M. Hollas, eds. (Kluwer Academic, Dordrecht, 1988), p. 353.

53. H. Schomberg, H. F. Döbele, and B. Rückle, *Appl. Phys. B* **B30**, 131 (1983).

54. K. G. H. Baldwin, J. P. Maranger, D. D. Burgess, and M. C. Gower, *Opt. Commun.* **52**, 351 (1985).

55. A. L. Harris and N. J. Levinos, *Appl. Opt.* **26**, 3996 (1987).

56. D. G. Sarkisyan, A. A. Badalyan, S. O. Sapondzhyan, and G. A. Torosyan, *Sov. J. Quantum Electron.* **16**, 571 (1986).

57. F. Zernike and J. E. Midwinter, *Applied Nonlinear Optics* (Academic, New York, 1973).

58. P. G. Harper and B. S. Wherrett (eds.), *Nonlinear Optics* (Academic, London, 1977).

59. G. D. Boyd, A. Ashkin, J. M. Dzeidzic, and D. A. Kleinman, *Phys. Rev.* **137**, 1305 (1965).

60. D. A. Kleinman, A. Ashkin, and G. D. Boyd, *Phys. Rev.* **145**, 338 (1966).

61. G. D. Boyd and D. A. Kleinman, *J. Appl. Phys.* **39**, 3957 (1968).

62. W. L. Glab and J. P. Hessler, *Appl. Opt.* **26**(16), 3181 (1987).

63. X. Zuyan, D. Daoqun, Z. Tienan, and S. Hansen, *Chin. Phys. Lett.* **5**, 389 (1988).

64. C. Chen, *Laser Focus World* **25**(11), 129 (1989).

65. D. S. Bethune and A. C. Luntz, *Appl. Phys. B* **B40**, 107 (1986).

66. D. S. Bethune, M. D. Williams, and A. C. Luntz, *J. Chem. Phys.* **88**, 3322 (1988).

67. S. Blit, E. G. Weaver, F. B. Dunning, and F. K. Tittel, *Opt. Lett.* **1**, 58 (1977).

68. Z. A. Tagiev, *Opt. Spectrosc.* **57**, 217 (1984).

69. R. Hilbig, G. Hilber, A. Lago, A. Timmerman, and R. Wallenstein, in *Laser Spectroscopy*, Vol. 7, T. W. Hansch and Y. R. Shen, eds. (Springer, Berlin, 1985), p. 181.

70. M. N. R. Ashfold, C. D. Heryet, J. D. Prince, and B. Tutcher, *Chem. Phys. Lett.* **131**, 291 (1986).

71. H. W. Messenger, *Laser Focus World* **26**(10), 30 (1990).

72. C. L. Tang, W. R. Busenberg, T. Ukachi, R. J. Lane, and L. K. Cheng, *Laser Focus World* **26**(10), 107 (1990).

73. R. L. Byer, in *Nonlinear Optics*, P. G. Harper and B. S. Wherrett, eds. (Academic, London, 1977).

74. I. I. Ashmarin, Yu. A. Bykovskii, V. A. Ukraintsev, A. A. Chistyakov, and L. V. Shishonkov, *Sov. J. Quantum Electron.* **14**, 1237 (1984).

75. R. L. Byer, R. L. Herbst, and R. N. Fleming, in *Laser Spectroscopy*, S. Haroche,

J. C. Pebay-Peyroula, T. W. Hansch, and S. E. Harris, eds. (Springer, Berlin, 1975), p. 207.

76. G. C. Catella, J. H. Bohn, and J. R. Luken, *IEEE J. Quantum Electron.* **QE-24,** 1201 (1988).

77. S. Leach, in *Frontiers of Laser Spectroscopy of Gases,* A. C. P. Alves, J. M. Brown, and J. M. Hollas, eds. (Kluwer Academic, Dordrecht 1988), p. 89.

78. A. Gouer, A. Friedman, and A. T. Drobot, *Laser Focus World* **26**(10), 95 (1990).

79. Bureau International des Poids et Mesures, *Le Système International d'Unites (SI),* 5th French and English Edition (BIPM, Sèvres, 1985).

80. E. R. Cohen and B. N. Taylor, *CODATA Bull.* **63,** 1 (1986).

81. K. M. Evenson, D. A. Jennings, F. R. Petersen, and J. S. Wells, in *Laser Spectroscopy,* Vol. 3, J. L. Hall and J. L. Carlsten, eds. (Springer, Berlin, 1977), p. 56.

82. C. R. Pollock, D. A. Jennings, F. R. Petersen, J. S. Wells, R. E. Drullinger, E. C. Beaty, and K. M. Evenson, *Opt. Lett.* **8,** 133 (1983).

83. D. J. E. Knight and P. T. Woods, *J. Phys. E* **9,** 898 (1976).

84. I. G. Nolt, G. D. Lonardo, K. M. Evenson, A. Hinz, D. A. Jennings, K. R. Leopold, M. D. Vanek, J. V. Radostitz, and L. R. Zink, *J. Mol. Spec.* **125,** 274 (1987).

85. S. Gerstenkorn and P. Luc, *Atlas du Spectre d'Absorption de la Molecular d'Iode 14800–20000 cm*$^{-1}$ (Laboratoire AIME-Cotton CNRS, Orsay, 1978).

86. D. K. Hsu, D. L. Monts, and R. N. Zare, *Spectral Atlas of Nitrogen Dioxide 5530Å to 6480Å* (Academic, New York, 1978).

87. J. R. Nestor, *Appl. Opt.* **21,** 4154 (1982).

88. J. A. Dyet, Ph.D. thesis, Heriot-Watt University, 1987.

89. M. H. Begemann and R. J. Saykally, *Opt. Commun.* **40,** 277 (1982).

90. G. A. Bickel and K. K. Innes, *Appl. Opt.* **24,** 3620 (1985).

91. M-C. Su, S. R. Ortiz, and D. L. Monts, *Opt. Commun.* **61,** 257 (1987).

92. B. R. Ready, P. Venkateswarlu, and M. C. George, *Opt. Commun.* **75,** 267 (1990).

93. N. J. Dovichi, D. S. Moore, and R. A. Keller, *Appl. Opt.* **21,** 1468 (1982).

94. L. J. Cotnoir, *Laser Focus World* **25**(4), 109 (1989).

95. J. J. Snyder and T. W. Hansch, *Laser Focus World* **26**(2), 69 (1990).

96. F. Dochio, F. P. Schafer, J. Jethwa, and J. Jasny, *J. Phys. E* **18,** 849 (1985).

97. J. J. L. Mulders, P. A. M. Steeman, A. H. Kemper, and L. W. G. Steenhuysen, *Opt. Laser Technol.* **17,** 193 (1985).

98. B. L. Bukovskii, Yu. F. Tomashevskii, V. V. Arkhipov, and B. A. Kiselev, *Sov. J. Opt. Technol.* **52,** 282 (1985).

99. J. Ishikawa, N. Ito, and K. Tanaka, *Appl. Opt.* **25,** 639 (1986).

100. M. I. Belovolov, E. M. Dianov, A. V. Kuznetsov, V. Kh. Pencheva, V. A. Sychugov, and T. V. Tulaikova, *Soc. Tech. Phys. Lett.* **12,** 544 (1986).

101. M-L. Junttila, B. Stahlberg, E. Kyro, T. Veijola, and J. Kauppinen, *Rev. Sci. Instrum.* **58,** 1180 (1987).

102. W. J. Evans and D. K. Lambert, *Appl. Opt.* **25,** 2867 (1986).

103. H. Lew, N. Marmet, M. D. Marshall, A. R. W. McKellar, and G. W. Nichols, *Appl. Phys. B* **B42**, 5 (1987).

104. P. Juncar and J. Pinard, *Rev. Sci. Instrum.* **53**, 939 (1982).

105. J. Kowalski, R. Neumann, S. Noehte, R. Schwarzwald, H. Suhr, and G. Z. Putlitz, *Opt. Commun.* **53**, 141 (1985).

106. C. Reiser, *Proc. SPIE* **912**, 214 (1988).

107. D. Rees and M. Wells, *J. Phys. E* **19**, 301 (1986).

108. C. J. White, T. L. Boyd, R. B. Miolne, and J. W. Keto, *Proc. SPIE* **912**, 234 (1988).

109. W. S. Gornall, *Wavelength Measurement of Pulsed Lasers,* presented at LEO '89 (Tokyo, 1989).

110. J. J. Snyder, in *Laser Spectroscopy,* Vol. 3, J. L. Hall and J. L. Carlsten, eds. (Springer-Verlag, Berlin, 1977), p. 419.

111. J. J. Snyder, *Proc. SPIE* **288**, 258 (1981).

112. J. J. Snyder, *Appl. Opt.* **19**, 1223 (1980).

113. M. B. Mons, T. J. McIlrath, and J. J. Snyder, *Appl. Opt.* **23**, 3862 (1984).

114. J. L. Gardener, *Appl. Opt.* **24**, 3570 (1985).

115. D. F. Gray, K. A. Smith, and F. B. Dunning, *Appl. Opt.* **25**, 1339 (1986).

116. J. L. Gardener, *Appl. Opt.* **25**, 3799 (1986).

117. C. Reiser and R. B. Lopert, *Appl. Opt.* **27**, 3656 (1988).

118. W. Kedzierski, R. W. Berends, J. B. Atkinson, and L. Krause, *J. Phys. E* **21**, 796 (1988).

119. S. Speiser and J. Jortner, *Chem. Phys. Lett.* **44**, 399 (1976).

120. *Basic Physics and Statistics of Photomultipliers,* Photodetection Information Service Publication R/P 063 (Thorn EMI Electron Tubes, Ruislip, 1984).

121. *A Comparison of the Performance of Photomultiplier Tubes and Silicon Photodiodes,* Photodetection Information Service Publication R/P 073 (Thorn EMI Electron Tubes, Ruislip, 1984).

122. D. M. Hul and A. Stewart, *Lasers and Applications* **4**(10), 75 (1985).

123. J. Fleischer, *Laser Focus World* **25**(4), 131 (1989).

124. T. F. Johnston, Jr., *Laser Focus World* **26**(5), 173 (1990).

125. G. Brost, P. D. Horn, and A. Abtahi, *Appl. Opt.* **24**, 38 (1985).

3
Electronic Photoabsorption Laser Spectroscopy

W. Demtröder

3.1. Introduction

The energies of electronic excited states lie for most atoms and molecules around a few electron volts above the ground state. Their excitation by absorption of one of several photons generally requires visible or ultraviolet photons. Electronic photoabsorption laser spectroscopy is, therefore, mainly based on wavelength-tunable lasers in the visible and ultraviolet region. Compared to infrared spectroscopy, the higher photon energies have the great experimental advantage that sensitive and fast detectors are available, such as photomultipliers and optical multichannel analyzers with image intensifiers, which reach quantum efficiencies of up to 80%. This allows the detection of single photons, emitted from excited atoms or molecules. For analytical applications this also implies in favorable cases the possibility of detecting single atoms.

There are three main goals of electronic photoabsorption spectroscopy. The first is of fundamental importance for basic research: From the line positions and line intensities of absorption spectra the energy levels and transition probabilities of atoms and molecules can be inferred. Although the rotational and vibrational level structure in the electronic ground state can generally be obtained with high accuracy from microwave and infrared spectroscopy, the applications of various techniques based on absorption of visible or UV lasers opens up many new possibilities for the detailed study of molecular structure in electronically excited states. This aspect of molecular structure determination requires a static or stationary spectroscopy.

Electronic excited states play an important role in collision dynamics and in chemical reactions. Many reactions are greatly enhanced by photoexcitation or may even not occur at all in the electronic ground state. The second goal of electronic photoabsorption spectroscopy, which forms a major part of the wide field of photochemistry, is a detailed understanding of the energetics, the dynamics, and the different reaction paths in photo-induced chemical reactions. This goal anticipates a thorough knowledge of the level structure of reactants and reaction products and mostly requires, besides spectral resolution, additional time resolution. Examples are the time-resolved spectroscopy of the transient species formed during chemical reactions or of short-lived isomeric conformations in biological processes.

A third, more applied goal of photoabsorption laser spectroscopy is determination of the abundances of certain species from measured absorption coefficients. This represents the major type of application in analytical chemistry and environmental surveys.

The amount of information that can be obtained from photoabsorption spectroscopy depends not only on the sensitivity of the applied technique but also on the spectral resolution. Here the development of various spectroscopic methods with sub-Doppler resolution has revolutionized classical absorption spectroscopy and allows, in combination with the cooling of molecules in supersonic molecular beams, rotational resolution of electronic transitions, even of complex molecules.

In this chapter several sensitive techniques of photoabsorption laser spectroscopy with pulsed and CW lasers are presented and their advantages and limitations are discussed. The most important methods of sub-Doppler linear and nonlinear absorption spectroscopy are briefly outlined in Section 3.4. Of particular importance for the assignment of molecular spectra are various kinds of optical double-resonance techniques, which are covered in Section 3.6. The chapter concludes with several examples that illustrate different applications of the techniques discussed earlier. Since such a short chapter, of course, cannot completely cover the extensive field of photoabsorption laser spectroscopy, the reader who is interested in more details is referred to reviews or books[1-8] and to original papers given as references at the end of the chapter.

3.2. Direct Absorption Measurements

When a parallel lightbeam of incident power P_0 and light frequency ω passes through a sample of absorbing species the transmitted power

$$P_t = P_0 \, e^{-\alpha(\omega)L} \tag{3.1}$$

is given by Beer's absorption law, where L is the absorption pathlength [m] and

$$\alpha(\omega) = \left[N_i - \left(\frac{g_i}{g_k} \right) N_k \right] \sigma_{ik}(\omega) \qquad (3.2)$$

is the absorption coefficient [m^{-1}], which depends on the absorption cross section $\sigma_{ik}(\omega)$ [m^2] and the number densities N_i, N_k of the lower and upper state of the absorbing transition $|i> \rightarrow |k>$. The numbers $g_i = J_i (J_i + 1)$ are the statistical weights of levels $|i>$ and $|k>$. In most cases, where absorption spectroscopy is performed at temperatures T with $kT << \Delta E = E_k - E_i$, the population N_k is very small compared to N_i and the second term in (3.2) may be neglected.

For small absorptions ($\alpha L << 1$) the exponential function in (3.1) can be expanded and yields the relative absorption:

$$\frac{\Delta P}{P_0} = \frac{P_0 - P_t}{P_0} = \alpha(\omega)L \qquad (3.3)$$

The detection limit for small concentrations N_i is set by the minimum power differences ΔP_m that can still be detected and is, according to (3.3), given by

$$N_i = \frac{\alpha}{\sigma_{ik}} \geq \frac{\Delta P_m}{P_0 L \, \sigma_{ik}} \qquad (3.4)$$

This illustrates that for the detection of small concentrations N_i of absorbing species the minimum detectable absorbed power ΔP_m should be as small as possible and that the incident power P_0, the absorption path length L, and the absorption cross section σ_{ik} should be as large as possible.

3.2.1. Comparison of Laser Absorption Spectroscopy with Classical Techniques

The advantages of absorption spectroscopy with narrowband tunable lasers over classical techniques using broadband continuous radiation sources may be summarized as follows (Figure 3.1):

1. While a broadband source requires wavelength-dispersing elements (spectrographs or interferometers) to obtain absorption spectra, a tunable laser combines the radiation source and wavelength selector, and no extra spectrograph is needed. If the laser bandwidth $\Delta\omega_L$ is small compared to the linewidth $\Delta\omega_a$ of the absorbing transitions, the spectral res-

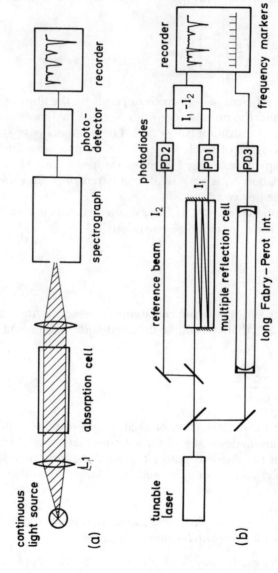

Figure 3.1 Comparison of (a) conventional absorption spectroscopy with a broadband incoherent light source and (b) absorption spectroscopy with a single-mode tunable laser, using multiple reflection cell and frequency markers from a Fabry–Perot interferometer.

olution is solely limited by the true absorption width $\Delta\omega_a$. In classical absorption spectroscopy the resolvable interval $\Delta\omega_{sp}$ of a spectrograph is generally much larger than $\Delta\omega_a$ (Figure 3.2). Besides a worse spectral resolution, this has the additional drawback of lower sensitivity because the relative absorption (3.3) monitored by the detector is reduced to

$$\frac{\Delta P}{P_0} \approx \alpha(\omega_0)\, L\, \frac{\Delta\omega_a}{\Delta\omega_{sp}} \tag{3.5}$$

With typical values of $\Delta\omega_a = 2\pi \times 10^9\,\text{s}^{-1}$ (Doppler-width) and $\Delta\omega_{sp} = 2\pi \times 5 \times 10^{10}\,\text{s}^{-1}$ (this corresponds to $\Delta\lambda = 0.5$ Å at $\lambda = 5{,}000$ Å) this already gives a reduction of the sensitivity by a factor of 50!

2. Because of the good collimation of laser beams multireflection absorption cells can be used where the effective path length L may be increased by up to a factor of 100 without spatial overlap of the beams (which would give rise to undesirable interference effects). According to (3.4) this increases the sensitivity by the same factor.

3. The minimum detectable absorbed power ΔP_m is limited by intensity fluctuations of the source. There are many ways of stabilizing the output power of CW lasers more effectively than that of broadband incoherent sources.

With all these improvements relative absorptions $\Delta P/P_0 = \alpha L \geq 10^{-6}$ may be still detectable. This means with typical values of $L_{\text{eff}} = 10$ m, $\sigma_{ik} = 10^{-20}$ m^2 that a minimum density $N_i \geq 10^{13}$ m^{-3} can be monitored with direct laser absorption spectroscopy, on strongly absorbing transitions.

Figure 3.2 Comparison of spectral resolution and sensitivity between conventional spectroscopy of molecules with absorption linewidths $\Delta\omega_a$, using a spectrometer with bandwidth $\Delta\omega_{sp}$, and laser spectroscopy with laser bandwidth $\Delta\omega_L$.

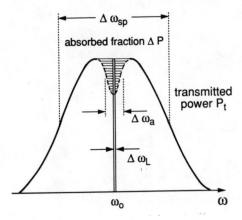

3.2.2. Frequency Modulation Techniques

In microwave spectroscopy it is well known that the sensitivity can be increased if the frequency ω of the microwave source is modulated at a frequency Ω while ω is tuned through the spectral interval of interest. This technique can be directly transferred to laser absorption spectroscopy. If the laser frequency is $\omega_0 + a \sin \Omega t$, where the center frequency ω_0 is tunable, the absorbed laser power can be obtained by expanding (3.3) into a Taylor series around ω_0. This yields

$$\frac{\Delta P(\omega)}{P_0} = L(\alpha(\omega_0) + \left(\frac{d\alpha}{d\omega}\right)_{\omega 0} a \sin \Omega t + \dots) \qquad (3.6)$$

Lock-in detection at the modulation frequency Ω gives the derivative $d\alpha/d\omega$ of the absorption profiles (Figure 3.3.), called the first-derivative spectrum.

This technique has the advantage that the detection can be transferred into a frequency range Ω, where the sum of all noise sources (laser intensity and frequency fluctuations, density fluctuations of the absorbing species, etc.) has a minimum. It is particularly useful in nonlinear spectroscopy, where small narrow resonances are detected against a large Doppler-broadened background (see Section 3.4). With this frequency modulation technique the minimum detectable absorbed power ΔP_m in equation (3.4) can be further reduced, using lock-in detection with a small bandwidth Δf around the modulation frequency Ω.

Figure 3.3 (a) Narrow resonance (Lamb peak) of the neon transition $1s_2-2p_2$ at $\lambda = 588.2$ nm on a Doppler-broadened background compared to (b) the derivative line profile $d\alpha/d\omega$.

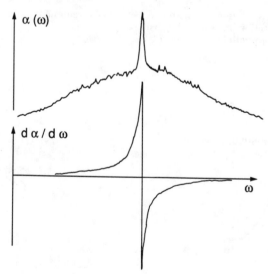

Narrow resonances in the presence of a broad background can be even more effectively filtered by the third derivative technique. Since frequency modulation of the source at a frequency Ω gives many higher harmonics $m\Omega$ ($m = 1, 2, 3, \ldots$) in the Fourier spectrum of the measured absorbed power $\Delta P(\omega)$, the lock-in can be tuned to 3Ω, which yields the spectrum $d^3\alpha/d\omega^3$ (see Ref. 7).

3.2.3. Intracavity Absorption

Placing the absorbing sample inside the resonator of a tunable laser (Figure 3.4) can increase the sensitivity of absorption spectroscopy by several orders of magnitude. This is due to several effects.

First, the laser cavity acts as a multipass cell. If the two mirrors of a linear laser resonator have reflectivities $R_1 = 1$ and $R_2 < 1$, a laser photon travels about $1/(1 - R_2)$ times back and forth between the two mirrors, thus increasing the effective absorption path lengths to $L_{\text{eff}} = L/(1 - R_2)$. With $R_2 = 0.99$ this means $L_{\text{eff}} = 100\,L$. The second effect that increases the sensitivity is based on the fact that the output power P_L of a CW laser operating

Figure 3.4 Intracavity absorption spectroscopy. (a) Either the laser-induced fluorescence (LIF) detector 1 or the laser output $P_L(\omega)$ can be monitored while the laser frequency ω is tuned. (b) Modulation of the laser at a frequency f and lock-in detection at $3f$ allows measurement of the third derivative spectrum.

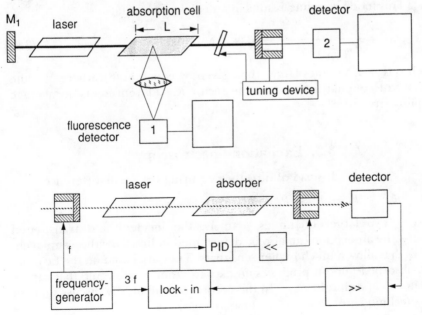

close above threshold is very sensitive to small changes of intracavity losses. The relative change of the output power is[9]:

$$\frac{\Delta P_L}{P_L} = \frac{G_0}{G_0 - \gamma} \frac{\Delta\gamma}{\Delta\gamma + \gamma} \tag{3.7}$$

where G_0 is the unsaturated gain per round trip, γ represents the losses without the absorbing sample, and $\Delta\gamma = 2\alpha L$ represents the additional losses due to absorption by the sample.

Comparing equation (3.7) with the single-pass absorption outside the laser cavity one sees that the sensitivity is enhanced by a factor

$$M = \frac{2G_0}{(G_0 - \gamma)(\gamma + \Delta\gamma)} \approx \frac{2\,G_0/\gamma}{G_0 - \gamma} \quad \text{for } \Delta\gamma << \gamma \tag{3.8}$$

Close above threshold $(G_0 - \gamma) << G_0$ and M may become a very large number. However, the closer the laser is operated above threshold, the larger its intensity fluctuation becomes. Therefore, an optimum value of M has to be chosen that depends on the power stability of the laser and that differs for different types of laser.

With broadband CW or pulsed dye lasers or color-center lasers a third effect can be utilized for intracavity absorption. Because of the coupling between different resonator modes within the homogeneous gain profile of a dye laser the attenuation $\Delta I(\omega_{ik})$ of the laser intensity at the frequency ω_{ik} of an absorbing transition results in an increase of the gain $G(\omega_{ik} \pm \Delta\omega)$ for laser modes oscillating at neighboring frequencies $\omega_{ik} \pm \Delta\omega$. Their intensity $I(\omega_{ik} \pm \Delta\omega)$ therefore increases, which further diminishes the gain at ω_{ik}. This mode-coupling results in an enhancement factor

$$M = \frac{2G_0}{(G_0 - \gamma)(\gamma + \Delta\gamma)} (1 + KN) \tag{3.9}$$

which is $(1 + KN)$ times larger than given by equation (8), where N is the number of coupled modes and the factor $K < 1$ represents an average coupling strength.[10]

3.3. Excitation Spectroscopy

Instead of directly measuring the small difference

$$\Delta P = P_0 - P_T = P_0\, \alpha(\omega)L$$

between two larger quantities—namely, the incident and transmitted power—the absorption can also be determined indirectly with several techniques that allow a much higher sensitivity. They are based on the fact that each absorbed photon produces an excited atom or molecule that can be monitored in different ways. In this section we will briefly discuss some of these techniques of *excitation spectroscopy*.

3.3.1. Fluorescence Detection

If a laser beam with incident power $P_0 = n_L \hbar\omega$ passes through an absorbing sample of length L, the rate of photon absorption associated with a transition $|i> \rightarrow |k$ with absorption cross-section σ_{ik} is given by

$$n_a = N_i \sigma_{ik} n_L L \tag{3.10}$$

where N_i is the number density of absorbing molecules in level $|i>$. The rate at which fluorescence photons are emitted from N_k excited molecules is

$$n_{Fl} = N_k A_k = n_a \eta k$$

where

$$A_k = \sum_m A_{km}$$

is the spontaneous transition probability into all lower levels $|m>$ and where the quantum efficiency defined by $\eta_k = A_k/(A_k + R_k) \leq 1$ is limited by the radiationless depopulation probability R_k (Figure 3.5). If the light-collecting optics covers a fraction $\delta < 1$ of all fluorescence photons emitted into the total solid angle of 4π steradians and the quantum efficiency of the photon detector is η_{ph}, we obtain the rate of measured photoelectrons (i.e., photon counts) as

$$n_{PE} = n_a \eta_k \eta_{ph} \delta = N_i \sigma_{ik} n_L \eta_k \eta_{ph} \delta \cdot L \tag{3.11}$$

An example will illustrate the application of these principles. With $\eta_k = 1$, $\eta_{ph} = 0.2$, $\delta = 0.5$, the ratio of detected photons to absorbed laser photons becomes $n_{PE}/n_a = 0.1$. With cooled photomultipliers a dark pulse rate $n_D < 10 \text{ s}^{-1}$ can be reached. If the minimum signal rate n_{PE} should be 10

Figure 3.5 Schematic diagram of excitation spectroscopy with detection of the laser-induced fluorescence or with ionization of the excited molecules by laser L2.

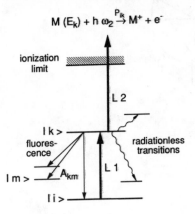

times higher, this implies that at least $n_a \geq 10^3$. With an incident laser power of 100 mW at $\lambda = 500$ nm we have $n_L = 3 \times 10^{18}$ s^{-1}. One can therefore still detect a relative absorption $n_a/n_L \geq 3 \times 10^{-15}$!

The sensitivity increases with η_{ph}, η_k, and δ. The collection efficiency δ can be optimized by special light-collecting systems, shown in Figure 3.6, which are particularly adapted to cases of small light-emitting volumes, such as the crossing volume of a collimated molecular beam with a perpendicular laser beam. If the focus of a parabolic mirror is placed into the center of this crossing volume, the parallel light beam reflected from the mirror can be imaged by a lens onto the cathode of a photomultiplier. An even higher collection efficiency is reached with an elliptical mirror, where the light source is placed at one of the two focal points and one end of an optical fiber bundle is at the other focus. The fiber collects more than 50% of the emitted fluorescence and guides it either directly to a photomultiplier or to the entrance slit of a monochromator. In the latter case the cross section of the fiber bundle end can be formed into a rectangular shape in order to match the geometry of the monochromator slit.

In the case of two-level systems (i.e., atoms, where only two levels are involved in the absorption/emission cycle) the excited atom returns after its excited state lifetime τ into the initial state $|i\rangle$ and can reabsorb another laser photon. With a mean time T of interaction between the atom and the laser field, up to $m = T/(2\tau)$ fluorescence photons (photon bursts) may be emitted from a single atom. For example, with a laser beam diameter of $d = 3$ mm and an atomic velocity of $v = 500$ m/s we obtain $T = 6 \times 10^{-6}$ s. For a lifetime $\tau = 10^{-8}$ s the photon burst has $m = 360$ photons.

Figure 3.6 Two experimental arrangements for fluorescence detection with high collection efficiency.

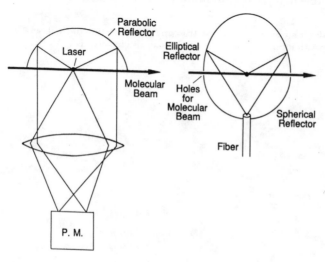

3.3.2. Two-Photon Ionization Spectroscopy

Instead of monitoring the absorption via the fluorescence from excited molecules, one may also photoionize these molecules with a second laser (Figure 3.5) with an ionization probability P_{kI} for the level $|k\rangle$. The ion rate for N_k excited molecules becomes

$$R_{\text{Ion}} = N_k\, P_{kI} = n_a \left(\frac{P_{kI}}{P_{kI} + A_k + R_k} \right)$$

$$= N_i\, n_L\, \sigma_{ik}\, L \left(\frac{P_{kI}}{P_{kI} + A_k + R_k} \right) \quad (3.12)$$

To reach a high ionization efficiency, the ionization probability P_{kI} has to be large compared to the spontaneous decay probability A_k or the collision-induced deactivation probability R_k. This case can be readily achieved with pulsed lasers, if their pulse duration is shorter than the lifetime of level $|k\rangle$. Since all ions produced in a small ionization volume can be collected by an electric field, accelerated onto an open multiplier and detected with 100% efficiency, every excited molecule can be monitored during the laser pulse. This implies that every atom or molecule in the initial state $|i\rangle$ within the overlap region of the two laser beams can be detected (single-atom detection[11–13]), provided the first laser is sufficiently strong to saturate the transition $|i\rangle \rightarrow |k\rangle$.

However, there are serious drawbacks if pulsed lasers are used. The spectral resolution is lower, because pulsed lasers have a larger bandwidth than single-mode CW lasers. Furthermore, the atoms are only detected during the time of the laser pulse. In the case of a continuous molecular flow (e.g., in a continuous molecular beam) most of the molecules escape detection during the long time intervals between successive laser pulses. With a pulse repetition rate of 100 Hz and a laser pulse duration of 10^{-8} s, only 10^{-6} of all molecules in a continuous flow can be monitored even if the detection efficiency during the laser pulses reaches 100%!

The situation can be improved with CW lasers. Since the laser power is down by several orders of magnitude compared to pulsed lasers, one has to focus the two laser beams to reach sufficient intensities for the ionizing step. Both laser beams must be perfectly superimposed, because the flight time of a molecule between excitation and ionization must be shorter than the lifetime τ_k; otherwise the excited molecules decay into other lower levels and can no longer be ionized.

A possible experimental realization for resonant two-photon ionization in molecular beams which is used in our labs employs optical fibers to bring the dye laser beam to the experiment (Figure 3.7). The divergent output from the single-mode fiber is collimated by a spherical lens into a parallel beam, the argon laser beam is superimposed by a dichroic mirror, and both beams are then focused by a cylindrical lens into a focal line with a cross section of about 20 μm × 1 mm. All molecules in the molecular beam pass

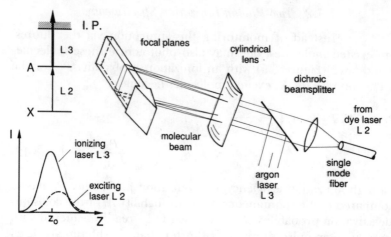

Figure 3.7 Experimental setup for two-color resonant two-photon ionization with CW lasers in a molecular beam.

through the two thin "sheets of light," with a Gaussian intensity profile in the z-direction of the molecular beam, which overlap within a micrometer (see insert). If the first laser L1 (λ_1) is tuned through the spectral interval of interest and λ_2 is kept constant, the ion rate $R_{Ion}(\lambda_1)$ gives the absorption spectrum of the first laser L1. With this arrangement, a very high detection sensitivity can be achieved, which exceeds that of fluorescence detection (Figure 3.8).

Figure 3.8 Section of the excitation spectrum of Cs_2 (bandhead of the 0–0 band in the $C^1\Pi_u \leftarrow X^1\Sigma_g$ system) monitored with sub-Doppler resolution by two-photon resonant ionization in a collimated molecular beam.

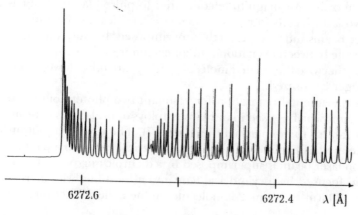

3.3.3. Combination with Mass Spectrometry

Often different atomic or molecular species with overlapping absorption spectra are present. These may be several isotopomers of the same molecule, or different molecules. In gas discharges, for instance, or in chemical reactions, fragmentation occurs, and besides the reactants and reaction products many transient species may be present. In seeded molecular beams, van der Waals complexes or clusters are formed with a wide mass distribution. In such cases it is difficult to separate the spectral contributions of the different species and to assign their spectra without further information. A combination of laser spectroscopy and mass selection is very helpful for solving these problems. (See also Chapter 9.)

When pulsed lasers are used, time-of-flight mass spectrometers are the best choice. The ions are formed during a short time interval ($\Delta t \approx 10^{-8}$ s) by resonant two-photon ionization, extracted by a small electric field, accelerated by a voltage V, and reach the detector after a flight path L at times

$$T(m) = \frac{L}{v(m)} = L \sqrt{\frac{m}{2 \text{ eV}}} \tag{3.13}$$

Time-resolved detection (e.g., with a boxcar or a transient digitizer) allows mass-selective detection (Figure 3.9). If laser L1 is tuned through the spectral interval of interest, simultaneous recording of the different ion rates $R_{Ion}(m, \lambda_1)$ allows the simultaneous and mass-selective measurement of otherwise overlapping absorption spectra.

Figure 3.9 Mass-selective photoabsorption spectroscopy with pulsed lasers using resonant two-photon ionization and a time-of-flight mass spectrometer.

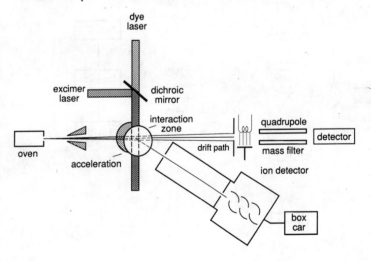

With CW lasers a quadrupole mass filter is a good choice. Two-photon ionization is performed with the arrangement of Figure 3.7, where the crossing volume of molecular beam and the laser beams is placed inside the ion source of the mass spectrometer. Such a system has been used in our labs for the mass-selective spectroscopy of small alkali clusters in cold molecular beams, where the molecular beam is sent coaxially into the quadrupole mass spectrometer (Figure 3.10).

3.3.4. Depletion Spectroscopy

For the spectroscopy of excited states, absorption transitions from an excited level $|k>$ to still higher levels $|j>$ have to be monitored (Figure 3.11). Since the spontaneous transition probability decreases strongly with increasing principal quantum number, detection of the fluorescence from levels $|j>$ is often not feasible. Photoionization of the levels $|j>$ would require a third laser if a second photon from laser L1, used for pumping the transition $|i> \rightarrow |k>$, cannot be used. In this case the *depletion* of level $|k>$, caused by the absorption of laser L2 on a transition $|k> \rightarrow |j>$, can be monitored by the corresponding decrease of decay fluorescence from $|k>$ into lower levels $|m>$.

This depletion spectroscopy is not restricted to transitions between excited states but is also very helpful for assigning complex molecular absorption spectra of molecules in their electronic ground state. If the pump

Figure 3.10 Mass-selective sub-Doppler resonance ionization spectroscopy of Na_n-clusters in a cold supersonic Ar/Na beam with a quadrupole mass spectrometer. The first crossing point of Ll allows LIF observation.

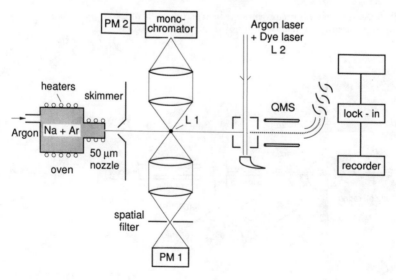

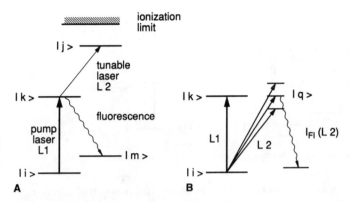

Figure 3.11 Depletion spectroscopy.

laser, tuned to a transition $|i>\rightarrow|k>$, is chopped at a frequency f, the population N_i of level $|i>$ will be accordingly modulated due to saturation by the pump laser. When a second "probe laser" L2 is tuned through the spectral interval of interest, its laser-induced fluorescence will be modulated at the chopping frequency f whenever λ_2 coincides with a transition $|i> \rightarrow |q>$ from the level $|i>$ labeled by the pump laser L1. When the laser-induced fluorescence signal is recorded through a lock-in amplifier, only those lines that start from level $|i>$ appear in the spectrum. If level $|i>$ has been assigned, the upper levels $|q>$ are more readily identified because of the known selection rules for optical transitions. One may regard depletion spectroscopy as a special case of the optical–optical double-resonance techniques that are discussed in Section 3.6.

3.3.5. Optogalvanic Spectroscopy

The spectroscopy of electronic transitions in gas discharges can give valuable information on transitions between the highly excited levels of atoms, molecules, radicals, and ions, which are populated by electron impact in the discharge. The emission spectroscopy of discharges is an important field of classical spectroscopy. Optogalvanic spectroscopy does not use the optical emission as such, but is based on the fact that the electrical impedance of the discharge depends on the population distribution of the species in the discharge. Its basic principle may be understood in a simple model.

When a laser beam passes through a discharge (Figure 3.12) and the laser wavelength λ_L is tuned to a transition $|i> \rightarrow |k>$, the population of the two levels changes by $\Delta N_i < 0$, $\Delta N_k > 0$, because of saturation of the transition. The ionization probability P_I for electron impact or collisional ionization generally differs for the different levels. Since the discharge

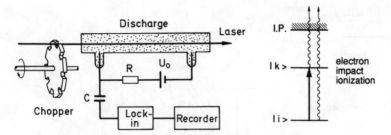

Figure 3.12 Optogalvanic spectroscopy in a gas discharge with a tunable laser.

impedance depends on the density of electrons and ions, the discharge current I at a constant voltage V will change according to

$$\Delta I = a[\Delta N_k\, P_I(E_k) + \Delta N_i\, P_I(E_i)] \tag{3.14}$$

when the laser induces the population changes $\Delta N_i = -\Delta N_k$. If the laser beam is chopped at a frequency f, the voltage $V = RI$ across a ballast resistor R will be modulated at the frequency f with a modulation amplitude

$$\Delta V = R\,\Delta I \approx aR\,\Delta N_k\,[P_I(E_k) - P_I(E_i)] \tag{3.15}$$

which can be fed through a capacitor C into a lock-in detector.

The real situation is more complex because the population changes ΔN_i, ΔN_k also affect the electronic temperature and, therefore, the electron excitation and ionization rates. This may also change the population densities of other levels. The observed signals $\Delta V(\lambda_L)$ can be positive as well as negative (Figure 3.13). According to (3.15) the sign depends mainly on the

Figure 3.13 Optogalvanic spectrum of neon at a discharge current of 1 mA and a pressure of 0.8 torr. (Reproduced with permission from Ref. 14.)

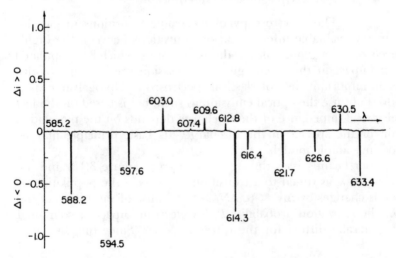

difference $P_I(E_k) - P_I(E_i)$, but is also affected by other processes not included in our simple model.

With hollow-cathode discharges, nonvolatile species such as high-temperature metals (tungsten, uranium, etc.) can be vaporized by sputtering from the cathode surface and are thus made accessible to optogalvanic spectroscopy. The spectroscopy of molecules and their fragments produced in the discharge has brought much information on intermediate, short-lived radicals.

In summary, the technique is relatively simple, does not need any sophisticated equipment except a tunable laser, and has, therefore, found a wide range of applications.[14]

3.4. Sub-Doppler Absorption Spectroscopy

The last section has covered a number of techniques that allow a *high detection sensitivity*. We will now discuss methods for achieving a *high spectral resolution*. They are based either on linear laser spectroscopy in collimated molecular beams, where the transverse velocity distribution of the molecules is reduced by collimating apertures, or on nonlinear techniques in gas cells or beams, where only those molecules are detected that have small Doppler shifts of their absorption lines.

3.4.1. Linear Laser Spectroscopy in Collimated Molecular Beams

When molecules effuse from a reservoir through a small aperture A into a vacuum (Figure 3.14), only those molecules can pass a slit B (with a width b in the x-direction) that have transverse velocity components $v_x < v \tan \varepsilon$, where the theoretical collimation ratio $\tan \varepsilon = b/2d$ is given by the geometrical arrangement. Because of collisions between molecules directly behind the nozzle A, where the density is still sufficiently high, the effective collimation ratio is somewhat smaller.

Figure 3.14 Sub-Doppler excitation spectroscopy in a collimated molecular beam with reduced transverse velocity components.

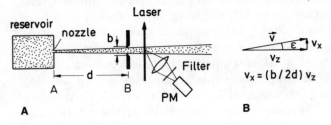

A

B

$$v_x = (b/2d)\, v_z$$

If the beam of a tunable single-mode laser is perpendicularly crossed with the molecular beam, the absorption profile

$$\alpha(\omega) = \int \left(\int_{x1}^{x_2} n(v_x, x) \, \sigma(\omega, v_x) \, dx \right) dv_x \qquad (3.16)$$

corresponds to the convolution of a Lorentzian profile, because of the homogeneous absorption cross section $\sigma(\omega, v_x)$ of the absorbing molecules, and a Gaussian profile with a reduced Doppler width

$$\Delta\omega_r = \Delta\omega_D \sin \varepsilon$$

which depends on the collimation ratio $\sin \varepsilon \approx \tan \varepsilon = b/2d$. For example, with $d = 100$ mm and $b = 1$ mm, $\tan \varepsilon = 5 \times 10^{-3}$. The residual Doppler width becomes $\Delta\omega_r \approx 5 \times 10^{-3} \Delta\omega_D$ and, therefore, in many cases reaches the natural linewidth of a molecular transition.

The absorption is monitored either by laser-induced fluorescence (Section 3.3.1) or by resonant two-photon ionization (Section 3.3.2). By way of illustration, Figure 3.8 shows a section of the 0–0 band in the $C^1\Pi_u \leftarrow X^1\Sigma_u$ system of Cs_2 molecules obtained with a CW dye laser for excitation and a CW argon laser for ionization using the arrangement of Figure 3.7.[15] The density of molecules in the absorbing levels was about 10^{12} m^{-3} and the absorption pathlength 0.2 cm, demonstrating the sensitivity of the technique.

When molecules M_x are mixed in the reservoir with a noble gas G at sufficiently high carrier pressures ($p(G) \approx$ 1–10 bar; $p(M_x)/p(G) \approx 0.01 - 0.05$), a supersonic seeded beam is obtained. Adiabatic expansion of the mixture through the nozzle A results in a cooling of the molecules.[16,17] Their internal energy is partly transferred into directional flow energy by inelastic collisions with the cold noble gas atoms. Although no real thermal equilibrium is reached, one often describes the internal energy distributions of the molecules after their expansion by a rotational temperature T_{rot} and a vibrational temperature T_{vib}. For example, the expansion of 0.1 bar sodium vapor mixed with 10-bar argon through a nozzle with 50-μm diameter results in formation of cold Na_2 molecules with $T_{rot} \approx 5$ K and $T_{vib} \approx 50$ K.

In supersonic cold beams the population distributions $N(E_{rot}, E_{vib})$ of molecules is compressed into the lowest rotational and vibrational levels. This has the great advantage of greatly reducing the number of absorbing levels and of drastically simplifying the absorption spectra.[18] This is illustrated by Figure 3.15, which shows a section of the NO_2 absorption spectrum measured (a) in a cell at room temperature, (b) in a collimated effusive NO_2 beam, and (c) in a supersonic cold argon beam seeded with 5% NO_2, where a rotational temperature of $T_{rot} = 10$ K was reached. Although in the cell spectrum the rotational lines are not resolved, the sub-Doppler

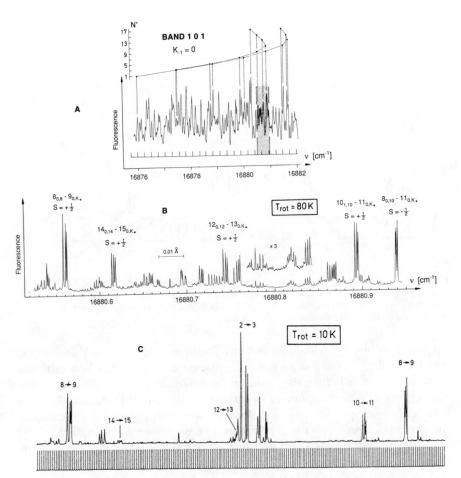

Figure 3.15 Section of the NO_2 visible absorption spectrum measured with a single-mode CW dye laser: (a) In a cell at $p = 10^{-5}$ bar, $T = 300$ K (with Doppler-limited resolution); (b) In a collimated pure NO_2 beam; (c) In a cold supersonic argon beam, seeded with 5% NO_2.

spectrum in (b) even shows the hyperfine structure of the rotational transitions and (c) demonstrates the reduction of the line density due to rotational cooling of NO_2. The spectral resolution is illustrated by Figure 3.16, which shows the fine and hyperfine structure of the $N' = 1 \leftarrow N'' = 0$ rotational line in the NO_2 spectrum.

Cold collimated molecular beams are of particular importance for accurate measurements of the collision-free lifetimes of individual excited rovibronic levels. Because of the reduction of the Doppler width and of the number of absorbing levels, the overlap of neighboring absorption lines can be greatly reduced and a selective excitation into single levels becomes possible even with molecules that have otherwise complex spectra. In the

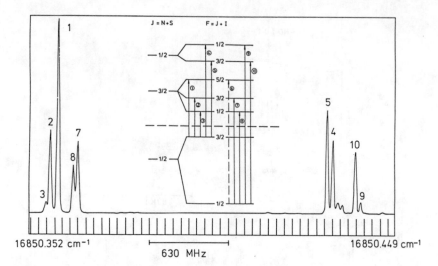

Figure 3.16 Fine and hyperfine structure of the rotational transition $N' = 1 \leftarrow N'' = 0$ in a vibronic band of NO_2, measured in a collimated beam of argon seeded with 5% NO_2. The linewidth of 15 MHz is due to the residual Doppler width caused by the finite collimation of the molecular beam.

case of long-lived excited levels with lifetimes τ the excited molecules, which travel at thermal velocities v a distance $\Delta z = \tau/v$, may leave the observation volume before they emit a fluorescence photon. Although this often leads to systematic errors in the determination of lifetimes when gas cells with isotropic velocity distributions are used, the one-dimensional movements of molecules in a collimated beam allows a collimation of the fluorescence light with a collection efficiency that is independent[19] of the point z of emission over a wide range of z.

3.4.2. Saturation Spectroscopy

Another Doppler-free spectroscopic technique that can be applied to atoms or molecules in the gas phase is based on the saturation of absorbing transitions.[20,21] If a single-mode CW laser with intensity I is tuned to a transition $|i> \rightarrow |k>$ with $N_k << N_i$, the stationary population density N_i of the absorbing molecules in level $|i>$ is obtained from the rate equation

$$\frac{dN_i}{dt} = -N_i \left(\sigma_{ik} n_L + \sum_m R_{im} \right) + \sum_m R_{mi} N_m + R = 0 \qquad (3.17)$$

where $n_L = I/\hbar\omega_{ik}$ is the incident photon flux density, σ_{ik} = the absorption cross section, and R_{im} = the relaxation rate for transitions $|i> \rightarrow |m>$ that may be due to collisions or fluorescence. The last term R in (3.17) represents the refilling rate caused by the diffusion of unpumped molecules into the excitation volume. The solution of (3.17) gives the saturated population density

$$N_i = \frac{R + \sum_m R_{mi} N_m}{\sigma_{ik} n_L + \sum_m R_{im}} = \frac{N_{i0}}{1 + S} \tag{3.18}$$

where $N_{i0} = N_i (n_L = 0)$ is the unsaturated population density and

$$S = \frac{\sigma_{ik} I/\hbar\omega_{ik}}{\sum_m R_{im}} \tag{3.19}$$

is the saturation parameter that equals the ratio of the induced absorption rate $\sigma_{ik} n_L$ to the total rate of depopulation without incident light, $\sum_m R_{im}$.

The absorbed power density

$$dP(\omega_{ik}) = N_i \sigma_{ik} I = \frac{N_{i0} \sigma_{ik} I}{1 + aI} \tag{3.20}$$

with $a = S/I$, depends in a nonlinear way on the incident intensity I. Any spectroscopic technique that utilizes the saturation of molecular transitions is, therefore, a "nonlinear spectroscopy."

When a monochromatic wave $E = E_0 \cos(\omega t - kz)$ passes through a sample, the absorption cross section of a molecule moving with a velocity component v_z becomes

$$\sigma_{ik}(\omega, v_z) = \sigma_0 \frac{(\gamma/2)^2}{(\omega - \omega_0 + kv_z)^2 + (\gamma/2)^2} \tag{3.21}$$

which shows that essentially only those molecules can absorb that are Doppler-shifted into resonance with the laser field. This means that their population density $N_i(v_z)$ is decreased around $v_z = (\omega - \omega_0)/k$, when the laser saturates the transition $|i> \rightarrow |k>$. The monochromatic laser burns a narrow "hole" with a width γ into the Doppler-broadened absorption profile $\alpha(\omega)$ with a corresponding peak in the population distribution $N_k(v_z)$ in the upper level, as shown in Figure 3.17a.

If the laser beam is split into two parts ("pump" and "probe" beams) that pass in opposite directions $\pm z$ through the sample, two different velocity groups $v_z = \pm(\omega - \omega_0)/k$ can absorb for $\omega \neq \omega_0$, whereas only molecules with $v_z = 0 \pm \gamma$ absorb for $\omega = \omega_0$. These molecules see for $\omega = \omega_0$ twice the intensity as for $\omega \neq \omega_0$, and their transition is, therefore, more strongly saturated. This means that the absorption coefficient $\alpha(\omega)$ of a Doppler-broadened transition for a sample traversed by two opposite beams from the same laser exhibits a dip at the center where $\omega = \omega_0$ (Lamb dip). The dip width γ equals the homogeneous width (natural linewidth + transit-time broadening + pressure-broadening), which is generally two orders of magnitude smaller than the Doppler width, Figure 3.17c. These narrow

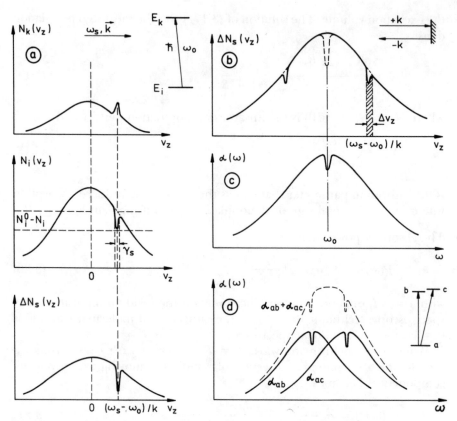

Figure 3.17 Selective saturation of a Doppler-broadened absorption line profile.
(a) Saturation hole in the velocity distribution of molecules in the lower level with the
corresponding population peak in the upper level; (b) saturation holes in a monochromatic
standing wave with $a \neq \omega$; (c) Lamb dip in the absorption profile;
(d) Lamb dips in two closely spaced transitions with a common lower level.

resonances can be used for spectroscopy with sub-Doppler resolution. Where two closely spaced transitions overlap within their Doppler widths, Figure 3.17d, their narrow Lamb dips, however, can be clearly separated.

A possible experimental arrangement is shown in Figure 3.18. The Lamb dips can be monitored either by observation of the transmitted intensity $I_T(\omega)$ of the probe beam or through the laser-induced fluorescence $I_{Fl}(\omega)$ (detector $D2$). If the saturating pump beam is chopped at a frequency f, lock-in detection of $I_T(\omega)$ yields the difference of the unsaturated absorption $\alpha_0(\omega)$ (pump beam off) and the saturated absorption $\alpha_s(\omega)$ (pump beam on). The difference eliminates the Doppler-broadened background and leaves only the narrow Lamp-dip profiles at the center ω_0 of an absorption line.

When I_{Fl} is monitored, the contributions to the LIF from the pump and probe beams cannot be spatially separated because both beams overlap inside the sample. Here an intermodulated technique[22] is useful, in which the pump and probe beams are chopped at different frequencies f_1 and f_2, respectively and the fluorescence $I_{Fl}(\omega, f_1 + f_2)$ is monitored at the sum-frequency $(f_1 + f_2)$. If $I_{L1} \gg I_{L2}$, the total fluorescence intensity can be written as

$$I_{Fl} \propto N_i(I_{L1} + I_{L2})$$

$$= N_{i0}(1 - aI_{10}\cos 2\pi f_1 t)(I_{10}\cos 2\pi f_1 t + I_{20}\cos 2\pi f_2 t)$$

$$= -\frac{N_{i0}}{2}aI_{10}^2 + N_{i0}I_{10}\cos 2\pi f_1 t - \frac{N_{i0}}{2}aI_{10}^2\cos 4\pi f_1 t$$

$$- \frac{N_{i0}}{2}aI_{10}I_{20}[\cos 2\pi(f_1 + f_2)t + \cos 2\pi(f_1 - f_2)t]$$

$$(3.22)$$

Figure 3.18 Experimental arrangement for saturation spectroscopy in a gas cell. The spectra are monitored through either the transmitted probe intensity or the laser-induced fluorescence.

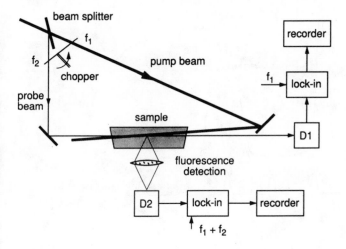

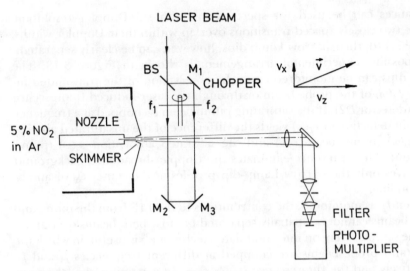

Figure 3.19 Saturation spectroscopy in a collimated molecular beam. Either the pump beam alone is chopped at frequency f_1 or the probe beam is also chopped at f_2.

Thus I_{F1} contains a term with cos 2π $(f_1 + f_2)t$ that is proportional to the product $I_{10}I_{20}$ of the two laser intensities. Lock-in detection at $(f_1 + f_2)$ therefore filters the nonlinear part and rejects the linear, Doppler-broadened profiles of $I_{F1}(\omega)$.

Saturation spectroscopy can also be performed in molecular beams. Because of the absence of collisions the refilling of the depleted lower level is only due to the flow of unpumped molecules into the excitation region.

Figure 3.20 Lamb-dip spectrum of the NO_2 transition $l_{1,0} \leftarrow 0_{0,0}$ shown in Figure 3.16.

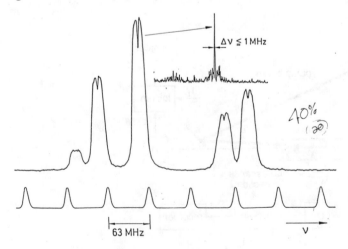

The saturation intensity is therefore low and laser powers of less than 1 mW may be sufficient to saturate a transition. A possible experimental arrangement is shown in Figure 3.19, where the incident beam is split by a beamsplitter (BS). The two beams cross the molecular beam perpendicularly. If the fluorescence they induce can be resolved spatially, chopping of the pump beam at frequency f_1 is sufficient to suppress the linear background with a reduced Doppler width. If, however, the fluorescence from both crossing points overlaps on the detector, again it is necessary to adopt the technique of intermodulated fluorescence, in which both beams are chopped at different frequencies f_1, f_2 in order to eliminate the linear background. This is illustrated by Figure 3.20, where at the lock-in frequency f_1, narrow Lamb dips $\Delta\nu = 1$ MHz are observed at the center of NO$_2$ lines with a residual Doppler width of 15 MHz, because of the finite collimation of the molecular beam.[23] Monitoring $I_{F1}(\omega)$ at the sum-frequency $(f_1 + f_2)$ yields the narrow structure shown in the insert. In this case the Lamb dip linewidth of 1 MHz is solely due to frequency fluctuations of the CW dye laser, while the natural linewidth of the NO$_2$ transitions is less than 10 KHz.

3.4.3. Polarization Spectroscopy

Whereas saturation spectroscopy is based on a change in absorption due to saturation of a molecular transition by a pump laser, polarization spectroscopy utilizes the change of the complex index of refraction that is caused by a population change and also a partial orientation of molecules induced by a polarized pump laser.[24]

The experimental arrangement is shown in Figure 3.21. The beam of a laser L1 is split into a strong pump and a weak probe beam, which pass in opposite $\pm$ z-directions through a sample cell placed between two crossed linear polarizers P1 and P2. Without the pump beam the sample is isotropic

Figure 3.21 Experimental arrangement for polarization spectroscopy. Laser L2 is only necessary for optical double-resonance experiments. DM = dichroic mirror, WH = Wood's horn, BS = beamsplitter.

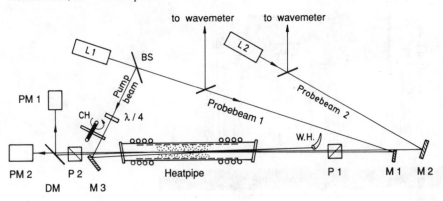

and does not affect the polarization state of the linearly polarized probe laser. The transmitted probe beam intensity is, therefore, strongly attenuated by a factor of about 10^7 by the crossed polarizer P2. The pump beam, circularly polarized by a quarter-wave plate, induces molecular transitions $\Delta M = +1$ (M is the quantum number of the projection of the angular momentum onto the propagation axis of the pump laser) and induces by M-selective saturation a partial orientation of the molecules. This causes a birefringence of the sample. If probe and pump beam can interact with the same molecules, then the plane of polarization of the probe beam is slightly turned by an angle β and the polarizer P2 transmits the fraction $I_2 \sin \beta$ of the probe beam intensity I_2.

Since the pump and probe beams with wave vectors $\pm \mathbf{k}$ and frequency ω travel in opposite directions, a molecule moving with a velocity component $v_z \neq 0$ experiences opposite Doppler shifts $\Delta \omega = \pm \mathbf{k} \cdot \mathbf{v}_z$ for the two beams and cannot be simultaneously in resonance with both beams unless $\mathbf{k} \cdot \mathbf{v}_z < \gamma$ (i.e., the Doppler shift becomes smaller than the homogeneous width γ of the molecular transition). This shows that the signals $I_t(\omega)$ transmitted by the polarizer P2 have Doppler-free line profiles, located at the center frequency ω_0 of the Doppler-broadened absorption lines. In the case of dense molecular spectra, where many lines overlap within their Doppler width, such a Doppler-free technique is essential for the resolution of individual lines.[25] This is illustrated by Figure 3.22, which shows a section of the polarization spectrum of the Cs_2 molecule, where rotational lines of different vibronic bands overlap. Note that the Doppler width $\Delta \omega_D$ is much larger than the spacings of the lines.

Figure 3.22 Section of the polarization spectrum of the Cs_2 ($CB^1\Pi_u \leftarrow X^1\Sigma_g^+$) system. (Reproduced from Ref. 26.)

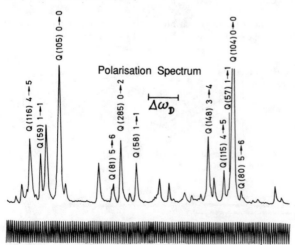

The advantages of polarization spectroscopy compared to saturation spectroscopy may be summarized as follows:

1. Because of the crossed polarizers the background is essentially suppressed. This helps to increase the signal-to-noise ratio. With extinction ratios of 10^{-7} through the crossed polarizers, small angles β can be detected.

2. Since the turning angle β depends on the phase shift between the σ^+ and σ^- circularly polarized components that add up to the linear polarization state of the probe wave, small changes $\Delta n = (n^+ - n^-)$ of the refractive index, caused by the M-selective pump saturation, result in a large signal. Polarization spectroscopy is, therefore, more sensitive than saturation spectroscopy.

3. The line profile $I_t(\omega)$ of the transmitted probe intensity depends on the change $\Delta J = J' - J''$ of rotational quantum numbers during the molecular transition. It is, therefore, possible to distinguish between $P(\Delta J = -1)$, $Q(\Delta J = +0)$, and $R(\Delta J = +1)$ transitions. This considerably facilitates the assignment of molecular spectra.

3.5. Laser Spectroscopy of Ionized Species

Atomic and molecular ions play an important role in gas discharges, plasmas, chemical reactions, and stars and interstellar clouds. Nevertheless their spectroscopy is far less advanced than that of neutral species. The main reason is the fact that in normal experimental environments their abundances are smaller than those of neutral molecules. In recent years increased efforts have been made to gain more information on ions by spectroscopic techniques.[27]

3.5.1. Velocity-Modulation Spectroscopy

The spectroscopy of molecular ions in gas discharges meets the problem that besides the parent molecules and ions, many neutral and ionized fragments are present. The spectra of many of these species may overlap, and it is often extremely difficult or even impossible to assign spectral lines to a special species if little is known about its structure. Here the technique of velocity modulation developed by Saykally and his group is very useful.[28] It is based on the fact that ions in a discharge are accelerated by the electric field, whereas the velocity of neutrals is affected only indirectly by collisions with ions. In a discharge driven by an ac voltage at a frequency f, the ions suffer periodic velocity changes $v = v_0 \cos 2\pi ft$. If the beam of a narrowband tunable laser with frequency ω_L is sent through the discharge, the absorption frequency

$$\omega_a = \omega_0 + \mathbf{v} \cdot \mathbf{k} = \omega_0 \left(1 + (v_0/c) \cos 2\pi ft\right) \tag{3.23}$$

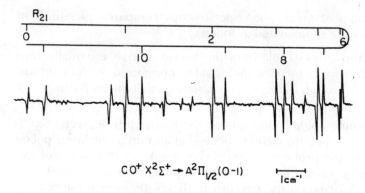

$$CO^+ \; X^2\Sigma^+ \to A^2\Pi_{1/2}(0\text{-}1)$$

Figure 3.23 The R_{21} bandhead of the $X^2\Sigma^+ \to A^2\Pi(0,1)$ system of CO^+ near 420 nm measured by velocity modulation laser absorption spectroscopy. (Reproduced with permission from Ref. 29.)

of ionic species is periodically Doppler-shifted and the transmitted laser intensity

$$I_t = I_0[1 - \alpha(\omega_L)L]$$

detected through a lock-in at the modulation frequency f exhibits a dispersion profile $d\alpha/d\omega$, as discussed in Section 3.2.2. Such a spectrum is shown in Figure 3.23.

3.5.2. *Spectroscopy in Ion Beams*

The problem of overlapping spectra from different species, often present in gas discharges, can be avoided in ion beams where the ions are produced in an ion source by electron impact or by a microwave discharge. The ions are accelerated through a voltage V, are mass-separated by a magnetic or electric field, and then interact along a distance L with a coaxial laser beam (Figure 3.24). This coaxial arrangement has

Figure 3.24 Laser spectroscopy of ions in a coaxial arrangement.

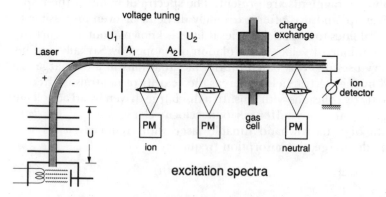

several advantages. First, the interaction length L can be made much larger than in a crossed-beam geometry. The laser absorption can be monitored via the fluorescence or, in the case of photodissociation, by detecting the photofragments. Second, the spectral resolution can be much higher, because of an effect called acceleration cooling.[30]

Suppose that two ions with mass m and velocities v_{10}, v_{20} leave the ion source, where thermal equilibrium at a temperature T is assumed. After acceleration by a voltage U their kinetic energies are

$$W_1 = \frac{m}{2}v_{10} + eU = \frac{m}{2}v_1^2 \qquad (3.24a)$$

$$W_2 = \frac{m}{2}v_{20} + eU = \frac{m}{2}v_1^2 \qquad (3.24b)$$

Subtracting (3.24b) from (3.24a) yields

$$\Delta v = \Delta v_0 \frac{\bar{v}_0}{\bar{v}} \approx \Delta v_0 \sqrt{\frac{kT}{2eU}}$$

where $\Delta v = v_1 - v_2$; $\Delta v_0 = v_{10} - v_{20}$;

$$\bar{v} = \frac{v_1 + v_2}{2} \approx \left(\frac{2eU}{m}\right)^{1/2}; \qquad \bar{v}_0 = \frac{v_{10} + v_{20}}{2} \approx \left(\frac{kT}{m}\right)^{1/2} \qquad (3.25)$$

The initial velocity spread Δv_0 is therefore reduced by a factor $(kT/2eU)^{1/2} \ll 1$.

For example, with $T = 1,000$ K, $U = 10^4$ V we obtain $\Delta v = 2 \times 10^{-3}\Delta v_0$. The velocity spread is therefore reduced by a factor of 500. The corresponding Doppler width of absorption lines for the collinear arrangement therefore decreases by the same factor, provided the voltage U is sufficiently stable. This shows that in fast ion beams sub-Doppler spectroscopy is possible. Either a single-mode tunable laser is tuned through the spectral interval of interest while the voltage V is kept constant or a fixed-wavelength laser can be used, and the absorption lines are Doppler-shifted across the laser wavelength λ_L by tuning the accelerating voltage U and therefore the ion velocity v.[31]

Nonlinear saturation spectroscopy in a collinear beam arrangement can be performed with a single laser beam, using two interaction zones (see Figure 3.24), where the ion velocity can be independently varied in the second zone. The laser is tuned to an absorption line and burns a hole into the velocity distribution of the ions in the first zone. When the LIF in the second zone is observed as a function of the voltage $\Delta U = U_2 - U_1$ between the apertures A_1 and A_2 in Figure 3.24b, a Lamb dip is observed with a linewidth that depends on the transit time and the excited state lifetime.[32,33]

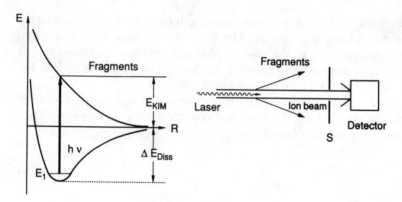

Figure 3.25 Measurements of molecular ion fragmentation.

The ions may be neutralized by charge-exchange collisions. The neutral atoms are generally produced in highly excited states. Since the velocity spread is not appreciably increased by these collisions with large impact parameters, the narrow velocity distribution is preserved and high-resolution spectroscopy of highly excited atoms becomes possible.[34]

Of particular interest are investigations of photon-induced fragmentation of molecular ions. If the laser excites molecular ions from a bound level E_1 into predissociating states or directly dissociating states with a repulsive potential, the total kinetic energy of the fragments in the center-of-mass system is

$$E_{kin} = h\nu + E_1 - \Delta E_{Diss}$$

Because most fragments have transverse velocity components in the laboratory system, laser-induced dissociation decreases the number of ions reaching the detector behind a collimating aperture (Figure 3.25). The measured ion current I_{ion} (λ_L) gives information about the dissociation probability as a function of laser wavelength.[35] More detailed information about the photofragmentation channels can be extracted if the fragment ions are mass-selected and their energies are measured. This allows determination of the repulsive potential curve $V(R)$ and can be performed with mass- and energy-selective ion optics in an apparatus, as shown in Figure 3.26.[36]

3.5.3. Ion Production and Spectroscopy in Cold Neutral Beams

For the high-resolution photoabsorption spectroscopy of more complex molecular ions it is desirable to cool the ions to reduce the number of absorbing levels. This can be achieved as for neutral mole-

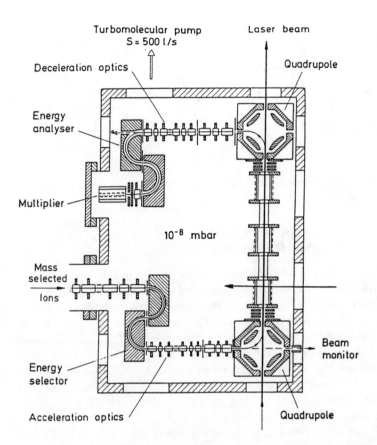

Figure 3.26 Apparatus for photofragmentation with determination of fragment masses and energies. (Reproduced with permission from Ref. 36.)

cules, by adiabatic cooling in neutral beams seeded with ions. Several experimental realizations have been published so far.[37,38] Two examples are given in Figure 3.27. The first arrangement utilizes an electric discharge, in a carrier noble gas seeded with molecules, between a thin tungsten wire in the glass nozzle and an anode on the low-pressure side of the nozzle expansion. The discharge produces ions and molecular fragments that are cooled by collisions with cold noble gas atoms.

The second method is based on ionization of cold molecules by electron impact. The electrons are emitted from a hot cathode and can produce rotationally cold ions since the light electrons do not appreciably change the angular momentum of the cold neutral molecules. The vibrational temperature depends on the Franck–Condon factors for transitions from the vibrational ground state of the neutral molecule to the vibrational levels of the ion ground state.

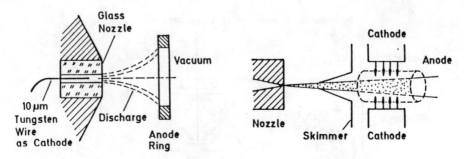

Figure 3.27 Two different arrangements for producing cold molecular ions in neutral beams.

3.6. Optical–Optical Double-Resonance Techniques

Optical–optical double resonance (OODR) means the simultaneous interaction of atoms or molecules with two optical fields that are tuned to two molecular transitions sharing a common level. There are three possible level schemes, depicted in Figure 3.28. For all three OODR schemes two lasers are necessary. One is a "pump laser" L1, which is tuned to a selected transition $|i> \rightarrow |k>$ and changes the populations from their unsaturated values N_{i0}, N_{k0} to the saturated values N_{is}, N_{ks}. The second is a "probe laser" L2, which is tuned over the spectral range of interest and which monitors those transitions that start from the levels $|i>$ or $|k>$ "labeled" by the pump laser L1.

In the first scheme both lasers share the lower level $|i>$. When L1 is chopped at a frequency f_1, the population density N_i is accordingly modulated between N_{i0} with L1 off and N_{is} with L1 on, provided that the time interval $\Delta t = 1/f_1$ is longer than the refilling time of N_i due to collisions or fluorescence. When L2 is tuned and its absorption is monitored through a lock-in amplifier at the chopping frequency f_1, only those transitions are detected that start from level $|i>$ (depletion spectroscopy). This allows one to filter out of a complex and dense absorption spectrum just a few lines, representing transitions from $|i>$ to upper levels $|m>$, connected to $|i>$ by allowed dipole transitions. If level $|i>$ has been assigned, the identification

Figure 3.28 Possible schemes of OODR: (a) Depletion spectroscopy; (b) Stepwise excitation; (c) Stimulated resonance Raman transitions.

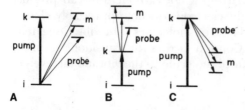

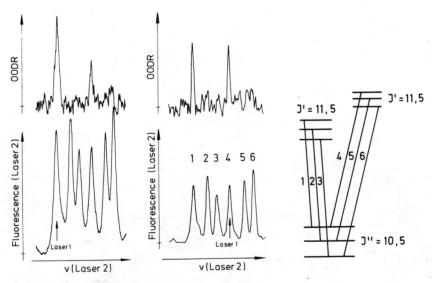

Figure 3.29 OODR-spectroscopy of NO_2.

of these upper levels is generally straightforward, since selection rules limit the possible rotational quantum numbers and the symmetry of the vibronic levels. This is illustrated by Figure 3.29, which shows a small section of the complex visible absorption spectrum of NO_2 in which the upper levels are heavily perturbed. Depletion spectroscopy allows the assignment of pairs of rotational transitions, starting from a common lower level (v'', J'') with 3-

Figure 3.30 Section of the 0–0 band in the sub-Doppler absorption spectrum of the $A \leftarrow X$ system of Na_3.

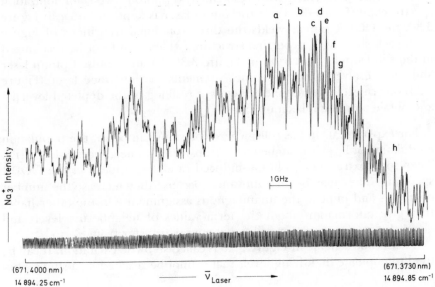

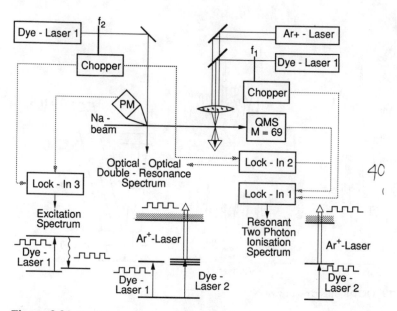

Figure 3.31 OODR spectroscopy with two-photon two-color detection in a cold molecular beam.

hfs components, to pairs of mutually perturbed levels of two different electronic states.[38,39] For very complex molecular spectra, such as the Na_3 spectrum of Figure 3.30, where many lines overlap even at sub-Doppler resolution, OODR often provides the only means of unambigious assignment.[40]

Very sensitive OODR depletion spectroscopy in a collimated molecular beam can be performed with resonant two-photon, two-color ionization spectroscopy (Figure 3.31). If the pump laser is kept on line *a* in Figure 3.30, the OODR signals yields the three rotational transitions of Figure 3.32, each showing a hyperfine structure, which could not be recognized in the sub-Doppler spectrum of Figure 3.30.[41] With a pulsed pump laser and time-resolved absorption measurements of the probe laser (Figure 3.33) one can follow the time-dependent refilling of the depleted level $|i>$ and obtain absolute relaxation rates $R (\sum_m |m> \rightarrow |i>)$.

For experiments in gas cells, where collisions can occur, the population modulation of level $|i>$ can be partly transferred to neighboring levels $|j>$. Therefore, extra lines (collision-induced satellites) appear in the OODR spectrum. They may be an annoyance, because they increase the number of signals and impede the unambiguous assignment. On the other hand, they give information about the term values of neighboring levels and about the cross sections for collisional energy transfer from $|j>$ to $|i>$.

The second scheme in Figure 3.28 is called stepwise excitation. It can be used to populate high-lying levels that cannot be reached by a single pho-

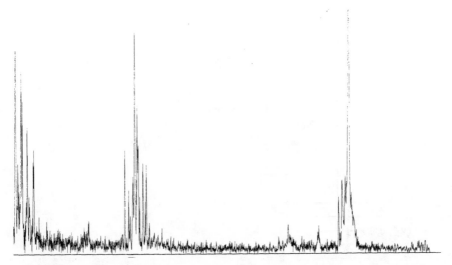

Figure 3.32 OODR signals of NA_3, obtained with the pump on line *a* of Figure 3.30.

ton from the ground state. Most of the high-resolution spectroscopy of atomic and molecular Rydberg states has been performed in this way. The absorption of the second laser L2 can be monitored either by the fluorescence emitted from levels $|j>$,[42] by the *decrease* of the fluorescence from level $|k>$ (depletion spectroscopy), or by photoionization of levels $|j>$ with a third photon.[43] In case of high-lying Rydberg levels, close below the ionization limit, field ionization is a very effective way to monitor the population of the Rydberg levels.[44,45] Collisional ionization in a thermionic heatpipe[46] also represents a sensitive technique for detecting the excitation of Rydberg levels.

The third scheme of Figure 3.28 turns out to be very useful for the study of high-lying rovibronic levels of ground-state molecules, which cannot be reached by direct absorption from thermally populated levels in the ground state. The probe laser transfers the molecules pumped by L1 into an excited level $|k>$ by stimulated emission into levels $|m>$. Laser L1 is kept on a transition $|i> \rightarrow |k>$, while L2 is tuned, and the resonances $|k> \rightarrow |m>$ can be again monitored by the *decrease* of fluorescence from $|k>$ if it is observed perpendicularly to the laser beam direction of L2.[47,48]

Another detection scheme is based on OODR polarization spectroscopy, where the oriented molecules, optically pumped by L1, influence the polarization characteristics of the probe transition L2. The experimental arrangement for this technique is shown in Figure 3.21. The Λ-type OODR spectroscopy may be regarded as a resonance-stimulated Raman process[49,50] and is also called stimulated emission pumping.[50]

In the case of single-mode lasers the linewidth Γ of the OODR signal depends only on the sum of the homogeneous level widths $(\gamma_i + \gamma_m)$. The width γ_k contributes to the signal only with a fraction $(1 - \lambda_1/\lambda_2) \gamma_k$. For

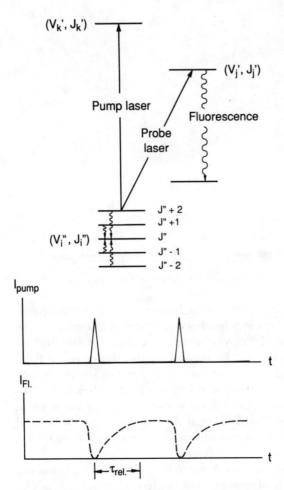

Figure 3.33 Time-resolved OODR spectroscopy for determination of inelastic collision cross sections.

long-lived levels $|i\rangle$ and $|_m\rangle$ and a short-lived level $|k\rangle$ the width Γ of the OODR signal may therefore be even smaller than the *natural linewidth* of the transition $|i\rangle \rightarrow |k\rangle$ (subnatural linewidth spectroscopy).[7,51]

3.7. Applications of Absorption Spectroscopy

Electronic photoabsorption spectroscopy has found many useful applications. They range from analytical chemistry, where the highly sensitive techniques discussed in Section 3.2 are utilized to monitor low concentrations of absorbing atoms and molecules, to various applications in photochemistry or photobiology, where the excitation energy transferred by the absorbed photon to molecules initiates chemical reactions. In medicine photoabsorption is used for diagnostics and the therapy of tumors. Examples of technical applications are afforded by studies of combustion processes by optogalvanic spectroscopy or by LIF. We will give some specific examples for illustration.

3.7.1. Analytical Applications

A very important branch of environmental studies is the measurement of gaseous pollutants, such as NO_x, SO_x, and O_3, in our atmosphere. Since the concentrations of such pollutants are generally within the ppb–ppm range (i.e., $10^{-9} - 10^{-6}$), sensitive detection techniques are necessary. Often remote sensing is required, which means that the *chemical* analysis of samples is not possible. Here spectroscopic techniques become superior. They are based either on infrared absorption on vibrational–rotational transitions (the fingerprint spectral region of molecular identification) or on electronic transitions in the visible or UV. Because of the larger transition probability and the higher sensitivity of detectors in the visible and UV, electronic photoabsorption spectroscopy is generally superior to infrared absorption techniques in all cases where tunable lasers can be applied.

A widely used spectroscopic technique for atmospheric research is LIDAR (light detection and ranging).[52] The output pulse of a pulsed laser is sent through a telescope into the atmosphere. The backscattered light, scattered by dust particles or small water droplets (Mie-scattering) at a distance d is collected by the telescope and is detected through an electronic gate at a time $t = 2d/c$ after the pulse emission (Figure 3.34). The observed signal is:

$$S(x,t) = P_0 \, N \, \sigma_{scatt} \frac{D^2}{d^2} \exp[-2 \int a(x,\lambda) \, dx] \qquad (3.26)$$

where P_0 is the power of the emitted laser pulse, N is the number density of scattering particles, σ_{scatt} is the scattering cross section, and a_t ($= \alpha_x + \alpha_s$) is the total attenuation coefficient, due to absorption and scattering.

If the laser wavelength λ is periodically switched from a value λ_1, where the molecules under investigation absorb, to a nearby value λ_2, where absorption is negligible, the ratio

$$R(t) = \frac{S(\lambda_1,t)}{S(\lambda_2,t)} = \exp[+2 \int_0^d \left(a(\lambda_2) - a(\lambda_1) \right) dx]$$

$$\approx \exp[-2 \int_0^d \alpha_a(\lambda_1) \, dx] \qquad (3.27)$$

of the received signals yields the absorption at λ_1, integrated over the light path $2d$, provided the scattering cross section does not change. From the ratio

$$\frac{R(t + \Delta t)}{R(t)} = e^{-\alpha(\lambda_1)\Delta d} \approx 1 - \alpha(\lambda_1) \, \Delta d$$

$$= 1 - N_a \sigma_a(\lambda_1) \, \Delta d \qquad (3.28)$$

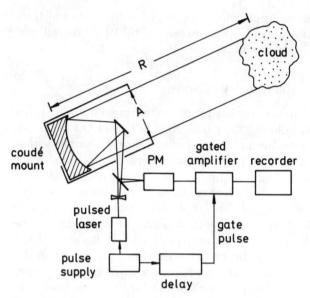

Figure 3.34 Experimental set up for LIDAR.

the density N_a of absorbing molecules with absorption cross section $\sigma_a(\lambda_1)$ at a distance $d = ct/2$ in a spatial interval $\Delta d = c\,\Delta t/2$ can be obtained.

One example for the application of this differential LIDAR technique is the determination of ozone concentration, its dependence on altitude and latitude, and its seasonal variations.[53] Here the two wavelengths are provided by an XeCl excimer laser at $\lambda_1 = 308$ nm, where ozone absorbs, and at $\lambda_2 = 353$ nm, where O_3 does not appreciably absorb, produced by Raman shift in a hydrogen high-pressure cell.

Another example of the analytical applications of photoabsorption spectroscopy is the spatially and time-resolved determination of radical concentrations in flames and combustions. If the absorption is monitored by laser-induced fluorescence (LIF), the quantum yield depends on pressure and temperature because collisional quenching plays an important role. For a quantitative determination of the concentrations of absorbing molecules the excitation of predissociating upper levels (with a predissociation rate that is fast compared to the quenching rate) can solve this problem. Although now only a small fraction η of all molecules contributes to the fluorescence, the value of η is independent of pressure and temperature.

With this "predissociation-limited" LIF technique, time- and space-resolved concentrations of OH, NO, CO, and other radicals in the combustion chamber of a car engine have been measured.[54–56] The laser-induced fluorescence is imaged through spectral filters onto a video camera, and a computer transforms the information into a density profile of the radicals under investigation. With a variable delay between the ignition time of the discharge and the probe laser pulse, a time-resolved picture of the combustion front (where the radicals are formed) can be obtained.

As a last example of analytical applications the measurement of laser ablation and desorption can be mentioned (Figure 3.35). A laser pulse impinging onto a solid surface vaporizes atoms, molecules or ions. A tunable probe laser can detect these gaseous species either by LIF or by multiphoton ionization with subsequent mass-selective detection. If the electronic spectra of the absorbing atoms or molecules are known, the internal energy distribution of the vaporized particles can be determined by tuning the probe laser to different transitions. If these transitions can be saturated, the resulting LIF or ion signals are directly proportional to the density of molecules in the absorbing level. These measurements allow a more detailed investigation of the mechanisms of laser ablation of solids and of desorption of molecules absorbed at a surface.[57] Sputtering processes induced by ion bombardment of surfaces can be studied in a similar way.[58,59]

3.7.2. Applications in Chemistry

The absorption of photons by molecules can enhance or induce chemical reactions by different mechanisms (Figure 3.36). If the electronic transition induced by photoabsorption results in photodissociation of molecules, the fragments are often radicals, which may react with added reactants more readily than the parent molecules. With one-photon absorption or resonant two-photon absorption the desired dissociation channel can generally be selectively opened by choosing the proper wavelength. In such cases laser-induced chemical reactions are possible. The photolysis of chloroethene

$$C_2H_3Cl + h\nu_1 \rightarrow C_2H_3 + Cl \qquad\qquad (a)$$

$$C_2H_3Cl + h\nu_2 \rightarrow C_2H_2 + HCl \qquad\qquad (b)$$

Figure 3.35 Detection of the mass distribution and internal energy distribution of particles, sputtered from a surface by laser irradiation or ion bombardment.

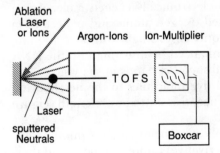

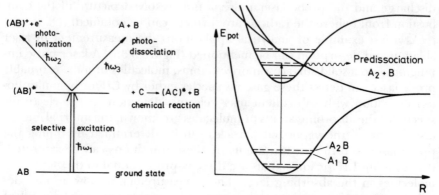

Figure 3.36 Different ways of inducing chemical reactions by photoabsorption.

may serve as an example, where the dissociation channel (a) or (b) depends on the laser wavelengths and on the temperature of the reaction cell.[60]

If bound excited states of polyatomic molecules are selectively populated by photoabsorption, the excitation energy may be distributed rapidly among many degrees of freedom because of intramolecular couplings between vibrational modes in different electronic states or by collisions. The selectivity of laser-induced chemical reactions therefore depends on the time scale of excitation and chemical reaction. Excitation of molecules with sufficiently short pulses at high pressures may lead to specific reaction channels if the time between excitation and reactive collision becomes shorter than the internal energy redistribution time in the excited molecule. Many more details on the applications of lasers in Chemistry can be found in Refs. 61 and 62.

3.7.3. Biological and Medical Applications

Electronic photoabsorption laser spectroscopy plays an increasing role in investigations of photoinduced biological processes. Examples are the very fast primary steps in the visual process, where photons are absorbed by the photoactive molecule retinal contained in the protein rhodopsin, the sunburn effect where epidermal cells are affected by UV light, the energy transfer processes in electronically excited molecules in DNA bases, and the reactions of excited oxygen atoms and molecules in living cells. Although DNA molecules only absorb light with $\lambda < 400$ nm, the absorption of visible light can be enhanced by inserting into the DNA helix dye molecules that have specific absorption bands in the visible. The energy transfer from the photoexcited dye molecules to the neighboring bases can be monitored by the fluorescence with shifted wavelength (Figure 3.37).[63]

A particular interesting medical application of photoabsorption by dye molecules is HPD (hematoporphyrin derivative) tumor diagnosis and ther-

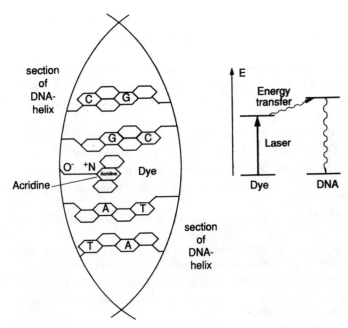

Figure 3.37 Energy transfer from a laser-excited dye molecule to the DNA.

apy.[64,65] If HPD dye is injected into the blood, it is distributed all over the body. Although normal cells release the dye after about two days, tumor cells keep it for a longer time. This HPD dye absorbs dye laser radiation between 600 and 700 nm. The LIF from photoexcited HPD can be used for diagnosis of tumor tissue. With sufficiently intense pulsed dye lasers sufficient energy is specifically absorbed in the tumor cells to destroy them. The mechanism of this destruction is not yet completely clear, but the formation of excited singlet oxygen by spin exchange with triplet dye molecules probably plays a crucial role.

REFERENCES

1. A. Corney, *Atomic and Laser Spectroscopy* (Clarendon, Oxford, 1977).

2. D. C. Hanna, M. A. Yuratich, and D. Cotter, *Nonlinear Optics of Free Atoms and Molecules* (Springer, Berlin, 1979).

3. D. S. Kliger, ed. *Ultrasensitive Laser Spectroscopy* (Academic, New York, 1983).

4. B. A. Garetz and J. R. Lombardy, eds. *Advances in Laser Spectroscopy* Vols. 1 and 2 (Heyden, London, 1982/83).

5. Y. Prior, A. Ben-Reuven, and M. Rosenbluth, eds. *Methods of Laser Spectroscopy* (Plenum, New York, 1986).

6. W. Demtröder and M. Inguscio, eds. *Applied Laser Spectroscopy* (Plenum, New York, 1991).

7. W. Demtröder, *Laser Spectroscopy*, 2nd ed. (Springer, Heidelberg, 1991).

8. S. Svanberg, *Atomic and Molecular Spectroscopy* (Springer, Heidelberg, 1991).

9. T. D. Harris, in *Ultrasensitive Laser Spectroscopy*. D. S. Kliger, ed. (Academic, New York, 1983), p. 343.

10. H. Atmanspacher and H. Scheingraber, *Phys. Rev.* **A32,** 254 (1985).

11. V. S. Letokhov, *Laser Photoionization Spectroscopy* (Academic, New York, 1987).

12. G. S. Hurst and M. G. Payne, eds. *Resonance Ionization Spectroscopy* (Institute of Physics, Bristol, 1984).

13. G. S. Hurst, M. G. Payne, and S. P. Kramer, *Rev. Mod. Phys.* **51,** 767 (1979).

14. B. Barbieri, N. Beverini, and A. Sasso, *Rev. Mod. Phys.* **62,** 603 (1990).

15. U. Diemer, J. Gress, and W. Demtröder, *Chem. Phys. Lett.* **146,** 330 (1991).

16. G. Scoles, ed., *Atomic and Molecular Beam Methods* (Oxford University Press, New York, Vol. 1, 1989; Vol. 2, 1991).

17. P. P. Wegener, ed., *Molecular Beams and Low Density Gas Dynamics* (Dekker, New York, 1974).

18. D. H. Levy, C. Wharton, and R. E. Smalley, in *Chemical and Biochemical Applications of Lasers*, Vol. 2, C. B. Moore, ed. (Academic, New York, 1977).

19. G. Persch, H. J. Vedder, and W. Demtröder, *Chem. Phys.* **105,** 471 (1986).

20. M. D. Levenson and S. S. Kano, *Introduction to Nonlinear Laser Spectroscopy,* 2nd ed. (Academic, New York, 1986).

21. V. S. Letokhov and V. P. Chebotayev, *Nonlinear Laser Spectroscopy* (Springer, Berlin, 1977).

22. M. S. Sorem and A. L. Schawlow, Opt. Commun. **5,** 148 (1972).

23. F. Bylicki, G. Persch, E. Mehdizadeh and W. Demtröder, *Chem. Phys.* **135,** 255 (1989).

24. R. E. Teets, F. V. Kowalski, W. T. Hill, N. Carlson, and T. W. Hansch, *Proc. SPIE* **113,** 80 (1977).

25. M. Raab, G. Höning, W. Demtröder, and C. R. Vidal, *J. Chem. Phys.* **76** 4370 (1987).

26. U. Diemer, R. Duchowicz, M. Ertel, E. Mehdizadeh, and W. Demtröder, *Chem. Phys. Lett.* **164,** 419 (1989).

27. T. A. Miller and V. E. Bondybey, *Molecular Ions: Spectroscopy, Structure and Chemistry* (North-Holland, Amsterdam, 1983).

28. Ch. S. Gudeman and R. J. Saykally, *Velocity Modulation Infrared Laser Spectroscopy of Molecular Ions, Ann. Rev. Phys. Chem.* **35,** 387 (1984).

29. Ch. S. Gudeman, C. C. Martner, and R. Saykally, *Chem. Phys. Lett.* **122,** 108 (1985).

30. S. L. Kaufman, *Opt. Commun.* **17,** 309 (1976).

31. R. A. Holt, R. Carré, S. Abed, M. Larzillière, J. Lermé, and M. G. Gaillard, *Opt. Commun.* **48,** 403 (1984).

32. D. Poulsen, in *Atomic Physics*, Vol. 8, I. Lindgren, S. Svanberg, and A. Rosen, eds. (Plenum, New York, 1983), p. 485.

33. D. Poulsen, Ph.D. thesis, University of Aarhus, Denmark, 1984.

34. D. Schulze-Hagenest, H. Harde, W. Brandt, and W. Demtröder, *Z. Phys.* **A282,** 149 (1977).

35. S. Abed, M. Broyer, M. Carræ, M. L. Gaillard, and M. Larzillière, *Chem. Phys.* **74,** 97 (1983).

36. L. Andrics, H. Bissantz, E. Solarte, and F. Linder, *Z. Phys.* **D8,** 371 (1988).

37. M. A. Johnson, R. N. Zare, J. Rostas, and S. Leach, *J. Chem. Phys.* **80,** 2407 (1984).

38. D. Klapstein, S. Leutwyler, J. P. Maier, C. Cossart-Magos, D. Cossart, and S. Leach, *Mol. Phys.* **51,** 413 (1984).

39. W. Demtröder, D. Eisel, H.-J. Foth, G. Höning, M. Raab, H. J. Vedder, and D. Zevgolis, *J. Mol. Struct.* **59,** 291 (1980).

40. H.-J. Foth, J. Gress, W. Demtröder, *Z. Phys.* **Dxx,** 000 (1991).

41. H. A. Eckel, H. J. Gress and W. Demtröder, *J. Chem. Phys.,* to be published.

42. R. A. Gottscho, P. S. Weiss, and R. W. Field, *J. Mol. Spec.* **82,** 283 (1980).

43. M. Broyer, J. Chevaleyre, G. Delacretaz, S. Martin, and L. Wöste, *Chem. Phys. Lett.* **99,** 206 (1983).

44. R. F. Stebbings and F. B. B. Dunnings, *Rydberg States of Atoms and Molecules* (Cambridge University Press, Cambridge, 1983).

45. J. C. Gallas, G. Leuchs, H. Walther, and H. Figger, *Adv. Atom. Mol. Phys.* **20,** 414 (1988).

46. H. K. Weber and K. Niemax, *Z. Phys.* **A312,** 339 (1983).

47. R. W. Field and D. H. Katayama, *J. Chem. Phys.* **75,** 2056 (1981).

48. C. H. Hamilton, J. L. Kinsey, and R. W. Field, *Ann. Rev. Phys. Chem.* **37,** 493 (1986).

49. G. Z. He, A. Kuhn, S. Schiemann, and K. Bergmann, *J. Opt. Soc. Am.* **B7,** 1960 (1990).

50. Hai-Lung Dai, *J. Opt. Soc. Am.* **B7,** 1802 (1990).

51. H. Weickenmeier, U. Diemer, W. Demtröder, and M. Broyer, *Chem. Phys. Lett.* **124,** 470 (1986).

52. R. M. Measure, *Laser Remote Chemical Analysis* (Wiley, New York, 1988).

53. W. Steinbrecht, K. W. Rothe, and H. Walther, *Appl. Opt.* **28,** 3636 (1988).

54. A. M. Wodtke, L. Hüwel, H. Schlüter, H. Voges, G. Meijer, and P. Andresen, *Opt. Lett.* **13,** 910 (1988).

55. P. Andresen et al. Report 11 (1989).

56. R. Suntz, H. Becker, P. Monkhouse, and J. Wolfrum, *Appl. Phys.* **B47,** 287 (1988).

57. R. W. Dreyfus, R. E. Walkup, and R. Kelly, *Appl. Phys. Lett.* **49,** 1478 (1986).

58. H. L. Bay, *Nucl. Instr. Meth. Phys. Res.* **B18,** 430 (1987).

59. R. De Jonge, *Commun. Atom. Mol. Phys.* **22,** 1 (1988).

60. M. Schneider and J. Wolfrum, *Ber. Bunsenges. Phys. Chem.* **90,** 1058 (1986).

61. D. L. Andrews, *Lasers in Chemistry,* 2nd ed. (Springer, Berlin, 1990).

62. K. L. Kompa and J. Warner, *Laser Applications in Chemistry* (Plenum, New York, 1984).

63. A. Anders, *Appl. Phys.* **18,** 373 (1979).

64. T. J. Dougherty, J. E. Kaufmann, and A. Goldfarb, *Cancer Res.* **38,** 2628 (1978).

65. T. J. Dougherty, J. E. Kaufmann, and A. Goldfarb, *Cancer Res.* **41,** 4606 (1981).

4 *Laser-Induced Fluorescence Spectroscopy*

J. Pfab

4.1. Introduction

Laser-induced fluorescence (LIF) spectroscopy was introduced about 20 years ago and has developed into a standard technique for monitoring many atomic, neutral, and ionic molecular species by optical means (i.e., via their emitted light).[1-5] A tunable source of laser light with narrow spectral bandwidth is brought into resonance with a spectroscopic transition to an *electronically excited level* whose spontaneous radiative decay provides the signal to be observed. The detection of fluorescence against a background of zero or of little scattered laser light is intrinsically much more sensitive than the direct measurement of absorbed light. Various alternative sensitive techniques are available for detecting absorbed light indirectly[2] (e.g., photoinization or photothermal deflection[3]). Disadvantages of fluorescence as a detection method are the limited photon collection and detection efficiencies (usually less than one in 20 emitted photons will be detected) and the requirement that species to be probed fluoresce with a finite quantum yield.

Among the most impressive advantages of the technique are its incisive power to unravel details of chemical, photochemical, and scattering dynamics and its ability to achieve ultrahigh sensitivity. The power of LIF for probing hostile environments like flames and discharge plasmas with high selectivity and spatial as well as temporal resolution is also valuable in applied, technologically relevant research. Fluorescence detection in the optical region is usually straightforward and can be achieved with simple, rugged, and cost-effective equipment. The cost of LIF instruments is therefore generally dominated by the laser source unless it is necessary to exploit the latest developments in low-level light detection or time resolution. Current tunable laser techniques routinely cover the 190–1,000-nm wavelength range, which is most interesting for standard LIF spectroscopy.

We are currently witnessing a dramatic expansion in the use and the range of applications of LIF spectroscopy, fueled by continuing improvements in the reliability, wavelength coverage, and performance of laser sources. The relative ease with which the detection of fluorescence can be implemented in laboratory experiments or field measurements contributes to the increasing popularity and widespread use of what has undoubtedly already become one of the leading and most universal laser spectroscopic techniques.

The emergence of the LIF technique as a key tool in modern research in chemistry and physics stems from its ability to accommodate a wide range of applications. A diverse range of spectroscopic, structural, and dynamical problems can be studied by LIF with spectroscopic resolution down to MHz, time resolution down into the subpicosecond domain, and spatial resolution less than 100 μm, often in combination with exceedingly high specificity for particular chemical species. Figure 4.1 shows a comparison of the γ (0,0) (i.e., the origin band of the $A\ ^2\Sigma^+ \leftarrow X\ ^2\Pi_i$ electronic spectrum of gaseous nitric oxide (NO) recorded at room temperature with 100 cm^{-1} ($\sim$0.5 nm) and 0.2 cm^{-1} spectral bandwidth, where the rotational structure of the band is readily resolved. The two easily distinguished pairs of rotational band heads merge to form two overlapping humps at low resolution. They arise from the 123 cm^{-1} separation between the upper $^2\Pi_{3/2}$ (F_2) and lower $^2\Pi_{1/2}$ (F_1) spin–orbit states that is characteristic for the L–S interaction of an unpaired electron in a molecular orbital of Π symmetry, with their intensity difference reflecting the difference in the Boltzmann occupancy. The advantage of a narrowband excitation source for studying systems with discrete and narrow level structure are readily apparent from the comparison.

LIF can be applied to studies of gas, liquid, and solid-state problems, but its most impressive applications in terms of information content generally use the narrow spectral bandwidth of tunable laser sources. This advantage is best exploited by choice of sample conditions, where discrete, well-defined energy levels are available for the substrate to be studied. This explains the wide use of the technique in gas-phase studies, in research utilizing supersonic expansion cooling and molecular beam techniques, and in work on low-temperature matrices and single crystals. There are, however, fascinating uses of time-resolved LIF that do not require quantum state selection or resolution to unravel dynamic behavior (e.g., complex molec-

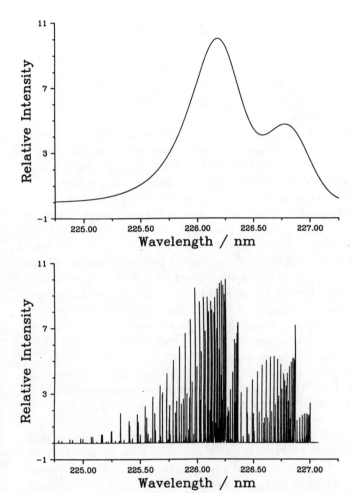

Figure 4.1 Comparison of (top) the room-temperature electronic absorption spectrum of gaseous NO recorded with a spectral bandwidth of 0.2 nm, with (below) a spectrum of the γ (0, 0) band simulated with a linewidth of 0.2 cm^{-1}, a temperature of 300 K, and the spectroscopic constants and line-strength factors reported in the literature.

ular systems in fluid solution). With conformational changes of biological systems or the segmental mobility of polymers in solution, the time resolution of LIF becomes essential and the spectral resolution is of minor interest.

With this chapter an attempt will be made to provide a broad overview of the current use of LIF with the main emphasis on gas-phase studies and the interface between chemistry and physics. Some applications in engineering and environmental or analytical applications can only be dealt with briefly. Applications in biological, medical, and forensic diagnostics are important but have not been included.

4.2. Principles and Basic Theory

The power of the LIF technique can be exploited most fully in applications involving diatomic or small polyatomic molecules, whose internal states are to be interrogated at low pressure in the gas phase. The species of interest is detected by tuning a pulsed or continuous-wave (CW) laser across a rovibronically resolved electronic band system. When the tuned narrow-bandwidth laser light comes into resonance with a rovibronic transition, the species absorbs a small fraction of the incident light, which the pumped level of the electronically excited state reemits by spontaneous radiative decay i.e., fluorescence. A fraction of this resonance fluorescence is collected and imaged onto the detector, providing the fluorescence signal in proportion to the excited state number density. As the laser scans across a more or less resolved electronic band an excitation spectrum can be recorded that essentially mirrors an *absorption spectrum with the optical resolution of the laser.* The most dramatic difference between absorption and fluorescence excitation spectra lies in their sensitivity. Although it is difficult to distinguish small dips in the amount of transmitted light, one records a positive fluorescence signal on a zero background, enabling fractions of absorbed to incident light of less than 10^{-7} to be routinely measured by LIF. Thus, the LIF technique represents the laser variant of the long-established fluorescence excitation method.

LIF experiments designed well and with attention to detail can provide a sensitivity approaching 10^{12} molecules m^{-3} and per quantum state for many diatomics, such as CN, NO, CO, and OH in gas-phase studies. It is possible to achieve *single-atom detection* using optically saturated LIF schemes, but in general it is easier to achieve single-species detection with a multiphoton ionization (MPI) scheme using single-ion detection techniques, since ions can be both collected and detected with 100% efficiency, at least in principle. This advantage is, however, balanced by the drawback of having to perform the MPI analysis in the small volume of a tightly focused laser beam, whereas LIF can probe rather large volumes. Thus, when detection sensitivity is measured in units of concentration or number density rather than species number, there is less to choose between the two techniques.

Clearly, only species with bound electronically excited states that can be accessed within the available range of tunable laser radiation are amenable to LIF detection. On the one hand, the excited state must be sufficiently stable to allow emission to compete during its lifetime; on the other hand, it should have a reasonably short radiative lifetime, preferably less than 1 μs. At any rate, an appreciable fluorescence quantum yield, preferably not depending on the internal state, is mandatory. These are serious constraints that limit the utility of the technique more for molecules than for atoms, where suitable excitation schemes can be located more readily and nonradiative processes are less important. The range of atomic, diatomic, and light polyatomic species amenable to LIF detection is, however, impressive and still keeps growing rapidly as the wavelength range of pulsed tunable laser devices keeps on extending into the far ultraviolet (UV) and vacuum (VUV) ultraviolet. It includes most of the small species important in combustion, plasma, and atmospheric chemistry and physics, such as OH, NO, CN, CH, SiH, CCl, O_2, CO, CH_2, CF_2, SiH_2, CCl_2, and NO_2. LIF is, of course, also suitable for the spectroscopy of a wide range of inorganic and organic species, including large biomolecules. Here its utility as a detection and interrogation technique is more limited because of the congestion of quantum states and concomitant dilution of number densities per state.

4.2.1. Basic Theory

The underlying concepts of the theory describing the interaction of laser light with absorbing systems are outlined in Chapters 1 and 3. Detailed material on optical transition rates in LIF, the effects of lineshapes and anisotropy or polarization, and optical saturation can be found elsewhere.[6–9] Consider LIF in a collision-free isotropic environment in order to focus on the most elementary principles involved. For a two-level system the rate of excitation to the upper level u by a laser tuned to the $l \rightarrow u$ transition is given by

$$\frac{dn_u}{dt} = B_{lu} I \, n_l \tag{4.1}$$

where n_l and n_u are the number densities in m^{-3} in the lower (l) and upper level (u), I stands for the intensity of the laser light in Wm^{-2}, and B_{lu} stands for the Einstein absorption coefficient. Let us neglect that B_{lu} and I would need to be convoluted by

$$B_{lu} I = \int B_{lu}(v) I(v) \, dv \tag{4.2}$$

to take into account the lineshapes of the laser light and the spectroscopic transition. The Einstein coefficients B_{lu} for absorption of *unpolarized isotropic* light and the rate constant for the total emission A_{ul} are related by

$$A_{ul} = 4h\bar{v}^3 B_{lu} \tag{4.3}$$

where the wavenumber is $\bar{v} = v/c$, and the radiative lifetime of the excited level τ is the inverse of radiative rate constant and Einstein coefficient for total emission:

$$(\tau)^{-1} = k_{\mathrm{rad}} = A_{ul} \tag{4.4}$$

For a typical *diatomic* molecular system spontaneous (resonance) emission from the vibronic level v' to the lower levels v'' will proceed with a rate constant

$$k_{v'} = (\tau_{v'})^{-1} = \sum_{v''} A_{v'v''} \tag{4.5}$$

Here the Einstein coefficients for the spontaneous emission of individual vibronic bands emitted to the ground electronic state may be approximated by

$$A_{v'v''} = \frac{64\pi^4}{3h} v^3_{v'v''} q_{v'v''} \mid R_e \mid^2 \tag{4.6}$$

Here v is the frequency, q is the Franck–Condon factor and $\mid R_e \mid^2$ is the electronic transition moment whose dependence on the internuclear distance cannot necessarily be neglected and may have to be determined experimentally from lifetime measurements or from calculations.

For atomic and bound–bound transitions in diatomic species the fluorescence quantum yield Φ_f is usually unity in the absence of collisional quenching. For some diatomic species (e.g., the $A\ ^2\Sigma^+$ states of OH and SH predissociation can lead to a level-dependent fluorescence yield. As v' increases, the rate of predissociation by crossing to a lower repulsive state increases, rapidly reducing Φ_f. Since the measured total rate of decay is

$$k = \tau^{-1} = k_{\text{rad}} + k_{\text{nonrad}} \tag{4.7}$$

it is feasible to estimate Φ_f as follows:

$$\Phi_f = k_{\text{rad}}\,\tau = \frac{k_{\text{rad}}}{k_{\text{rad}} + k_{\text{nonrad}}} \tag{4.8}$$

where k_{nonrad} stands for the rate constant associated with the nonradiative decay. For polyatomic molecules nonradiative pathways other than predissociation by curve-crossing become more important with increasing complexity. Here other unimolecular decay processes leading to loss of fluorescence are intersystem crossing (ISC), internal conversion (IC), and molecular rearrangements, and the observed fluorescence decay is often dominated by the rates of nonradiative processes. These processes, in addition to the dilution of number densities per state, reduce the sensitivity of LIF detection for larger polyatomic molecules, but they form the basis and object of time-resolved fluorescence studies where pulsed lasers compete as excitation sources with pulsed lamps and synchrotrons.

When the fluorescence from the excited state is spectrally and temporally integrated under collision-free conditions, a fluorescence signal FS is obtained:

$$\text{FS} = Cn_f \sum_{v''} A_{v'v''} = Cn_f/\tau \tag{4.9}$$

Here the apparatus constant C comprises all factors for geometry, optical collection and transmission, detection efficiency and electronic gain. Note that C may vary significantly over wider wavelength ranges.

4.2.2. Product-State Distributions and Alignment

One of the most elegant applications of LIF aims at dynamical measurements of state-resolved elementary chemical reactions, dissociation, and scattering processes. Here LIF provides, in conjunction with appropriate theoretical models, a microscopically detailed insight into the dynamics of such elementary processes. It constitutes the principal optical technique for interrogating the quantum states of atoms and diatomics in the ground or low-lying electronic states. Both scalar and vector attributes of the probed species are accessible. Applications to beam-surface scattering, the probing of surface chemistry and desorption processes will also be considered later.

All these applications demand the extraction of internal state populations from relative intensities of LIF excitation spectra. Here only a brief summary needs to be given, since the theoretical basis has been described in detail elsewhere.[1,6-8] An important requirement for the use of LIF detection is the need to understand and assign the relevant electronic band system. This may require spectral simulations or band contour calculations. Exothermic chemical reactions and photodissociation processes often produce rotationally and vibrationally highly excited nascent products where the spectroscopic data available in the literature are not applicable or cannot be extrapolated. The most important data required once a spectrum is assigned are the Franck–Condon factors $q_{v'v''}$ and the rotational line-strength factors. These are either listed in the literature or have to be calculated using quantum mechanics and, in the latter case, complex angular momentum algebra. Equations available for calculating the rotational line-strength factors applicable for LIF detection are based on the Breit formula and are often factored into the Hönl–London factors appropriate for isotropic excitation and emission, and a geometry factor that depends on the angles between the directions of linear polarization of analysis beam ε_a and the detector ε_d in a laboratory-fixed axis.[7,8,10,11]

In the standard LIF detection method the fluorescence is collected and detected without selection of the polarization and rotational branch of the emitted resonance light. In this case the geometry or angular factor mainly affects the ratio of line-strengths of P/R versus Q branches for high-J and changes the intensities for low-J transitions. For axially propagating polarized laser beams the Hönl–London factors S_J cannot predict line-strengths correctly in theory.[6-8] In practice, however, the Hönl–London formulas[8-10] are often adequate under these conditions except for low values of J. Thus, the simple expression

$$I_{rel}(vJ) = C'\tilde{v}\left(\frac{S_J}{(2J + 1)}\right) q_{v'v''}n_{vJ} \tag{4.10}$$

can be used to evaluate population distributions from different vibronic bands provided that the sample is isotropic. Here C' comprises the product of the apparatus constant defined previously (Section 4.2.1) and the laser power, $\tilde{v}$ is the wavenumber of the light, S_J are the isotropic (Hönl–London) line-strength factors, $q_{v'v''}$ are the Franck–Condon factors that permit the comparison of different vibronic bands, and n_{vJ} stands for the populations. Note that the validity of equation (4.10) does not extend to experiments with aligned samples such as nascent photofragments produced by a photolysis laser pulse preceding the LIF analysis pulse. Equation (4.10) also fails under conditions where optical saturation of the electronic transition becomes important, and much more elaborate analysis is required as described by Hefter and Bergmann.[6]

A simple test of the validity of (4.10) under a given set of experimental conditions can be carried out when the species to be probed can be rotationally thermalized by collisions with a bath gas. For analysis within a single vibronic band (4.10) can be transformed to

$$I_{rel}(J) = C''S_J \exp[-hcBJ(J + 1)/kT_{rot}] \tag{4.11}$$

Here the dependence on Franck–Condon factor and wavenumber has been dropped and the population n_J has been expressed through a Boltzmann distribution characterized by a rotational constant B for the probed species and a Boltzmann temperature T_{rot} in thermal equilibrium with the surroundings. A graph of $\ln(I_{rel}/S_J)$ versus $E_{rot} = hcBJ(J + 1)$ gives a linear plot with a slope equal to $-1/kT_{rot}$, and T_{rot} is identical to the thermal equilibrium temperature of the gas if there are no deviations of the line-strengths from the isotropic Hönl–London factors. The test is particularly stringent if data are taken and compared from different branches to eliminate further the possibility of alignment and optical saturation effects. A more simple test for the latter is a comparison of the observed and calculated isolated lines originating from the same ground-state level but differing in the selection rule by one ($\Delta J = \pm 1$). In this case we get directly from (4.10):

$$I^{R/P}(J)/I^{Q}(J) = S_J^{R/P}/S_J^{Q} \tag{4.12}$$

Equations (4.10)–(4.12) clearly do not hold for anisotropic samples—for example, where significant alignment is present, and when excitation conditions approach optical saturation.[6-9] Analysis of the alignment of the angular momentum vector **j** by polarized LIF can provide incisive information on the dynamics of formation of the species to be probed. The alignment factor $A_0^{(2)}$ is detected and measured by the characteristic redistribution of intensities between the rotational branches of an LIF excitation spectrum that occurs when the polarization of the probe beam is rotated by a Fresnel-Rhomb or half-wave retarder. Optical saturation effects are excluded by reducing probe beam intensities until the relative intensities between transitions with low and high S no longer change. This can be a more sensitive experimental test than power broadening where the Doppler widths are already significant. Note that any dependence of LIF spectra on orientation and alignment must be dealt with properly even if only scalar quantities such as populations or velocities are desired. Although m-resolved level populations can be measured by LIF,[6] it proves more practical for most purposes to measure the moments of the m-state distribution using the concept of multipoles, as elegantly presented by Greene and Zare.[7]

The axially symmetric geometries most relevant for beam-gas experiments have been considered in detail, and relationships usable by experimentalists are available,[7] as are their extensions to optical saturation effects.[6,9] In general, only the even moments in the form of the populations $A_0^{(0)}$ and the axially symmetric quadrupole and hexadecapole moments $A_0^{(2)}$ and $A_0^{(4)}$ are readily accessible. The alignment factor $A_0^{(2)}$ takes values in the interval $(-\frac{2}{5}, +\frac{4}{5})$ and represents the expectation value $A_0^{(2)} = 2 < P_2\,(\mathbf{j \cdot z}) >$ for perpendicular and parallel alignment with respect to the symmetry axis z in the laboratory frame ($A_0^{(2)} = -0.4$ for $\mathbf{j} \perp \mathbf{z}$ and $+0.8$ for $\mathbf{j} \parallel \mathbf{z}$). The clearest examples of alignment are found in the fragments of direct photodissociation processes where the parent molecules selected by the electric-vector of the photolysis beam $\boldsymbol{\varepsilon}_p$ produce an aligned ensemble of fragments whose $\mathbf{j}$ vectors retain memory of the original orientation of the molecule. An example from the author's laboratory is shown in Figure 4.2 for NO ($v = 0$) from the 560-nm dye laser dissociation of CH_3SNO in a supersonic jet.[12] The average $A_0^{(2)}$ of $+0.75$ is close to the limiting case for parallel alignment (i.e., $\mathbf{j}$ is oriented parallel with the electric-vector $\boldsymbol{\varepsilon}_p$ of the dissociating beam. The transition dipole moment $\boldsymbol{\mu}$ must be oriented perpendicularly to the planar CH_3SNO framework as expected for an $^1A'' \leftarrow {}^1A'$ (n, π^*) electronic transition.

The most reliable method for evaluating the alignment is to compare integrated peak intensities I as a function of branch and geometry[7]:

$$I = CSG \left(b_0 + \frac{5}{4} b_1 A_0^{(2)} \right) \tag{4.13}$$

where C is proportional to the population, S is the line-strength, and G accounts for different experimental geometries (e.g., laser beam overlap): b_0 and b_1 are calculated constants for a given rovibrational transition and geometry. Once the alignment $A_0^{(2)}$ has been evaluated from the LIF spectra all the scalar attributes (i.e., state-resolved population distributions over the internal degrees of freedom of the probed fragment) become available.

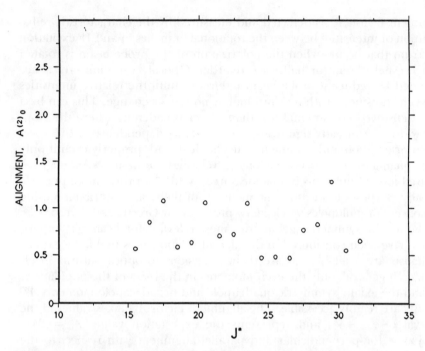

Figure 4.2 Alignment $A_0^{(2)}$ of NO ($v = 0$) from the polarized photodissociation of jet-cooled CH_3SNO at 560 nm as a function of the angular momentum of the photofragment. The average alignment parameter of $+0.75 \pm 0.24$ is close to the theoretical maximum ($+0.8$).

This includes fine-structure populations such as Λ-doublet, spin–rotation doublet, and spin–orbit ratios that can contain very subtle dynamical information.[13]

4.2.3. *Doppler Profiles, Velocity Distributions, and Photofragment Vector Correlations*

The Doppler profiles of atomic or molecular LIF transitions provide information on the translational energies and velocity distributions of the absorbing species. The Doppler shift induced by the motion of an absorbing species is

$$v = v_0 \left(1 \pm \frac{v}{c} \right) \tag{4.14}$$

where the positive shift relates to motion toward the light source, and the negative shift reflects motion away from the source. A Maxwell–Boltzmann velocity distribution characterized by a temperature T gives rise to a Gaussian spectral line profile described by

$$I(v) = I_0 \exp\left[-\left(\frac{c(v - v_0)}{v_0 v}\right)^2\right] \tag{4.15}$$

where $v = (2kT/m)^{1/2}$ is the most probable speed and the Doppler width [i.e., full width at half maximum (FWHM)] is

$$\text{FWHM} = 2 \, \Delta v_D = \frac{2v_0}{c} \left(\frac{2kT \ln 2}{m}\right)^{1/2} \tag{4.16}$$

This expression yields $2 \, \Delta v_D = 7.16 \times 10^{-7} \, v_0 \, (T/M)^{1/2}$ in s^{-1}, with T in Kelvin and M in units of g mol^{-1}.[14]

Direct or prompt indirect photodissociation processes induced by polarized laser beams prepare an anisotropic ensemble of fragments. As outlined earlier, polarized LIF excitation spectra can reveal the alignment [i.e., the (j,μ), or more precisely, (j, ε_p) correlation]. Dixon and others have shown how the analysis of Doppler profiles of isolated LIF lines of recoiling molecular fragments can reveal new information on the anisotropy, the kinetic energy release, and subtle vector correlations characterizing the dynamics of the dissociation process.[15–18] By recording LIF spectra with line-narrowed analysis beams either coaxial or perpendicular to the photolysis beam, and with polarizations set either parallel $(\varepsilon_a \parallel \varepsilon_p)$ or perpendicular $(\varepsilon_a \perp \varepsilon_p)$, fascinating new information can be gleaned on the correlations between several vector properties.[19] The $(\mathbf{v}, \boldsymbol{\mu})$ correlation is measured by the anisotropy parameter β that characterizes the angular distribution function for the recoiling fragments

$$I(\theta) = \frac{1}{4\pi} [1 + \beta P_2(\cos\theta)] \tag{4.17}$$

where $\cos\theta = \mathbf{v} \cdot \boldsymbol{\varepsilon}_p$. For fast dissociation the limiting cases are $\beta = +2$ when $(\mathbf{v} \parallel \boldsymbol{\mu})$ or $\beta = -1$ for $(\mathbf{v} \perp \boldsymbol{\mu})$, where $\mathbf{v}$ and $\boldsymbol{\mu}$ are the velocity and transition dipole moment vectors. The Doppler profiles also contain information on the alignment $A_0^{(2)}$ or $(\mathbf{j}, \boldsymbol{\mu})$ correlation, the correlation between the recoil velocity vector $\mathbf{v}$ and $\mathbf{j}$, and the triple vector correlation between $\mathbf{v}$, $\mathbf{j}$, and $\boldsymbol{\mu}$. Dixon's lineshape analysis treats the Doppler profile of an isolated rovibronic line with the lineshape function

$$g(\chi) = \frac{I}{2\Delta v_D} [1 + \beta_{\text{eff}} P_2(\cos\theta) P_2(\chi)] \tag{4.18}$$

where $\Delta v_D = v_0 \, v/c$ is the maximum Doppler shift, ω_0 is the line center, and χ is the ratio of the Doppler shift to its maximum shift, $\chi = (v - v_0)/\Delta v_D$: I is proportional to the integrated intensity of the transition, $P_2(x) = (3x^2 - 1)/2$ is the second Legendre polynomial, and θ stands for the angle between ε_p and $\mathbf{k}_a$ (i.e., the photolysis polarization and probe laser propagation vectors). The parameter β_{eff}, which represents the effective anisotropy parameter, is a function of four bipolar moments β_0^k (k_1, k_2) with coefficients $b_0 \cdots b_4$ that can be calculated for a given rovibrational transition and experimental geometry.[15] Since the alignment $A_0^{(2)} = \frac{4}{5} \beta_0^2$ (02) is measured separately as outlined earlier, the three remaining bipolar moments have to be determined from the analysis of the Doppler profiles. This requires recording LIF lines for different dissociation, probe, and detector geometries with isolated profiles from different branches to obtain a reliable set of moments.

The most interesting cases where such a detailed analysis can be justified is in the photodissociation of triatomic molecules, since the counter fragment is monoatomic and the probed fragment can be taken to have a single center of mass recoil velocity. Analysis of the profiles is usually more rewarding for light fragments recoiling from heavy counter fragments due to the increased Doppler width associated with their fast recoil.[16-18,20] The uncertainty of Doppler profile simulations increases when broad velocity distributions of heavy fragments and significant contributions from the thermal velocity distributions of parent molecules have to be included.

Figure 4.3 shows an example of a Doppler profile simulation for NO for

Figure 4.3 Doppler profile of LIF transition of nascent NO on the $J = 33.5$ line of the P_{11} branch of the $\gamma(0,0)$ transition. Nascent NO was formed by the 560-nm photodissociation of CH_3SNO in a supersonic expansion with He. Pump and LIF probe laser were aligned collinearly with $\varepsilon_a \perp \varepsilon_p$.

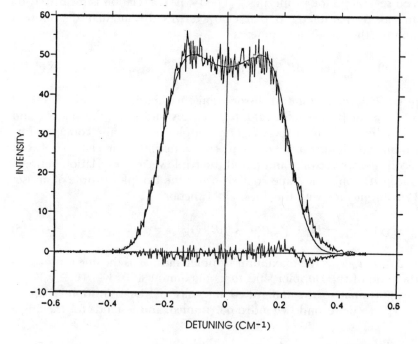

DETUNING (CM-1)

the 560-nm photodissociation of jet-cooled CH_3SNO, with a Doppler shift of $\Delta v_D = 0.210 \pm 0.003$ cm^{-1}, $\beta_{eff} = -0.45$ and a mean recoil velocity for the NO fragment of $1,450 \pm 100$ ms^{-1}. Docker et al. have recently reported an analysis of Doppler-broadened NO two-photon LIF lineshapes for the photodissociation of CH_3ONO at room temperature. Simplifying assumptions about the spread of translational energies have to be made to analyze the profiles in detail.[21]

4.2.4. Collisional Effects: Relaxation, Electronic Quenching, and Energy Transfer

An essential goal in applying LIF to dynamical studies such as those covered in Sections 4.2.2 and 4.2.3 is to obtain scalar and vector information on *nascent* products (i.e., the information carried by the species probed must not be scrambled or modified by collisions after the initial reactive or scattering event). The collisional relaxation of nascent fragments, reaction products, or scattered species is most readily avoided by using molecular beams or low-pressure gas samples under conditions where the average time between collisions is longer by several orders of magnitude than the time scale of the experiment. In pulsed experiments the time scale is defined as the time elapsed between the collision, half-collision, or scattering event and the time of LIF probing. The lifetime of the excited state accessed by LIF can also play a role under conditions where collisional redistribution of upper-state levels or state-dependent predissociation or electronic quenching occurs. These effects become more important in dynamical experiments for species with long fluorescence lifetimes and when collision partners with large quenching or energy transfer cross sections are present.

Collisional effects in free-jet expansions are commonly encountered in nanosecond dissociation-probe experiments, particularly with pulsed expansions and rare gases other than He. For rare gases as collision partners, the rate of collisional relaxation generally increases with the gas kinetic collision cross section in the order *E, V, R, T* (electronic, vibrational, rotational, translational). Thus, the Doppler profiles of LIF transitions of nascent fast, and light product species are usually the most sensitive indicators of relaxation effects.

Collision rates in supersonic beams or jets decrease with the axial distance from the orifice and can be calculated by the formulas of Lubman et al.[22] Close to the nozzle orifice the rates of collisional relaxation are obviously quite high. Careful studies of the effect of nozzle-laser distance on relative rotational intensities or band contours or on Doppler line profiles are necessary to exclude collisional relaxation or reorientation effects. Collisions of electronically excited states may result in quenching of the fluorescence, in nonradiative intramolecular transitions (i.e., predissociation, curve crossing), reorientation of **j**, and *V, R, T* energy transfer populating other rovibronic levels within the excited state manifold. This obviously changes the intensity, polarization, and spectral distribution of the emitted light. These general principles form the basis of the important field of electronic quenching and energy transfer,[23] which includes very detailed state-to-state and beam vector correlation studies.[24-27] These processes are important in molecular lasers, plasma discharges, combustion, and atmospheric physics. From a fundamental perspective, studies of inelastic (energy transfer) processes serve as test beds for the interaction potentials predicted by theory.[26,27]

While quenching leads to loss of fluorescence, collisional energy transfer redistributes population from the optically pumped level to neighboring levels, whose emission will be detected as loss only if it falls outside the spectral bandwidth of the detection system. Collisions with light rare gases, such as He or Ne, which cannot quench, rapidly transfer population to nearby levels and ultimately lead to rotational thermalization in the upper as well as the lower state. The relative rates of quenching and $V/R - T$ energy transfer need to be considered where LIF serves as a temperature or number density gauge at intermediate and high pressures.

Electronic quenching of fluorescence in the gas phase plays an important role in LIF, not to mention its wider significance in molecular gas lasers, discharges, combustion diagnostics, and atmospheric physics and chemistry. Collisional quenching by definition leads to loss of fluorescence. This can occur as a consequence of chemical reaction, resonant (Förster type) energy transfer, charge transfer with ion pair formation and straightforward $E - V/R/T$ energy transfer between the collision partners.

The kinetics of electronic quenching is described by the well-known Stern–Volmer relationship. The rate constant for the decay of fluorescence can be expressed as

$$k = k_f + \sum_i k_i Q_i \qquad (4.19)$$

where $k_f = (\tau_f)^{-1}$ is the fluorescence rate constant in the absence of a quencher as defined by equation (4.7), k_i are the bimolecular quenching rate constants for collision partners i, and Q_i are their pressures or concentrations, respectively. Remembering $k = \tau^{-1}$, we can express (4.19) in terms of fluorescence lifetimes for a simple case where only one quencher is present:

$$\frac{\tau_0}{\tau} = 1 + k_Q \tau_0 Q \qquad (4.20)$$

where τ is the measured lifetime, τ_0 is the fluorescence lifetime in the absence of a quencher, k_Q is the quenching rate constant, and Q is the partial pressure of the quencher. Self-quenching is a very common phenomenon in both gas and condensed phases.

The fluorescence quantum yield ϕ_f previously expressed by (4.8) changes in the presence of a quencher to

$$\phi_f = \frac{k_{rad}}{k_f + \sum_i k_i Q_i} \tag{4.21}$$

where $k_f = k_{rad} + k_{nonrad}$. In practice, conditions often prevail where $\sum_i k_i Q_i \gg k_f$—for example, electronic quenching by atmospheric O_2 at normal ambient pressure and where quenching by other species can be neglected; here

$$\phi_f \approx \frac{k_f}{k_q Q} \tag{4.22}$$

The most commonly encountered form of the Stern–Volmer relationship is

$$\frac{I_0}{I_Q} = 1 + k_Q \tau_0 Q \tag{4.23}$$

It is easily seen that k_Q can be evaluated from Stern–Volmer plots [i.e., from the linear relationships (4.20) or (4.23)], by measuring fluorescence lifetimes or steady-state emission intensities as a function of the pressure of the quencher from slopes $k_Q \tau_0$ and τ_0. Equations (4.19)–(4.23) summarize the practically important fact that electronic quenching shortens the lifetime of the excited states and decreases the fluorescence efficiency.

Efficiencies of electronic quenching and energy transfer are commonly expressed by the collision cross section σ and its ratio with the gas kinetic collision cross section $R^2 \pi$, where R is the Lennard–Jones collision diameter. The quantities usually measured are second-order rate constants commonly expressed in $\mu s^{-1} torr^{-1}$, $dm^3 mol^{-1}$, s^{-1} or cm^3 molecule$^{-1} s^{-1}$. The relationship between the energy transfer or quenching cross section and the conventional rate constant is

$$k = \sigma v \tag{4.24}$$

where v is the center-of-mass rms thermal velocity $v = (8kT/\pi\mu)^{1/2}$ with μ for the reduced mass of the collision pair.

Open-shell species such as O_2, NO, NO_2, and other radicals quench fluorescence as well as phosphorescence very efficiently. This is the main reason LIF techniques are not generally considered very promising under ambient atmospheric conditions. The main exception to this rule is a variant of LIF, laser-induced predissociation fluorescence (LIPF) spectroscopy[28,29] that uses the fact that certain rovibronic states of diatomics predissociate so rapidly that collisional quenching does not contribute significantly even at high pressures in a flame. Obviously, ϕ_f given by (4.21) and (4.8) becomes very small under these conditions since $k_f \gg k_{rad}$ if predissociation (k_{nonrad})

becomes very rapid. Powerful lasers are therefore needed to overcome the small fluorescence yields. The important application of this technique in combustion diagnostics and concentration imaging in flame will be summarized briefly elsewhere in this chapter (4.9.2). The main significance of the technique lies in the principle that the effects of collisional quenching can be suppressed.

4.3. Instrumentation and Ancillary Techniques

4.3.1. Tunable Laser Sources for LIF

CW and nanosecond pulsed dye lasers are the most widely used tunable sources for LIF, with argon ion and Nd:YAG or XeCl excimer lasers the most frequently employed pump sources. Very high spectroscopic resolution can now be achieved with CW single-mode, standing-wave ring dye lasers. Commercial equipment allows scanning of such systems over 20,000 GHz with an accuracy of the order of 50 MHz using I_2 or Te_2 lines as wavelength standards and confocal Fabry–Perot interferometers for frequency marking. Bandwidths less than 5 MHz are routinely achievable over shorter scans, provided appropriate stabilization techniques are employed. Such systems are well suited for intracavity frequency doubling and provide single-frequency radiation with <3-MHz bandwidth near 320 nm, where many larger aromatic molecules absorb.[30,31] Sub-Doppler resolution is needed for these molecules in order to resolve the rotational structure of electronic bands and to measure homogeneous linewidths in the MHz region.

The spectral region below 280 nm has not been explored much at this resolution, mainly because of the difficulties involved in frequency-doubling CW dye laser radiation. Another approach has to be adopted here and where high power is needed for nonlinear spectroscopy or frequency extension.[32] Excimer, Nd:YAG, or copper vapor lasers can be employed for pulse amplification of narrow linewidth CW radiation. This can provide pulsed laser radiation of very high peak power and close to Fourier-transform-limited bandwidth. Riedle et al., for example, have employed 5-ns pulsed radiation of 100-MHz bandwidth near 240 nm to explore the notorious channel 3 region of benzene.[32] Figure 4.4 shows a setup based on commercial equipment using excimer pumped pulsed dye amplification capable of producing 10 mJ per pulse with a bandwidth of 0.002 cm^{-1} (i.e., 60 MHz near 365 nm).[33] A similar setup has been described by Lee et al., who succeeded in using this source for measuring Doppler profiles of nascent H_2 in the VUV from the photoelimination of cyclohexadiene. Here the high power was required for (1 + 1) REMPI spectroscopy of H_2.[34,35] Pulse amplification schemes utilizing relatively long pump pulses with a

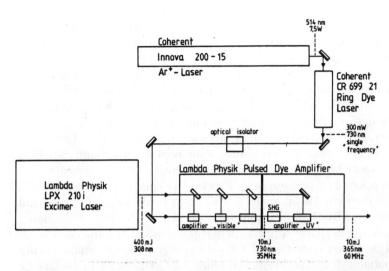

Figure 4.4 Setup based on commercial laser equipment for the production of tunable ~8–ns pulses with close to Fourier-transform limited bandwidth.[33] Similar pulse amplification schemes based on seeded Nd:YAG lasers have also been reported. (Reproduced with permission from Ref. 33.)

smooth temporal profile are clearly advantageous. The use of harmonics from injection-seeded Nd:YAG and long-pulse excimer lasers for such schemes opens up interesting new possibilities for high-resolution spectroscopy in the far-UV and VUV, as well as for two-photon and multiphoton spectroscopy.

Ion laser-pumped CW dye lasers can operate up to 1,000 nm, but the near-IR beyond 1 μm can be covered better by color-center, difference-frequency, and semiconductor diode lasers. Another interesting device for LIF in this region is the optical parametric oscillator (OPO). Very-high-resolution spectroscopy with bandwidths down to less than 1 MHz is possible in principle in the 650–1,600-nm region with properly stabilized commercial diodes used in fiber-optic information transmission.[36] Since the Einstein coefficients for spontaneous emission are much smaller and detection techniques become less straightforward, LIF spectroscopy in the near-IR is less developed than that in the visible and UV, although there are many atomic, diatomic, and triatomic species, such as SiN, HO_2, and C_2H with electronic transitions in this region.[37,38] A recent development for CW spectroscopy in the far red and near-IR is the argon ion pumped titanium sapphire laser, which uses a crystal rod rather than the notoriously toxic IR-dye solutions as the gain medium. Pulsed single-mode dye lasers capable of 150 MHz bandwidth have been available for a few years. They can be pumped by standard Nd:YAG and excimer lasers with diffraction coupled unstable resonator optics. For lower-resolution LIF spectroscopy in the near-IR a range of nonlinear frequency conversion techniques is available for use with standard pulsed dye, lasers as reviewed elsewhere.[13] (See also Chapter 2.)

Commercial pulsed dye lasers with grazing incidence oscillators, or angle- and pressure-tuned air-spaced intracavity etalons, provide line-widths close to 0.05 cm^{-1} or 1.5 GHz and permit convenient tuning over wide wavelength ranges. In conjunction with frequency doubling a wave-length coverage of 215–1,000 nm at this linewidth has become fairly routine. Typical performance levels of commercial excimer and Nd:YAG pumped systems have been summarized recently.[13] Two major developments have opened up new opportunities for LIF spectroscopy in the VUV, far-UV, and the near-IR—namely, the injection seeded Nd:YAG laser and β-barium-borate (BBO), which permit very efficient frequency-doubling, difference-frequency, and sum-frequency generation. The injection seeding technique allows operation of YAG lasers with monochromatic output of high spatial and temporal quality, while BBO combines high damage threshold and nonlinear coefficients with wide optical transparency for light from 190 nm to the IR. Thus, with Raman shifting and sum-frequency generation the range down to 190 nm can now be covered more routinely.

LIF studies in the VUV have also become easier following recent advances in the generation of tunable laser radiation in this difficult spectral region. The techniques available and a brief summary of achievable performances have been compiled elsewhere.[13] All available methods rely on highly nonlinear processes and require very powerful nanosecond pulsed dye lasers operating at peak efficiency and with output of high spectral and spatial quality. Briefly, the main methods available for the experimentalist are frequency-tripling or third harmonic generation (THG) in rare gases, two-photon resonant sum-frequency generation in metal vapors mixed with rare gases, and stimulated anti-Stokes Raman shifting in hydrogen. The latter technique covers wavelengths down to 180 nm, but the shorter wavelengths necessary to probe CO by one-photon LIF at high sensitivity and spectral resolution (i.e., 145 nm) requires the much more challenging four-wave mixing process.[39] Because of the difficulties involved in generating these very short wavelengths for one-photon spectroscopy, it is essential to examine two-photon options available for LIF interrogation of atoms and simple molecules.

Another high-power source for LIF studies is available with tunable, line-narrowed excimer lasers. These combine nearly diffraction-limited beam quality with extremely high spectral brightness and tunability of 0.2–0.6 nm with narrow linewidths of the order of 0.5 cm^{-1}. For ArF, for example, an optical resolving power of better than 100,000 can be obtained at 193 nm with a tuning range of 350 cm^{-1}. Both ArF and KrF lasers find important applications in combustion diagnostics by LIF, where concentrations of a variety of species, including O_2, NO, OH, and H_2O, can be imaged by laser-induced predissociation fluorescence (LIPF) spectroscopy without effects from collisional quenching.[28,29]

In summary, for LIF as for most laser spectroscopic techniques, the ideal source would combine sub-Doppler linewidth with the very high peak power of a nanosecond pulsed source to cover the wavelength range of 180–2,000 nm that can be achieved currently with well-established nonlinear frequency-doubling and mixing techniques. The physical limit that one might aim to approach is dictated by the Fourier transform of the laser pulse duration T given by $(4\pi T)^{-1}$. The Doppler width of a spectroscopic transition, however, can easily be reduced by jet cooling and some very simple experimental tricks. Obviously, small linewidths allow Doppler-free spectroscopy, but this cannot usually be achieved without CW lasers. In practice, it is therefore necessary to aim for a compromise, since CW dye lasers have a very limited wavelength range of 400–1,000 nm. The development of pulsed tunable dye lasers has slowed down markedly, but linewidths of 100–200 MHz at high power can already be achieved and will become available, we hope, at a lower price in the near future. Progress with semiconductor diode lasers, on the other hand, continues rapidly. These devices are small, cost very little, and are long-lived. In addition, they can be tuned and frequency-doubled very easily. The other area likely to benefit from recent progress in laser development is LIF spectroscopy in the VUV, where there is much scope for further applications.

4.3.2. Supersonic Beams and Jet-Cooling

The impact of supersonic expansion and jet-cooling techniques on advances in LIF spectroscopy has been dramatic, if not spectacular. Their most frequently exploited advantages are as a powerful preparation technique and as a method for removing spectral congestion to simplify electronic spectra and their analysis. Other advantages include the reduction or elimination of Doppler broadening of spectroscopic transitions and gains in sensitivity and specificity that can be valuable in analytical applications of LIF. Nozzle sources and the theory and practice of their application in molecular beam scattering have been summarized recently.[40]

The velocity and internal quantum state distribution of molecules in cells or bulbs under conditions of thermal equilibrium are characterized by Boltzmann statistics. Gas-phase experiments with polyatomic molecules other than light triatomic hydrides deal with very broad ensembles of species corresponding to Boltzmann distributions over translational, rotational, vibrational, and in some cases electronic degrees of freedom. By entraining species of interest in carrier gases and expanding the gas mixture through a small orifice into a vacuum adiabatic cooling takes place over a short axial distance by (V,R,T) energy transfer in collisions with the bath gas. Undirected translational and internal (E,V,R) energy of the sample is converted to axial translational energy of the carrier gas. The translational temperature of the gas as measured by the width of the velocity distribution in the axial direction from the orifice drops. Both T_{rot} and, with some delay, T_{vib} fall rapidly over distances of 5–10 times the nozzle diameter. Ultimate translational and rotational cooling can be achieved with He using very high reservoir pressures, but the heavier monatomic rare gases cool rotations and vibrations more efficiently and are preferred if their increased tendency to form van der Waals complexes and clusters with the seed gas can be tolerated.[41]

The effect on LIF spectra of reducing rotational and vibrational temperatures to less than 10 K is dramatic. Spectral congestion from overlapping vibronic bands disappears, and populations collapse to a few highly populated quantum states, giving rise to a large enhancement of sensitivity as laser excitation becomes more efficient. Figure 4.5 shows the LIF spectrum of the γ(0,0) vibronic band of NO, which may be compared with the calculated room temperature absorption spectrum of Figure 4.1 and the jet-cooled LIF excitation spectrum of Figure 4.6. Both LIF excitation spectra of Figures 4.5 and 4.6 were recorded with the same cell and observation/ detection system. The electronic spectrum of NO in the 225-nm region has been discussed briefly in connection with Figure 4.1.

The advantages of extreme cooling were already known from matrix isolation spectroscopy, where LIF has been used very successfully.[42] Supersonic jet expansions offer the advantage of permitting studies of isolated molecules free from complications arising from interactions with the matrix and from the occurrence of different matrix sites. Very small amounts of impurities such as O_2 and H_2O or chemical impurities already present

Figure 4.5 LIF spectrum of gaseous NO at 300 K on the $A^2\Sigma^+ \leftarrow X^2\Pi$,γ (0,0) transition.

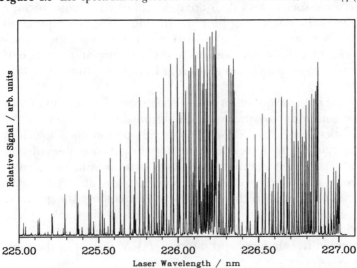

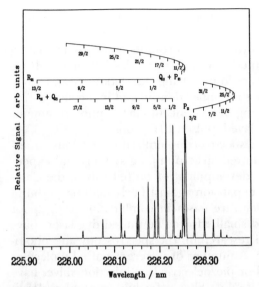

Figure 4.6 LIF spectrum of a few mtorr of NO in 400 torr N_2. The rotational temperature of the jet at a distance of 5 mm from the orifice is ~20 K. Only the F_1 rotational transitions from the $^2\Pi_{1/2}$ state are seen on this scale, since the population of the $^2\Pi_{3/2}$ higher-lying rotational manifold is drastically reduced due to the 123 cm^{-1} spin–orbit gap.

in the sample can have dramatic effects on the emission spectra of matrix-isolated species. These effects often bedevil matrix spectroscopic studies but are absent in supersonic jets.

For high-resolution LIF spectroscopy beams and jets also offer the possibility of reducing the Doppler widths of atomic and molecular transitions. Both the velocity components v_z in a direction perpendicular to the jet axis and the width of the axial velocity distribution decrease and become very small for well-collimated supersonic beams. In the most common configuration for Doppler-free LIF excitation spectroscopy the laser beam crosses a skimmed and collimated molecular beam at right angles. Majewski et al. achieved a resolution of 3 MHz around 300 nm in the UV by probing at a distance of 100 cm from the nozzle.[30,31] The standard way to obtain high resolution in freely expanding supersonic jets is to probe at large distances from the nozzle and to restrict the angle viewed by the collection optics such that only a small region on the jet axis is imaged onto the detector. The velocity spread Δv_z along the laser beam then decreases in proportion to the distance from the nozzle. Here resolution is limited by the residual Doppler width due to the angular divergence of the jet and by collisions upstream that scatter molecules into the observation zone. This technique can nevertheless be exploited for Doppler-free LIF spectroscopy without the need to produce a highly collimated molecular beam. With careful attention to experimental detail a practical limit around 50 MHz appears achievable in the visible. With nanosecond pulsed narrow-linewidth sources the best resolution achievable in the visible and UV is therefore limited by the laser.

LIF detection techniques are often used in conjunction with jet-cooling in experiments on molecular photodissociation dynamics. The main advantages of expansion cooling are here in preparing translationally and internally cold parent molecules at high number density and in a collision-free environment, which allows probing of nascent unrelaxed photofragments using pulsed dissociation and delayed, pulsed LIF probing.

Initial pioneering work on jet spectroscopy utilized continuous supersonic nozzle expansions and required large pumping systems and expensive CW lasers.[43] Since then the development of pulsed valves, the Campargue technique exploiting jet expansion against fairly high background pressure, and the use of heavier rare gases in combination with "poor man's" pumping stations[41,44] and small, affordable pulsed dye lasers have resulted in a rapid expansion of research on the spectroscopy and photophysics of jet-cooled molecules.[13] A number of different pulsed molecular beam sources based on solenoid or piezoelectric fuel injection valves have been described in the literature.[13,40,45,46] They offer higher number densities, lower gas and sample consumption, and better duty cycles in combination with pulsed lasers for LIF studies. Also, some of these sources are capable of delivering very short pulses down to 100 μs at repetition rates up to a few hundred hertz and can be constructed to permit operation at high and low temperature, and with corrosive vapors. Disadvantages in comparison to continuous sources are lower reliability, rapid mechanical and corrosive wear of moving seals, and the need to synchronize sample delivery with pulsed lasers. Pulse-to-pulse fluctuations in the sample supply can also deteriorate the signal-to-noise ratio.

4.3.3. Sample Handling and Preparation Techniques

The use of low sample pressures reduces self-absorption and self-quenching of the fluorescence and similar effects by other components of a mixture. As discussed previously (Section 4.2.4), high pressures of quenching gases can be tolerated for species with very short fluorescence lifetimes because of the large Einstein coefficients for spontaneous emission or rapid predissociation of the emitting levels.

Standard gas handling and mixing techniques appropriate for volatile species have to be modified for refractory inorganic and involatile organic compounds. Resistively heated cells and sample handling equipment based on the "heat pipe" principle can be used, which have a heated central section for establishing a sample-vapor equilibrium and cooled sections in front of the windows. In the presence of argon the diffusion of the sample vapor to the windows and scattering by deposits are prevented. Involatile samples can also be vaporized by heating, laser ablation, or nebulization of solutions and entrainment into a flowing rare gas. Many of these techniques have been developed to a high degree of refinement in connection with analytical atomic spectroscopy and mass spectrometry.[38]

The extension of these techniques to jet-cooling with pulsed or continuous nozzles is not necessarily straightforward. Grieman et al., for example, describe a high-temperature continuous free-jet expansion source that can be heated up to 1,000°C where substantial vapor pressures for quite refractory inorganic compounds are achieved.[47] This nozzle is shown in Figure 4.7 and has been used to record LIF excitation spectra of $NiCl_2$ cooled to a rotational temperature of 40 K with He. The standard technique to entrain thermally stable, large organic compounds into continuous supersonic jets for LIF spectroscopy consists of passing the rare gas over a heated bed of the finely divided solid and where necessary through a particle filter before expansion through the heated orifice.[30,31] High-temperature pulsed nozzles, where the choice of construction materials becomes quite critical, have been described by several groups.[48,49]

The jet-cooling of thermally unstable, involatile samples requires more elaborate equipment and can be effected by two techniques: (1) vaporization of solutions by "thermospray" methods and (2) laser desorption or vapourization. Rizzo et al. deposited the thermally labile amino acid tryptophan by the thermospray technique from its solution in the throat of a pulsed valve. Subsequent thermal desorption in vacuum and entrainment into the expanding jet permitted between one and three hours of pulsed jet spectroscopy, before the sample had to be replenished. The same group described a pulsed laser desorption technique for LIF of thermally sensitive large biomolecules.[50-52] Other techniques such as pulsed laser vaporization of solids within the throat or in an expansion channel bolted onto a

Figure 4.7 High-temperature supersonic beam source for LIF studies of involatile metal halides. Sample vapor from the heated alumina crucible is entrained by the carrier gas and expands through the circular orifice of the alumina nozzle. The source is surrounded by a light shield to minimize the detection of radiated light. (Reproduced with permission from Ref. 47.)

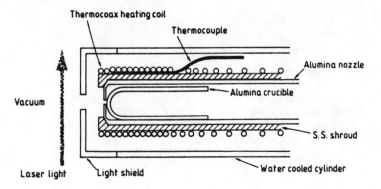

pulsed valve have also been described. These methods have been developed originally and independently by Dietz and Bondybey et al. and have become popular and important for the generation and spectroscopy of metal dimers and clusters as well as metal compounds.[53–55]

4.3.4. Preparation of Unstable Species

Metal vaporization techniques have become the workhorse for the in situ preparation of a variety of weakly bound metal dimers and small metal and semiconductor clusters.[56,57] Metal vaporization is achieved by focusing the beam of a pulsed laser onto a rotating rod made of the desired metal or onto pressed pellets of mixtures of metallic or semiconducting elements. The resulting vapor plume is entrained in He or Ar and is allowed to expand through a small orifice of 0.7–1.0-mm diameter into the evacuated observation region,[53–55,58,59] where tunable laser radiation excites the fluorescence of the jet-cooled species. Under conditions where the expansion is not supersonic, sufficient cooling of the hot plasma can be effected by external cooling of the vaporization fixture with liquid N_2. Nozzle expansion techniques also allow the production of a wide range of diatomic, triatomic, and large polyatomic van der Waals molecules for LIF spectroscopic studies.[60] Some representative examples are shown in Table 4.1.

LIF studies of free radicals and other reactive transients play a significant role and underpin progress in reaction kinetics and dynamics and re-

Table 4.1 Production of van der Waals Complexes for LIF Spectroscopy

Species	Method	Expansion Mode[a]	Backing Pressure	Ref.
MgAr	Laser vaporization	P	4 atm Ar	61
SiAr	Pulsed electric discharge through $Si(CH_3)_4$/ Ar mixture	P	2 atm Ar	62
XeCl	Pulsed electric discharge in Xe/HCl mixture	P		63
ArOH	Pulsed ArF laser photolysis of HNO_3	C	1 atm Ar	64
ArNO	Nozzle expansion of 5% NO in Ar	P	1 atm Ar	65
C_6H_6-Ar	Nozzle expansion of 2.5% C_6H_6 in Ar	P	2.6 atm Ar	66
Indole-S	S = H_2O, CH_3OH, formamide, acetamide	C	8 atm He	67
Tr-H_2O	Tr = tryptamine in He/H_2O	C	5 atm He	68
Cb-H_2O	Cb = carbazole in He/H_2O	C	5 atm He	69

[a]P: pulsed; C: continuous.

search on photophysics, energy transfer, and combustion processes. The classical techniques for the generation and detection of radicals in the gas phase are well developed and have been reviewed elsewhere in detail.[37,70] They include flow tube pyrolysis, radiofrequency, microwave, and glow discharges in low-pressure flowing gases, and photolytic and chemical reactions such as fast atom abstraction and combustion processes. Some of these continuous flow methods have been adapted successfully to jet-cooling and supersonic beams. Since all the preceding techniques produce "hot" (i.e., internally highly excited) species, very efficient cooling in the jet expansion is particularly advantageous. The marriage of radical production with jet-expansion techniques represents a dramatic advance and provides an opportunity to extend spectroscopic, dynamical, and kinetic studies to much larger and heavier chemical species relevant in atmospheric chemistry and low-temperature oxidation. Systematic studies using mixtures of different expansion gases are usually needed to find the best compromise between radical production and cooling for a given setup, while avoiding extensive formation of clusters with the rare gas. Table 4.2 summarizes a selection of examples from the recent literature. Clearly, both continuous and pulsed nozzle expansions can be coupled to a variety of methods for the production of radicals, as reviewed by Liu et al.[71]

In general, pulsed supersonic expansion techniques offer the potential of higher species concentrations and more efficient cooling, and there are clear advantages in combining pulsed production of the desired species with pulsed expansion cooling. Among photolysis techniques the pulsed laser method has become very popular, with ArF (193 nm) and KrF (248 nm) laser photolysis just below the nozzle orifice the most widely adopted variants. The target in the photolysis method is to achieve or at least to approach conditions of optical bleaching of the molecular parent as sum-

Table 4.2 Generation of Jet-Cooled Radicals and Unstable Molecules

Species	Production Method[a]	Expansion Mode	Ref.
CH, NH, OH, C_2	PLP of CH_3I, NH_3, HNO_3, $C_6H_5CH_3$	P	72–75
CH_2, HCO, CCl_2	PLP of CH_2CO, CH_3CHO, PED in CCl_4	P	76–78
HS, CN	PLP of H_2S, BrCN	C	79
CCl_2, CH_2S	CFP of $(CH_3)_3SiCCl_3$, cyclo-$(CH_2)_3S$	C	80, 81
HNO	Photosensitized reaction-Hg*/H_2/NO	C	82
NH_2	NH_3/F atoms from discharge in CF_4	C	83
CaX	PLV of Ca over HX (C_2H_2, NH_3, CN)	P	58, 59
Ca and CdX	PLV of Ca or Cd over RX (X = CH_3, cp)	P	58, 59
$C_6H_5CH_2$	PLP of benzyl chloride	P	84
C_5H_4CN	PLP of C_6H_5NCO or C_5H_5CN	P	85

[a]P: pulsed; C: continuous; PLP: pulsed laser photolysis; PED: pulsed electric discharge; CFP: continuous flow pyrolysis; PLV: pulsed laser vaporization of metal atoms followed by reaction with reagent vapor

marized by the relationship $I\sigma T > 1$, where I is the incident intensity, σ is the absorption cross section at the chosen wavelength, and T is the pulse duration. Since most molecules absorb strongly in the far-UV where excimer lasers deliver very high photon fluxes, ArF and KrF lasers have become the sources of choice for the generation of reactive species. Control over the irradiance is readily achieved by focusing or beam attenuation. Multiphoton dissociation of precursors right down to constituent diatomics and atoms commonly occurs under these conditions and can limit use of the technique for the generation of more complex radical species, where one relies on the primary process rather than on the photolysis of intermediates to produce the desired species. A technique with considerable potential for the generation of radicals is IR multiphoton dissociation (IRMPD), which can provide high local densities of a very wide range of radicals, such as CF_3, CHF, NCO, SiH_2, C_3, and many other species.[13,86,87]

For large molecular radicals and ions the review articles by Miller[79,88] on the spectroscopy of radicals provide an excellent survey. Here alternative techniques based on chemical reactions with metal atoms have recently been shown to offer significant advantages over photolytic techniques. Thus, Whitham et al. succeeded in using laser vaporization of Ca in the expansion region of a pulsed valve to prepare CaCCH (from HCCH), $CaNH_2$ (from NH_3), and CaCN (from CH_3CN) and obtain their LIF spectra in the visible.[58] A significant number of the vapourized Ca atoms are formed in metastable excited electronic states and react rapidly with the chosen precursor's abstracting H atoms. The radicals produced then combine with the ground state atoms and form the organometallic species. Miller et al. used a similar pulsed supersonic jet source with an excimer laser to vapourize Ca and Cd and simultaneously photolyze a precursor over the metal surface. The metal atoms and organic fragments then react, yielding the organometallic species. LIF spectra of CaH, $CaCH_3$, $CdCH_3$, and the monocyclopentadienyl derivatives of Ca and Cd were obtained in this way.[59]

Both continuous and pulsed electric discharge techniques offer advantages for preparing radicals in supersonic beams.[40,71] The combination of pulsed electric discharges with pulsed supersonic jet expansion has recently been applied to prepare van der Waals complexes as shown in Table 4.1. The carbenes CCl_2 and CBrF have also been studied using discharges through CCl_4 and $CFBr_3$ mixtures with argon at pressures of the order of 4 atm.[78,89] Two variants of the electric discharge/pulsed nozzle technique are of interest in conjunction with LIF spectroscopy. The first uses a pulsed high-voltage discharge in front of the nozzle orifice (i.e., at high pressure) whereas the second and simpler method employs a DC glow discharge that automatically ignites on opening the valve and extinguishes when the space between the electrodes is evacuated.[62,78,89] Both variants have also been described in connection with REMPI studies of the diatomic radicals CCl and NO.[90-92]

4.3.5. Cells, Fluorescence Collection, and Detection Systems

The main criteria to be applied in the design of LIF cells or chambers are efficient sample transport or containment, efficient collection of the fluorescence, and adequate suppression of scattered laser and ambient light. A number of carefully designed cells for LIF of room-temperature gas samples and molecular beams have been described in the literature.[6,93] The vacuum integrity of sealed low-pressure sample cells is of considerable practical concern where the fluorescence is quenched by O_2 and other atmospheric constituents, as discussed in Section 4.2.4. In flow cells the fluorescence excitation and observation volumes should be located at a point that is well swept by fresh sample gas. The distance between the excitation–observation volume and the observation window should be kept small to avoid attenuation of the fluorescence by sample vapor, a problem that frequently occurs in the far-UV and VUV.

Figure 4.8 shows a cell described by Hefter and Bergmann that illustrates many of the most important features.[6] Brewster angled entrance and exit windows for plane polarized beams minimize reflection losses and avoid backscattering of the laser light, and baffles and Wood's horns act as light traps. The sensitivity of LIF techniques is often dominated by scattered light, and such simple precautions as the painting of reflecting metal surfaces with absorbing coatings, the use of a stack of sharp-edged diaphragms carefully arranged with varying internal diameters in both en-

Figure 4.8 LIF chamber with laser light baffles, Wood's horns, and spatial filtering of the collected light to minimize the detection of scattered laser light. (Reproduced with permission from Ref. 6.)

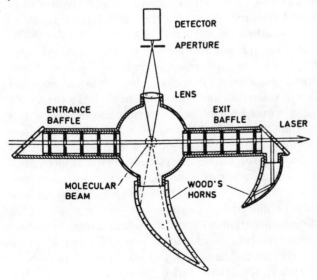

trance and exit baffle arms, or the use of a range of highly polished coni-
cally shaped light skimmers[94] are usually very effective in suppressing
scattered light. Placing an additional light trap on the axis of the observation system and spatially filtering the light by positioning small apertures
in front of the detector are often the only means to reduce scattered light
when fluorescence occurs in the same wavelength range as that of the laser
light.[95] Rejection of scattered laser light can usually be accomplished more
readily where the fluorescence occurs at shorter or significantly longer
wavelengths.

The choice of excitation and observation wavelengths is usually dominated by the ease of rejection of scattered light rather than by the Franck–
Condon factor of the exciting transition. The use of off-diagonal transitions for excitation and the diagonal Franck–Condon transition for
detection of the fluorescence allows much easier discrimination of fluorescence and laser light by spectral filtering. The excitation–observation sequence $v + 1 \leftarrow v$ followed by $v + 1 \rightarrow v + 1$ allows detection of the red-
shifted fluorescence and rejection of scattered laser light by a cutoff, interference, solution, or gas filter and is often advantageous. Classical cases of
diatomics where such a scheme has been employed are $OH(A \leftarrow X)$ and
$CN(B \leftarrow X)$. An example where the Franck–Condon principle favors emission into nondiagonal transitions is encountered with the $A \leftarrow X$ transition
of NO. Since the A state is a Rydberg state the internuclear distance decreases
and the potential curve for the upper state is shifted inward, favoring nondiagonal red-shifted fluorescence as shown schematically in Figure 4.9.

Efficient rejection of scattered laser light can be effected in cases where
the fluorescence occurs at much shorter wavelengths than the exciting
light. Examples of two-photon excited LIF and (1 + 1) two-color double-
resonance LIF with observation of the fluorescence to the ground electronic state or the sum or harmonic frequency. Here it is often possible to
combine suitable UV transmitting glass filters with solar blind or VUV
photomultipliers that have low quantum efficiency for the scattered light.
Where scattered laser light can be suppressed or entirely rejected by the
preceding measures and high sensitivity needs to be achieved, attention
should be payed to efficient collection and transmission of the fluorescence
onto the detector. Aspheric and Fresnel lenses permit the design of fast
collection systems with small f number. The use of achromatic antireflec-
tion-coated camera objectives and cemented doublets as field lenses offers
advantages for spectroscopy in the visible, particularly if the fluorescence
is to be focused onto the entrance slit of a monochromator for recording
dispersed fluorescence spectra. The fraction of light collected by a high-
speed lens system can be enhanced by positioning a spherical mirror on
the axis of the optical detection system, which functions as a retro-reflector.
Such measures may increase the collection efficiency to 30% from the more
usual 5% to 10%. More efficient collection of the fluorescence can only be
achieved by the use of spherical mirrors and light pipes or optical fiber

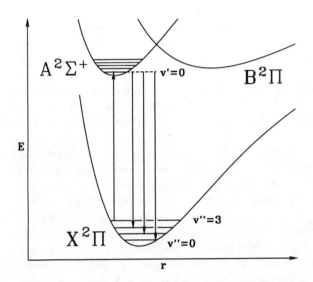

Figure 4.9 Schematic potential energy curves for NO showing excitation of the γ (0,3) hot band and blue-shifted fluorescence. The curve for the nearby *B* state is also included.

bundles, as described by Zare and Bergman et al.[6,93] Majewsky and Meerts have described a high-speed optical collection telescope based on the use of spherical mirrors and a fused silica condensor consisting of a planoconvex and biconvex doublet that collects the fluorescence from a skimmed molecular beam through a small hole drilled into one of the mirrors. Their design combines high collection efficiency (~50%) with spatial filtering of the fluorescence.[30]

Many of the preceding design principles can be applied to LIF spectroscopy in molecular beams and supersonic free-jet expansions. Continuous free-jet expansions are often operated in practice under conditions of fairly high background pressure unless very large and expensive pumping systems are available. The formation of a Mach disc enclosing the free-jet "zone of silence" is a well-known phenomenon under such conditions. Outside this zone collisions with the background gas result in a reheating of the jet-cooled molecules. It is therefore important to use spatial filtering that eliminates the imaging onto the detector of any fluorescence excited outside the zone of silence. Otherwise jet-cooled spectra of a species are observed on a background of warm spectra. Dunlop et al.[81] recorded LIF spectra of thioformaldehyde cooled by a free-jet expansion through a 0.15-mm-diameter nozzle at 2–3 atm pressure. They imaged the fluorescence onto a V-shaped mask to discriminate spatially against scattered laser light and emission of warm molecules excited outside the collision-free region of the jet. Sub-Doppler linewidths of 0.012–0.013 cm^{-1} were observed using a mask with a 1.5-mm slit parallel to the jet and perpendicular to the laser beam to observe fluorescence. The source of excitation was a < 1-MHz bandwidth single-mode ring dye laser that crossed the jet 30 nozzle diameters downstream of the 0.15-mm nozzle. With pulsed jets it is generally easier to avoid contributions to the observed LIF spectra from warm molecules, provided the valve is operated at a sufficiently small repetition rate and with short pulse duration to ensure proper supersonic expansion conditions by maintaining a background vacuum smaller than 1 mtorr.

The design of vacuum chambers for LIF spectroscopy of jet-cooled molecules can only be summarized briefly here, since detailed accounts are available elsewhere.[40] The size of the pumping system depends on the planned throughput. For continuous free-jet expansions a Roots blower followed by a sufficiently large two-stage rotary pump may be adequate for most purposes. Alternatively, a 23-cm oil diffusion pump with suitable backing pump could be used. Large booster oil diffusion pumps are only needed for extreme requirements. The minimum requirement for a pulsed jet setup is a 10-cm diffusion pump, which is adequate if a fast-acting valve with an opening time smaller than 0.5 ms and repetition rates less than 20 Hz can be employed at pressures less than 1 atm. A 15-cm pumping system offers sufficient flexibility for most requirements, and a 23-cm system would allow experiments at high repetition rates or expansion pressures higher than the usual requirement of <2–3 atm. Liquid-nitrogen-cooled traps are usually a good investment, where contamination by oil vapor and condensation on optical surfaces can cause problems, and the installation of a gate or quarter swing valve between the expansion–detection chamber and the trap may be advantageous where the chamber has to be opened frequently. In these circumstances the loss of pumping speed due to traps and valves has to be tolerated.

Figure 4.10 shows an experimental setup used in the author's laboratory for LIF spectroscopy in pulsed supersonic jets. The chamber, which is made of anodized aluminum, allows simultaneous observation of fluorescence in the visible and UV with separate collection and detection systems mounted on opposite faces of the cube. The system is used with a commercial pulsed valve at pulse durations of 0.5 ms, a repetition rate of 10 Hz, and expansion pressures of less than 1 atm and is pumped by a 10-cm trapped diffusion pump. The setup has been described in detail, and its performance has been demonstrated with two-photon spectra of NO and one-photon spectra of NO_2, CF_3NO, and other jet-cooled NO-containing molecules.[96,97]

The most widely employed optical detection systems for LIF spectroscopy are photomultipliers (PMTs) used in conjunction with phase sensitive or gated detection/boxcar averaging or transient digitizers and digital averaging systems. Multichannel detection systems such as intensified diode arrays or, for very low light levels, charge-coupled-device (CCD) detectors

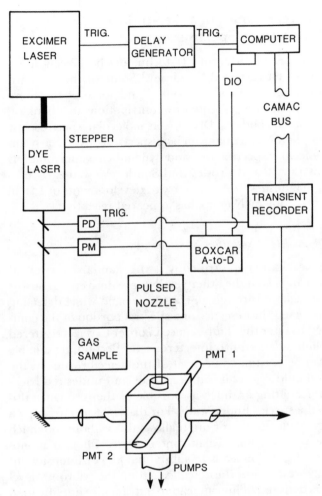

Figure 4.10 Schematic experimental setup for pulsed LIF and pulsed supersonic jet cooling showing timing, computer control, and signal acquisition systems. (Reproduced from Ref. 96.)

offer considerable advantages for LIF work when dispersed fluorescence spectra have to be recorded. Gated options for these multichannel devices are commercially available for use with pulsed excitation, but the high costs are justified only if the usual sources of noise have been eliminated. The basic principles involved in these optical detection systems are summarized elsewhere.[6,14,40] A good example of the use of multichannel detection in CW LIF studies is the recent work of Davis and Holtzclaw on energy transfer in IF.[98] Special problems with optical signal detection in LIF by the photon counting mode have been summarized by Bergmann.[6]

4.4. Techniques of Laser-Induced Fluorescence Spectroscopy

A number of different LIF techniques have been developed since the field was first reviewed by Kinsey.[1] Some of the standard methods are summarized in Demtröder's book,[14] and some of the more recent developments have been highlighted by contributions to a book on high-resolution spectroscopy, both by Dixon[99] on molecular spectroscopy and by Cagnac on two-photon and quantum beat spectroscopy.[100] Early applications of LIF to molecular spectroscopy and dynamics in supersonic jets have been reviewed by Levy,[43] and Crosley and Smith reviewed early uses of LIF in combustion diagnostics.[101] A summary of developments of LIF in analytical atomic spectroscopy by Niemax has appeared recently.[38]

4.4.1. LIF Excitation Spectroscopy

LIF excitation spectroscopy is the standard version of the technique from which most of the other versions are derived. It consists of scanning the laser radiation through an electronic band while detecting either spectrally undispersed fluorescence or a spectral portion of the emitted light. If it is desired to filter the fluorescence in order to reject scattered laser or background light, glass cutoff, interference, reflection, or suitable liquid or gas filters can be positioned in front of the detector. Ideally, the filter system selected should reject all scattered laser and undesired background light while transmitting as much as possible of the resonance fluorescence emitted by the species under study. For diatomic spectroscopy a small broadband filter monochromator often allows this without too much loss of intensity. The use of a narrowband interference filter or monochromator with higher dispersion leads to substantial loss of intensity but may ensure that only resonance fluorescence from the vibronic level pumped by the laser is detected. This can lead to a dramatic simplification of complex molecular spectra if nonradiative transitions transfer population from the optically pumped level to other nearby states that emit fluorescence in a different spectral region. Examples where molecular excitation spectra can be significantly simplified by this technique are NO_2[102] and BO_2.[103]

The effects of collisions in gas phase LIF experiments can also complicate spectra. This is possible even for diatomics such as NO, where excitation of certain C or D state levels can lead to emission from nearby levels of the A state that have a much longer fluorescence lifetime, a phenomenon known as cascading.[104–106] For polyatomic molecules where nonradiative processes become much more common, intramolecular vibrational energy redistribution in the excited electronic state can lead to changes in the excitation spectra as a function of the delay time between pulsed excitation and time-gated signal acquisition.

Complex excitation spectra can be simplified by using a variant that is commonly employed in condensed phase fluorescence, known as *synchronously scanned excitation spectroscopy*. Here both the excitation source and the spectral window selected by the emission monochromator are scanned synchronously. For gas phase spectra this method provides better discrimination between resonant and nonresonant fluorescence, and it enhances selectivity for a particular species where the induced emission is due to a mixture. This variant can offer advantages in spectral simplification for small as well as very large polyatomic molecules.[102,107]

The advantages of jet-cooling for simplifying complex excitation spectra have already been discussed in Sections 4.3.2–4.3.4. The effects of jet-cooling in achieving spectral simplification are particularly dramatic for molecules with low-frequency vibrations such as torsion and bending modes with high Franck–Condon factors. Examples from the author's laboratory are the fluorine-containing C-nitroso compounds CF_3NO, $CClF_2NO$, and CCl_2FNO.[97,108–110] An example is shown in Figure 4.11. The $\tilde{A}^1A'' \leftarrow \tilde{X}^1A'$ transition of CF_3NO shows a clear progression in the torsional mode 12 with a very weak Franck–Condon forbidden electronic origin associated with a transition from a minimum to a maximum in the periodic, threefold potential function for internal rotation of the CF_3 group against the bent $C-N{=}O$ chromophore. The analysis of the torsional structure yields an accurate excited state potential function for hindered internal rotation.[97]

In general, the aim of the jet-cooling technique is to achieve very low rotational and vibrational temperatures. Ideally, all the observed transitions originate from the lowest rovibronic level. The excitation spectrum then mirrors the excited-state manifold and provides little or no information on

Figure 4.11 Fluorescence excitation spectrum of CF_3NO cooled by pulsed supersonic expansion with 400 torr Ar. The spectrum is due to the $\tilde{A}^1A'' \leftarrow \tilde{X}^1A'$ (n, π^*) transition.

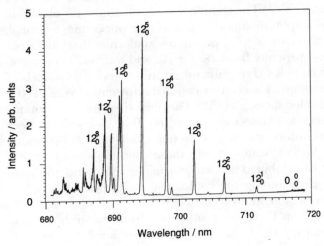

the ground state. Once the vibronic structure of the excited state is fully analyzed, however, warm spectra can often provide better-resolved information on the ground state through analysis of hot bands than dispersed fluorescence spectra. The rotational analysis of an electronic band is rarely satisfactory unless transitions of high J can be included. This holds for the analysis of rotationally resolved spectra as well as band contour synthesis. It is therefore often essential to record warm jet-expansion spectra by changing the expansion conditions or by using expansion gases that do not relax rotations as readily as the higher rare gases. The mounting of a short, 2-mm-diameter expansion channel underneath the orifice of a pulsed nozzle often achieves the same. Finally, it is worth noting that optical double-resonance techniques can often unravel difficult spectroscopic problems that cannot be solved by jet-cooling.

4.4.2. Dispersed LIF Spectroscopy

Dispersed LIF spectroscopy utilizes an experimental approach similar to Raman and resonance Raman techniques (i.e., excitation at a fixed wavelength, with collection and dispersion of light to be analyzed by a spectrometer/detector combination). By definition, however, it is the analysis of the spontaneously decaying electronic emission rather than scattered light that one is concerned with here. The distinction between short-lived fluorescence and resonantly enhanced Raman scattering is a moot issue. For gas-phase electronic spectroscopy the main use of the technique lies in determining Franck–Condon factors from observed intensities and in the vibrational analysis of the ground electronic state. The selection of single vibronic levels of the upper state manifold obviously simplifies spectroscopic analysis and requires collision-free conditions, a narrow source, and the elimination of spectral congestion by hot bands. Jet-cooling offers obvious advantages for single-vibronic-level (SVL) fluorescence studies of polyatomic molecules.[107,111–116] The technique is ideally suited for obtaining information on large-amplitude motions such as ring puckering, torsional, and other conformational motions of polyatomic molecules that can be obtained only with great difficulty from IR, far-IR, and microwave measurements.[107,113–115] Even complex tryptophan derivatives serving as models of large biomolecules have provided conformational information via SVL fluorescence of the jet-cooled molecules.[68,107] The limitations of the technique are worth noting, however: Molecules that do not fluoresce well are poor candidates, and only conformational motion that exhibits Franck–Condon activity is amenable to SVL study. Where these conditions are met, dispersed fluorescence in combination with jet-cooling represents the technique of choice for determining the conformational potential functions that often are of considerable interests for chemists.

An example of work on C-nitroso compounds from the author's laboratory is shown in Figure 4.12. As already noted in Section 4.4.1, the analysis of "warm" bands in the excitation spectra of incompletely cooled molecules can sometimes provide complementary, if not better, information on the electronic ground state. Rotationally resolved, dispersed emission spectra can provide very detailed information on level mixing in ground and excited states (e.g., in formaldehyde).[116] Finally, it is worth remembering that the emission spectra of diatomic molecules often provide new information on higher electronic states including ion-pair and unbound excited states giving rise to oscillatory continua, as shown by the work of Donovan et al. on I_2, for example.[117]

4.4.3. Two-Photon LIF

The main advantages of two-photon excited LIF are access to high-lying excited states without the need to produce laser radiation in inconvenient, short-wavelength regions, the existence of selection rules that are less restrictive or complementary to one-photon absorption, and the high spatial resolution associated with the use of focused beams. For high-resolution spectroscopy the potential to eliminate Doppler broaden-

Figure 4.12 Small portion of the dispersed SVL fluorescence spectrum of jet-cooled $CCIF_2NO$ excited on the 12_0^1 vibronic transition of the jet-cooled parent near 700 nm. The O_0^0 origin of this $\tilde{A} \leftarrow \tilde{X}$ (n, π^*) electronic transition is at 704.7 nm. The structure reveals transitions to low-frequency torsional and rocking vibrational levels of the electronic ground state. The normal mode 12 corresponds to the torsional vibration.

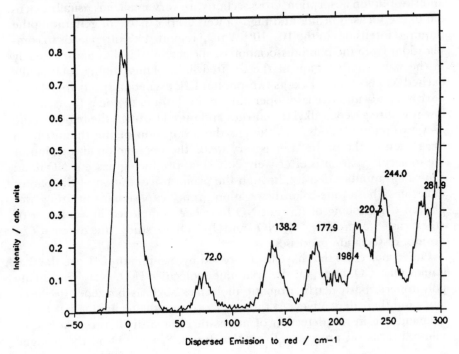

Intensity / arb. units

Dispersed Emission to red / cm−1

ing by cancellation of the velocity shift on simultaneous and coherent absorption of two photons of the same frequency is well known.[14,100] Although high-power pulsed laser radiation is generally required, because of the small two-photon absorption cross sections, the rejection of scattered light rarely presents a problem. A more serious limitation of the technique is its lower sensitivity than one-photon LIF, usually by a factor of 1,000–10,000. Since this is partially compensated by the 10- to 100-fold better collection efficiency of ions as compared to photons, and because ionization techniques are not restricted to fluorescent states, two-photon absorption has become much more favored in conjunction with REMPI (i.e., resonantly enhanced multiphoton ion detection, particularly the (2 + 1) scheme).

Figure 4.13 shows a comparison of observed and calculated two-photon LIF excitation spectra of a supersonic jet of argon seeded with ~1% NO. The spectrum is carried by the (0,0) band of the $A-X$ or γ-band system of NO, which is excited here by coherent two-photon absorption near 450 nm. Fluorescence of the γ-bands in the 225–295-nm region is readily discriminated from scattered blue laser light by a blocking filter and by the use of a solar blind photomultiplier with negligible quantum yield for visible light. Note that this simple technique can be utilized to measure the rotational cooling efficiency of gases and that local rotational temperatures in supersonic jets or beams can readily be measured in this way.[96]

Since the observed signal depends on the square of the laser intensity and two-photon absorption, cross sections are very small, it is usually essential to use focused beams and high-power lasers capable of generating the required intensities in the 10^7–10^{12} W m^{-2} region. This often leads to complications from the photodissociation of species, even if they absorb weakly in the wavelength region of the focused light. Thus, highly rotationally excited NO is observed via its two-photon LIF excitation spectrum when a narrow bandwidth dye laser operating in the 450-nm region is focused into low-pressure gaseous alkyl thionitrites and tuned through the region of the $A-X$ two-photon bands.[118] Although the absorption of the thionitrite in the region of the probe laser is very weak, the one-photon absorption of the molecular parent is effectively saturated, producing nascent NO at an equivalent number density through the prompt dissociation of the parent molecule. These "one-color dissociation-probe" experiments utilizing two-photon LIF probing of nascent NO have been employed in a number of other cases (e.g., NO_2,[96] ClNO, and BrNO) to study the dynamics of prompt dissociation processes.

Other molecules that have been probed by two-photon LIF are the diatomics OH, CO, and PO; the triatomic molecules H_2O, NH_3; and many polyatomics, particularly aromatic molecules such as benzene, naphthalene, and their derivatives. At the other extreme the technique has found frequent use in the detection of atoms in plasma discharges, flames, or photodissociation processes. Examples are Si, S, and O and the halogen

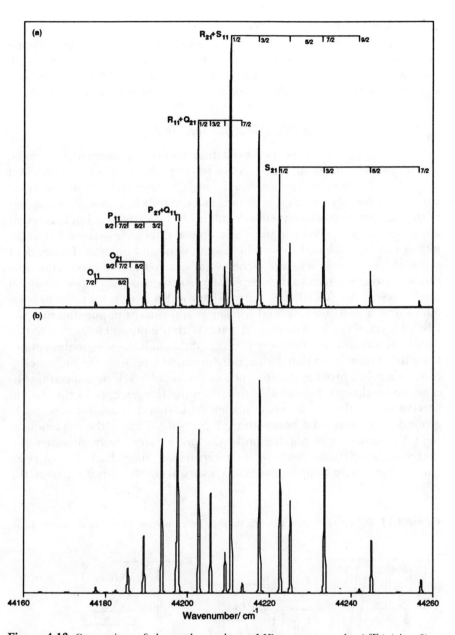

Figure 4.13 Comparison of observed two-photon LIF spectrum on the $A\,^2\Sigma^+$ ($v' = 0$) $\leftarrow \mathrm{X}^2\Pi_{1/2}$ ($v'' = 0$) transition, that is, the $\gamma(0,0)$ band of jet-cooled NO and simulated two-photon spectrum ($T_{\mathrm{rot}} = 20$ K).

atoms Cl, Br, and I. In favorable circumstances the sensitivity that can be achieved[119,120] is between 10^{15} and 10^{16} atoms m^{-3}. Two-photon LIF has been used for the detection of many metal atoms and has found a well-established application in the trace detection and analysis of metal species.[38] Further examples of two-photon excitation are discussed in Chapter 8.

4.4.4. Double-Resonance Techniques

Fluorescence excitation spectra of molecules at room temperature or at the higher effective temperature that is often required to produce unstable species are usually highly congested. Yet precise spectroscopic constants are often required, that can only be obtained by analysis of the rotational structure up to very high J or by analysis of hot bands that are absent in jet-cooled spectra. Here optical double-resonance LIF techniques can be exploited to simplify assignment and analysis. Figure 4.14 shows three possible schemes. Scheme (a) is a frequently used sequential double-resonance technique often described as two-color optical–optical double resonance (OODR). A narrow-bandwidth laser is tuned in to resonance with a well-characterized transition and pumps population into the labeled level of the intermediate A state. With the pump laser fixed to this transition a second tunable laser is scanned through the $B \leftarrow A$ absorption. Only those transitions that connect the pumped level with the higher electronic state will provide fluorescence in the suitably selected spectral window. The technique is most sensitive when the fluorescence of the state to be characterized (i.e., the state labeled B) is detected against zero background. For short-lived nonemitting high-lying B states the depletion of the intermediate state population by the probe laser can be detected as a decrease in the fluorescence of the intermediate state. Each $B \leftarrow A$ resonance gives rise to a dip in the fluorescence from the A state. Unlike the

Figure 4.14 Double resonance LIF schemes: pu = pump laser; pr = probe laser; du = dump laser.

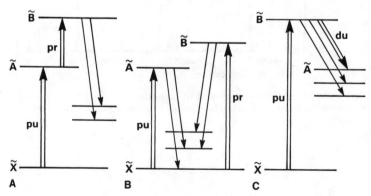

previous detection scheme, resonances are detected against a high background, and the sensitivity necessarily decreases.

Donovan et al. have demonstrated the advantages of sequential OODR spectroscopy in the study of high-lying ion pair states of I_2 and their collision dynamics in the presence of Xe. Using two dye lasers pumped by the same excimer laser, they succeeded in selectively populating specific rovibronic levels from $v' = 0$ to $v' = 107$ of the f ion pair state, and they recorded dispersed fluorescence spectra of a number of selected levels in this range.[117] Shibuya et al. used the same approach to characterize the $2\,^2B_2$ state of NO_2 with the well-known, notorious $\tilde{A}$ state in the visible as intermediate. In their case the line-narrowed pump laser (0.04 cm^{-1}) was tuned to a rovibronic transition in the visible absorption of NO_2 and the probe laser of 0.3 cm^{-1} bandwidth accessed the $2\,^2B_2$ state. Fluorescence from this state down to the ground state was detected in the UV.[121]

Scheme (b) of Figure 4.14 provides an opportunity to label and thereby assign a single-band system by identifying pairs of transitions with a common level. In this case competitive excitation, with the pump transition partially saturated, is used to detect probe resonances with the labeled ground state level. To detect the probe resonances the strong fluorescence from the pumped level needs to be suppressed by a suitable blocking filter. Johnson et al. took advantage of this population depletion technique to simplify the congested $C \leftarrow X$ band system of BaI, where the pump and probe lasers were tuned to the two components of this spin-orbit split system near 561 and 538 nm, respectively.[122]

Variation (c) of Figure 4.14 represents a folded variant of the sequential double resonance technique and is better known as the stimulated emission pumping (SEP) technique. This OODR scheme is particularly advantageous where high-lying levels are to be accessed that cannot be reached directly or that remain buried in highly congested or severely overlapping regions of direct absorption spectra. Unlike dispersed fluorescence, the nonstimulated equivalent of the SEP technique that utilizes spontaneous emission of the selected level, the losses inherent in the use of a monochromator are avoided. As the probe laser scans through the resonances with the selected level, it stimulates emission in the direction of the probe laser and thereby depletes the intensity of the spontaneous fluorescence. Thus, resonances are detected by depletion of the fluorescence or by the gain experienced by the probe laser.

The SEP techniques has been particularly effective in the exploration of high-lying ground state levels (i.e., where the levels designated as belonging to the A system in scheme (c) of Figure 4.14 actually represent high levels of the electronic ground state. The technique relies on pumping a sizable fraction of the population of the selected levels upward as well as downward. Thus, powerful, pulsed dye lasers are needed for the pump as well as the probe, which is fittingly described as the "dump" laser. Since the difference in frequency between the pump and dump photons can easily be tuned to match the final level, vibrational spectra spanning a range from zero to tens of thousands of cm^{-1} may be recorded. This enables regions of potential energy surfaces to be explored that are not accessible by conventional IR and UV-visible techniques, and it provides important information on molecules close to their thresholds for unimolecular dissociation or rearrangements. Field, Kinsey et al. have used the SEP method to characterize high levels of ground-state H_2CO and C_2H_2, finding level patterns that are strongly affected by anharmonic and Coriolis coupling.[123–125] Since the first demonstration of SEP it has been used widely for many systems, including molecules above the dissociation barrier such as HFCO,[126] radicals such as CH_3O and CH_2, small atomic clusters like C_3 and Na_3, and larger van der Waals complexes like Ar-glyoxal. Here vibrational levels in the frequency range 10–50 cm^{-1} are mapped by OODR in the visible rather than the far-IR, a notoriously difficult spectral region.[127]

Since SEP appears to be a widely applicable if not universal technique, a cautionary remark is appropriate. The detection scheme invariably requires a comparison of signal and reference fluorescence and differential amplification of the difference signal. This involves duplication of detection systems as well as the use of two optically synchronized pulsed dye lasers. The detection of gain using polarization suppression of the dump laser is possible and may overcome some of these problems.

4.4.5. Sub-Doppler LIF Spectroscopy

For large organic or very heavy species the density of rotational transitions exceeds the width of the individual Doppler profile and presents a fundamental limit on conventional high-resolution spectroscopy. This problem can be overcome by some well-established techniques reviewed elsewhere.[14,128] The three main approaches utilizing LIF are two-photon LIF, spectroscopy in molecular beams with LIF detection, and the OODR LIF techniques discussed earlier. All three techniques require tunable laser sources with a linewidth narrower than the Doppler width for the species under the particular conditions employed. Riedle, Neusser et al. have applied Doppler-free two-photon LIF to benzene vapor at room temperature and have demonstrated the impressive power of this approach for the rotational analysis, for emission spectroscopy, and for homogeneous linewidth measurements in congested spectral regions.[129] Another example is their recent study of the $\tilde{A}$–$\tilde{X}$ origin band of difluorodiazirine, F_2CN_2.[130]

The alternative method for reducing Doppler widths and rotational congestion is to use LIF spectroscopy in freely expanding jets or skimmed and well-collimated supersonic molecular beams. This has now become almost a standard technique of high-resolution spectroscopy, and there are many examples in the recent literature, some of which have been mentioned in Section 4.3.2. The main requirement for this approach is a narrow-linewidth tunable source in the spectral region of interest and good collimation of the beam.[95,131,132] Sub-Doppler resolution spectroscopy can also be used to measure homogeneous linewidths in molecular electronic spectra. These linewidths carry dynamical information on the lifetime of the levels involved and can provide rate constants for nonradiative processes and their dependence on energy, quantum number, and symmetry.[13]

Figure 4.15 shows a comparison of (b) an observed LIF spectrum of uranium atoms in a beam source, revealing rich hyperfine structure of the 17,362 cm^{-1} line, with (a) the line profile of the same transition in a hollow cathode lamp. Clearly, in (a) the nuclear hyperfine structure is almost completely obscured by the Doppler motion of the atoms.[133] The increasing availability of narrow-linewidth tunable diode lasers in the near-IR in conjunction with Doppler-free two-photon techniques is likely to lead to new applications in analytical atomic spectroscopy.[38]

Figure 4.15 Lineshape of the 17,362 cm^{-1} atomic uranium line from (a) a hollow cathode lamp where the hyperfine structure, indicated by a stick diagram, is obscured by the Doppler effect; (b) an atomic beam source. (Reproduced with permission from Ref. 133.)

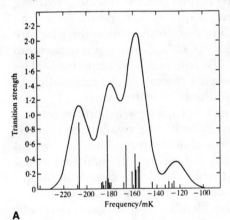

A

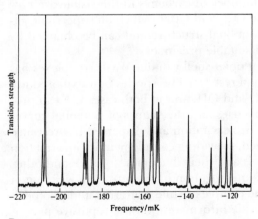

B

4.5. Applications in Molecular Spectroscopy

LIF techniques have played a major role in the renaissance of electronic spectroscopy that accompanied the emergence and development of the dye laser and associated nonlinear wavelength extension techniques. Reviews of high-resolution spectroscopy are available that survey this rapidly expanding area of contemporary research up to 1983.[128,134] Here only a few topics can be highlighted, and both the spectroscopy of radicals and unstable species and the conformational analysis of molecules with large-amplitude motions are areas where progress appears to be particularly rapid.

4.5.1. LIF Spectroscopy of Unstable Species

Techniques for producing transient species for LIF spectroscopy have been covered in Section 4.3.4, where many examples from the recent literature have been cited. While REMPI spectra of C-centered aliphatic and halogen-substituted methyl radicals revealing Rydberg transitions have been recorded for several species, LIF spectra do not appear to be known. In contrast, considerable information on the structure and photophysics of alkoxy and thioalkoxy radicals has recently been obtained from their LIF spectra in the UV and near-UV, respectively. The LIF spectra of benzyl and cyclopentadienyl radicals have been studied in considerable detail.[88] Ishida et al. have shown recently that the LIF spectrum of the species previously attributed to triplet phenylnitrene actually arises from the cyano-cyclopentadienyl radical. The rotationally resolved LIF spectrum of the jet-cooled radical could not be reproduced by spectral synthesis with a phenylnitrene structure, but was readily simulated with a cyanogen-substituted cyclopentadienyl structure. This example illustrates the power of the combination of jet-cooling with LIF spectroscopy and rotational band contour synthesis in the identification of gas-phase radicals.[135] LIF spectra of a sizable number of carbenes in the visible are now known. Recent examples include CCl_2 and CBrF, which on supersonic jet-cooling reveal well-developed rotational structure and can be obtained by electric discharge or pyrolysis of suitable precursors.[78,80,89]

The electronic spectroscopy of open-shell van der Waals complexes has recently attracted considerable interest.[64,65] The species best explored are complexes of rare gases with NO and OH, and in both cases LIF can provide useful information on geometries and the shapes of potential curves for these species. Both fluorescence excitation and dispersed fluorescence spectra can sometimes be recorded, providing complementary information on ground and excited state levels and their dynamics.[64,136] LIF and other spectroscopic techniques permit interrogation of the attractive well region of the interaction potential of the complex, while scattering and photodissociation experiments often provide information on the repulsive part of the potential via LIF spectra of the scattered free NO or OH.[137,138]

Numerous homonuclear and heteronuclear rare gas complexes in their electronically excited states have been characterized by one-photon spectroscopy in the VUV.[139] LIF and REMPI spectroscopy of supersonic gas expansions are both valuable techniques for the spectroscopy of excimers and exciplexes providing data on spectroscopic constants, potential curves, dissociation limits, and electronic-state symmetries. Similar information has recently also been obtained on metal rare gas van der Waals complexes, such as ZnAr, MgAr, or SiAr[60–62] using laser vaporization or electric discharge techniques, as discussed in Section 4.3.4. The spectroscopy of metal dimers and small metal clusters largely relies on laser vaporization techniques for sample preparation, and LIF excitation and detection are frequently employed.[53–55] A newly emerging area of spectroscopic research may well be seen in the LIF spectroscopy of jet-cooled organometallic free radicals like $CaNH_2$, CaCN, CaC_2H, and the Ca and Cd cyclopentadienides.[58,59]

4.5.2. LIF Spectroscopy and Conformational Analysis

The chemical reactivity and many physical properties of molecules in the gas phase and in solution depend critically on their conformational behavior. Consequently, a wide range of techniques has been developed to deduce information on the relative stabilities, rates of interchange, and spectroscopic differences between conformational isomers.[140] Spectroscopic techniques that reveal information on conformational potential functions and energy levels of isolated gas-phase molecules have traditionally played an important role. Microwave, IR, and Raman spectroscopies have provided most of the available data on electronic ground states, with far-IR FT techniques often contributing substantial information. Among the laser techniques UV-Raman, resonance-enhanced Raman, and LIF-based techniques have revealed much promise for the analysis of periodic and other potential functions involving large-amplitude motions such as torsion, ring puckering, and inversion vibrations. LIF with single vibronic level selection and dispersed emission spectroscopy represents a powerful technique for mapping low-frequency levels of the lower electronic state. Unlike most of the other techniques mentioned, it is an indirect method and does not rely on thermal population of the levels to be probed, but on their Franck–Condon overlap with the optically pumped vibronic level. This offers the advantage that levels in the difficult $10–500$ cm^{-1} range can be accessed in the optical region of the spectrum, where atmospheric transmission, dispersion, and detection of the radiation do not present appreciable hurdles. Dispersed emission spectroscopy as a technique used in conjunction with LIF excitation has already been discussed in Section 4.4.2, and some examples have been referenced there.

Among examples from the author's laboratory, the spectroscopy of C-nitroso compounds has revealed striking cases,[97,108–110] where excitation spectra of jet-cooled species in the visible provide detailed information on potential functions for hindered internal rotation. The absorbing chromophore in these cases is the bent CNO (i.e., the C-nitroso group), which gives rise to a structured $\tilde{A}^1A''-\tilde{X}^1A'$ electronic transition of n,π^* type in the convenient 520–800-nm wavelength range. Many of these molecules fluoresce with a lifetime that is dominated by nonradiative processes, including photodissociation to alkyl radicals and the ground-state NO radical. Since the dissociation dynamics of these molecules can also be studied using state-selective excitation and LIF probing of the nascent NO fragments and fluorescence lifetime measurements can provide complementary information on rates of nonradiative processes, these species have provided unique model systems where spectroscopic and dynamical studies can be combined to unravel the photophysics and subsequent chemistry in detail.[110] The electronic spectroscopy of these molecules in the visible offers some intriguing features for conformational analysis, since electronic excitation in the visible is accompanied by a change in the dihedral conformation, from eclipsed in the ground to staggered in the excited state. Thus, the torsional potential curves in the ground and the lowest electronically excited state are shifted by 60° with respect to each other, giving rise to unusual intensity distributions and strong Franck–Condon activity of hot bands. This permits access to high-lying torsional levels of both ground and excited states and provides a test for models of internal rotation that is far more taxing than the microwave data usually used in conjunction with customary semirigid internal rotor models.

Figure 4.16 shows an example of an excitation spectrum of the molecule CCl_2FNO in a pulsed supersonic jet. The spectrum is dominated by progressions in the torsional and low-frequency rocking or bending modes of the excited state, and analysis of the torsional structure yields potential functions for the excited state. Interestingly, excitation spectra of incompletely cooled molecules of this type provide better information on the ground-state levels than dispersed single vibronic-level spectroscopy. The latter involves detection of weak dispersed emission in the difficult near-IR spectral region, as shown previously with the same molecule in Figure 4.12. In CF_3NO, the threefold periodicity of the two torsional potential functions involved and the change in their phase on electronic excitation results in the minimum-to-minimum transition being Franck–Condon forbidden and extremely weak. This results in the characteristic and most unusual intensity distribution shown in the spectrum of Figure 4.11,[97] in which the intensities increase rapidly with the upper torsional quantum number x along the 12_0^x progression until the top of the barrier to rotation in the excited state is reached.

The excitation spectra of jet-cooled acetone and acetaldehyde in the near-UV at wavelengths longer than 315 nm also reveal rich torsional structure because of the hindered internal rotation of the methyl group. The spectra have been analyzed to yield potential functions and barrier heights.[116,141] These molecules adopt pyramidally distorted $S_1(n, \pi^*)$ states,

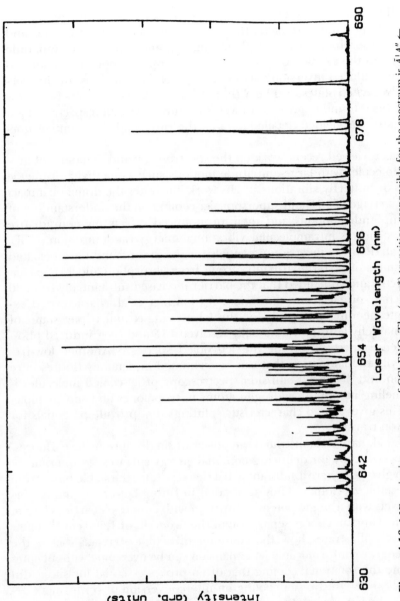

Figure 4.16 LIF excitation spectrum of CCl₂FNO. The (n, π^*) transition responsible for the spectrum is $\tilde{A}^1A'' \leftarrow X^1A'$.

and double-minimum potentials with inversion barriers of the order of 500 cm^{-1} are required to rationalize the observed vibronic structure. Baba and Hanazaki also observed out-of-plane wagging and other large-amplitude vibrations in the excitation spectra of cold cyclopentanone and cyclobutanone and deduced inversion barriers and potential functions for the electronically excited states.[142] The $S_1(n, \pi^*)$ state of 2-cyclopenten-1-one has recently been found to give rise to well-structured excitation spectra on jet-cooling and has yielded detailed data on the ring bending potential function.[143]

Ito has reviewed recent work on the spectroscopy and dynamics of aromatic molecules with large-amplitude motions such as methyl and phenyl torsions or butterfly vibrations.[144] His work illustrates the dramatic impact of the marriage of LIF and supersonic jet-cooling on the understanding of large-amplitude motions and their pronounced effects on nonradiative processes such as intramolecular vibrational energy randomization (IVR). Shin et al. deduced inversion potentials for the ground and lowest excited singlet states of 9,10-dihydroanthracene from absorption and LIF excitation spectra and concluded that the barrier increased substantially from 80 to 615 cm^{-1} on electronic excitation.[115] Clouthier et al. characterized excited-state bending potentials for the lowest singlet and triplet states of thioformaldehyde using rotationally resolved LIF and laser-induced phosphorescence spectroscopy of the jet-cooled species.[81] Another low-frequency vibrational motion that can be examined in detail by fluorescence excitation and dispersed emission spectroscopy of jet-cooled molecules is the tunneling motion of hydrogen atoms in tropolones and similar molecules. This gives rise to characteristic splittings that provide potentials for the proton transfer.[145]

Butz et al. have demonstrated the power of single vibronic-level fluorescence spectroscopy for deducing potential energy surfaces for internal rotation in glyoxal, where a substantial barrier separates the stable *trans*- from the less stable *cis*-isomer. They succeeded in fitting both *trans* and *cis* torsional levels with a single potential function with a *trans–cis* barrier close to 2,000 cm^{-1} and an energy separation of the two wells of 1,690 cm^{-1}. Their work nicely illustrates how the common difficulty of relaxation of the higher-energy conformer in a jet expansion can be overcome without compromising the rotational cooling that often proves essential for successful analysis.[146] The effect of jet-cooling on interconversion of conformers depends on the barrier heights separating the conformers and on the nature of the expansion gas, but in general relaxation of conformers appears to be inefficient if the barrier exceeds 400 cm^{-1}: He relaxes conformers much less than Ar or Kr, and in favorable cases high-energy rotamers can be frozen out with a nonrelaxing carrier gas.[147] Very-high-resolution LIF excitation spectra of jet-cooled 1- and 2-methylnaphthalene exhibit perturbations caused by the interaction of the methyl torsion with the overall rotation. The spectra have been analyzed in detail and have yielded V_3 barriers.[148]

Finally, it should be noted that REMPI spectra often also provide new insight into the conformational behavior of excited states. Thus, jet-cooled tetracene derivatives have yielded highly informative two-color, two-photon resonant ionization spetra using time-of-flight mass-resolved detection, and the spectra provide interesting data on phenyl torsions in the first excited singlet states of these large and highly fluorescent molecules.[48]

4.6. Photodissociation Dynamics

The recent advances in laser and sample preparation techniques have also revolutionized the study of photochemistry in both the condensed and gas phases and initiated major new developments in photodissociation dynamics and surface photochemistry.[13] One of the most exciting developments in this area has been the emergence of femtosecond pump–LIF probe techniques permitting the study of fast photodissociation or bimolecular collision processes in cold isolated molecules or via complexes in real time.[149–151] Both OH and CN appearance dynamics have been probed in this way on the femtosecond and picosecond time scale. Nanosecond pulsed laser photolysis and delayed LIF probing of simple fragments are currently evolving into a remarkably versatile tool for exploring photodissociation processes. The number of atomic and diatomic species that can be probed has significantly increased by the extension of wavelength regions for LIF into the VUV.[39] Supersonic jet-cooling of the parent molecules to be photolyzed, and polarization as well as Doppler lineshape techniques in the analysis of LIF spectra of nascent fragments permit and demand well-defined experimental conditions. Scalar as well as vector attributes of the dissociation process can now be unraveled in remarkable detail, as outlined in Sections 4.2.2 and 4.2.3.

In a few notable cases it has been possible to study the dynamics of individual quantum states.[19] Thus, the era of the venerable flash photolysis technique has evolved into the new era of "high-resolution photochemistry."[152,153] On the other hand, temporal rather than spectral resolution has also been extended into the femtosecond region, allowing LIF probing of the evolution of wavepackets and transition states into dissociation products in real time.[149–151] Ultrafast techniques are discussed in more detail in Chapter 10.

4.6.1. Triatomic Molecules

One of the first pulsed dissociation–LIF probe studies to be reported in the literature was the photolysis of NO_2 in the near-UV using a N_2-laser for dissociation and tunable delayed LIF of the NO photofragment on the $A–X$ γ-bands.[154] Vibrationally excited NO up to $v = 4$ with decidedly nonstatistical rotational distributions was observed and characterized. This seminal study indicated the potential of this pump-probe technique for exploring the details of dissociation processes. The near-UV photolysis of NO_2 plays an important role in atmospheric chemistry and continues to attract much attention. Recent work in our and other laboratories utilized jet-cooling of the parent and detailed LIF studies of nascent NO and $O(^3P_J)$ atoms close to the dissociation threshold near 398 nm and at shorter wavelengths. By scanning the dissociation laser while the probe laser resonantly excites a given J level of $NO(v = 0)$, the dissociation threshold and the pronounced structure in the NO yield spectra can be explored. The threshold for formation of ground-state O atoms coincides with that for formation of NO and disappearance of the fluorescence of the parent NO_2 molecule. Tsuchiya et al. have recently reported the same experiment using VUV LIF probing of the $O(^3P_J)$ atoms. Figure 4.17 shows a two-photon LIF spectrum of $O(^3P_J)$ atoms obtained by focusing the tuned UV probe light into jet-cooled NO_2 while monitoring atomic fluorescence to an intermediate level with a red-sensitive photomultiplier. The three lines seen are the $J = 0,1,2$ spin–orbit levels of $O(^3P)$. Their intensities reflect a statistical population distribution and correspond to a $T = \infty$ spin–orbit temperature, indicating that multiphoton absorption, ionization, and plasma production are involved in the formation of the O atoms seen. In contrast, photolysis in the near-UV yields much colder O atoms, as seen by delayed one-photon VUV and two-photon LIF probing near 225 nm.[155–157]

Figure 4.17 Two-photon LIF excitation spectrum of O^3P atoms obtained by monitoring the $3p^3P \rightarrow 3s^3S$ fluorescence at 844.6 nm as a function of the wavelength of the focused scanned UV laser. The sample consisted of 0.1 torr of NO_2 in 400 torr Ar jet-cooled by pulsed supersonic expansion through a 0.1-mm nozzle.

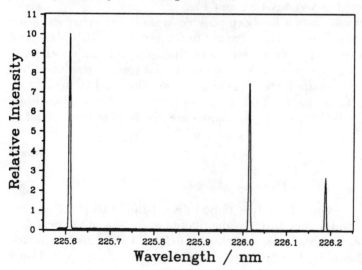

A celebrated example studied by Andresen et al. is the photodissociation of H_2O in the first absorption band at 157 nm using an F_2 excimer laser for photolysis and LIF probing of the OH fragments.[158] Each rotational state of an $OH(^2\Pi)$ fragment is split by the very small Λ-doubling into pairs of states differing in the orientation of the Π-lobe occupied by the unpaired electron with respect to the plane of nuclear rotation. The Λ-doublet states with the lobe perpendicular to the plane of rotation (electronic symmetry A'') are strongly favored in the dissociation of jet-cooled H_2O. This strong but j-dependent Λ-doublet selectivity is explained by the conservation of the electronic reflection symmetry during the dissociation process. Similar pronounced selectivity in the population of the Λ-doublets has been observed in many other examples of prompt dissociation processes from repulsive upper states producing OH, NO, or other $^2\Pi$ radicals in high J states. The Λ-doublet selectivity tends to be reduced by parent rotation and out-of-plane vibration. Thus, jet-cooling of the parent molecule should generally enhance the Λ-doublet selectivity. In most cases the ensemble of generated photofragments is aligned with respect to the laboratory frame through the photoselection involved in the absorption of polarized light by the parent molecule. In the inversion of the observed LIF fragment spectra to population distributions this alignment must be taken into account before population ratios for the Λ-doublets can be evaluated.

Very detailed studies of this type have recently been carried out by Reisler et al. for the nitrosyl halides using one-photon LIF of NO. A careful analysis of Doppler profiles of individual LIF transitions provides recoil anisotropy parameters, and this has led to a much improved understanding of the electronic transitions involved in the absorption spectrum of ClNO in the UV and visible.[153] Studies of the pulsed photolysis of cyanogen iodide (ICN) by LIF of the cyanogen radical have become more and more detailed and include alignment and fine structure effects, and the photolysis of carbonoxysulfide (SCO) with LIF probing of the CO has provided a striking example of vector correlation effects through analysis of Doppler broadened spectral profiles.[17]

Kable et al. recently succeeded in rationalizing in fascinating detail the dissociation dynamics of the radical HCO and its D isotopomer after excitation of the $\tilde{A}^2A''-\tilde{X}^2A'$ electronic transition in the 500–570-nm region. They used CO photofragment yield spectra of the jet-cooled radical, linewidth and hence predissociation lifetime measurements, and finally H as well as CO polarized sub-Doppler LIF spectroscopy to measure the quantum state dependence of the recoil anisotropy. The transition movement for the $\tilde{A} \leftarrow \tilde{X}$ transition of bent HCO is perpendicular to the molecular plane, yet both H and CO LIF Doppler profiles indicate that the recoil anisotropy β is positive, contrary to expectation. This unusual behavior has been explained quantitatively by the Renner–Teller interaction and the coupling of nuclear and electronic angular momenta associated with the crossing of the $\tilde{A}$ and $\tilde{X}$ surfaces.[77]

4.6.2. Four-Atom Molecules

Among four-atom species the photodissociation of nitrous acid (HONO) in the near-UV and that of nitrosyl cyanide (NCNO) in the visible have been studied particularly well and display very contrasting behavior. In each case both fragments from the primary dissociation differ and can be interrogated in detail by LIF in readily accessible regions of the spectrum.[159-161] The near-UV absorption spectrum of HONO shows well-developed vibrational structure but is rotationally diffuse because of rapid predissociation on the 100-fs time scale. The observed structure is due to resonances associated with a shallow well in the upper electronic state and represents a progression in the N=O stretching vibration. Both OH and NO have been probed for photolysis on different members of this progression. The OH fragment is born vibrationally cold and carries little rotational excitation, and its spin–orbit populations are not equilibrated. Polarization spectroscopy and LIF Doppler profiles reveal that it recoils at high velocity and is strongly aligned with a clear preference for the A' (i.e., the symmetric Λ-doublet). The dissociation proceeds promptly and is impulse dominated with the trajectories of the recoiling fragments preferentially lying in the plane of the transition state. Excitation of the parent N=O vibration in the upper state is not carried over into vibration of the OH fragment but is retained in the rotationally highly excited NO.[160] The reverse OH + NO reaction and the H + $NO_2 \rightarrow$ HO + NO reaction have received much attention by experimentalists and theoreticians interested in elementary reaction dynamics.[162,163] Detailed knowledge of the repulsive half of a full collision of OH with NO is valuable, and the work on HONO convincingly illustrates the benefits of such "half-collision" studies for chemical reaction dynamics.

In contrast to HONO, the photodissociation of NCNO from $S_1(n, \pi^*)$ is not impulse-dominated and provides a unique example for the decay dynamics of a longer-lived complex.[164-166] NCNO absorption spectra in the visible and near-IR reveal considerable vibrational and rotational structure due to an $\tilde{A}^1 A'' \leftarrow \tilde{X}^1 A' n, \pi^*$ transition, which is characteristic for C-nitroso compounds.[167] This has been examined by LIF excitation spectroscopy and jet-cooling.[168] When the photon energy exceeds $D_0 = 17{,}085$ cm^{-1}, the jet-cooled molecule predissociates within < 2 ns, as evidenced by the measurements of the appearance times of the CN radical by picosecond pump-probe techniques.[166] Unfortunately, dissociation only sets in 5,746 cm^{-1} above the electronic origin of the S_1–S_0 (n, π^*) transition where absorption features can no longer be assigned to specific rovibronic transitions, and the state dependence of the dynamics cannot be explored in a systematic fashion on the basis of spectroscopic assignments.

The cyanogen radical is well suited to LIF probing on the well-characterized violet bands associated with the $B \leftarrow X$ electronic transition near 380 nm. Both CN and NO fragments have been probed by LIF, revealing product state distributions over the internal *(E, V, R)* degrees of freedom of the fragments in great detail, and LIF Doppler profiles of the CN radical have been analyzed over a range of excess energies. These profiles indicate an isotropic distribution of recoil velocities, and the Λ-doublets of NO are equally populated except for minor deviations at high *J*. The observed product-state distributions conform well with phase space theory, provided it is modified to take into account the fact that energy in the excited complex cannot flow freely between all parent vibrations and rotations. The lifetimes of NCNO measured by Khundkar et al. for excess energies 0–600 cm^{-1} revealed 10 ps as the shortest lifetime. Extrapolations and theoretical estimates[165,166] yield lifetimes shorter than 1 ps for excess energies greater than 2,000 cm^{-1}. In conclusion, the observed energy disposal is statistical and energy randomization prior to dissociation appears to be complete except at high excess energy, where radiationless transition rates may become rate-limiting. Even there product state distributions are unaffected and should remain statistical, if intramolecular vibrational energy randomization (IVR) on the ground-state potential energy surface is complete prior to dissociation. The mechanism involved in the predissociation of NCNO according to these studies is internal conversion of the S_1 levels to vibrationally excited S_0 and unimolecular decomposition on the S_0 surface by vibrational predissociation.

It should, however, be noted that product state distributions are unfortunately not very sensitive to the details of the dynamics. The influence of other couplings and surface crossings, particularly intersystem crossing to and from T_1, may not be seen at all in product-state distributions. Nevertheless, the work on NCNO remains a landmark and a fascinating piece of research on the photophysics and unimolecular decay of a long-lived complex. It vividly illustrates the extraordinary power of LIF techniques for dynamical studies.

Currently, the best understood and most intensively studied small molecular system is formaldehyde,[20,169] where theory can provide complementary information through state-of-the-art *ab initio* techniques. Here, as in previous examples, LIF techniques for probing both the CO and the H_2 products of photolysis provide the key for detailed dynamical studies. The mechanism of molecular dissociation is well known. Electronically excited H_2CO internally converts to highly excited vibrational levels of the ground electronic state, where it dissociates over a steep barrier. Excitation of the jet-cooled molecule by a pulsed UV laser to specific rovibronic states of S_1 allows truly state-selective photochemistry to be studied with delayed LIF probing of CO and H_2 in the VUV. For photolysis energies near the barrier to dissociation, the CO product emerges highly rotationally excited with little vibrational energy, while the H_2 product vibrational distribution peaks at $v = 1$, and the rotational distribution is Boltzmann-like with T_{rot} decreasing with increasing v. Most of the available energy appears as translational energy of H_2. The rotation of the excited parent molecule during the

10–100-ns lifetime does not completely wash out the alignment of H_2 $(A_0^{(2)} = -0.31)$ deduced from polarized LIF spectroscopy, and the VUV–LIF Doppler profiles provide data on the H_2 average velocities and on vector correlations using Dixon's lineshape analysis outlined in Section 4.2.3. H_2 formed in high vibrational states is correlated with CO formed in low rotational states. A semiclassical model combining an impulsive force with the momenta of zero-point vibrations of the parent near the transition-state geometry reproduces all the general features of the product-state distributions and energy disposal near the barrier to dissociation.[169]

While formaldehyde will remain a unique example, other cases have provided outstanding advances in photochemistry. H_2O_2 is a particularly interesting case that has been studied by several groups. The LIF Doppler profiles of the OH fragment vary strongly with dissociation-probe configuration, polarization, and the branch probed, and their analysis provides detailed information on the vector correlations.[13,15,16,19] The H_2O_2 case is unique in producing two light fragments recoiling with their velocity and angular momentum vectors aligned parallel. Other interesting examples that have recently been studied by LIF with notable success are HNCO, HN_3, and $(CN)_2$. Such small systems offer the possibility of tractable calculations of potential energy surfaces connecting the initial nuclear configurations with the final products and have consequently attracted strong interest. A major source of excitement in many of these LIF product studies has been the fine structure branching ratios of open-shell products that are often nonstatistical and reflect strict symmetry correlations or the influence of nonadiabatic surface crossings during the separation of chemical bonds.[170]

4.6.3. Intermediate and Large Molecules

Here potential surfaces are rarely available, and very detailed studies are hardly justified, since information on the counterfragment will be sparse or absent. The LIF probe technique offers highly specific quantum state-resolved information on atomic and diatomic fragments, but there are serious practical difficulties in extending this approach to larger radicals because of the much greater dispersal of populations over the wider range of internal quantum states available. Thus, number densities per quantum state drop, and it becomes difficult to record spectra of sufficient quality to justify more than band contour and coarse vibronic analysis. Triatomic species such as CH_2, NH_2, NCO, NCS, and some larger radicals, such as CH_3O and CH_3S, are amenable to limited LIF analysis, at least in principle, as their spectra are understood in some detail.

Among a variety of molecular systems that can be studied, nitroso and carbonyl compounds are particularly interesting, since they present a large range of chemically interesting organic and inorganic species and produce

NO and CO in their electronic ground states on dissociation. Both fragments can be probed in detail by one- or two-photon LIF, and a significant number of systems has been studied recently.[13,77,169,171,172] Among the carbonyl compounds other than H_2CO dealt with earlier, the work by Bitto et al. on ketene ($H_2C{=}C{=}O$) yielding singlet CH_2 and CO, which can both be probed by LIF, is particularly noteworthy. Using tunable dissociation and LIF of ^1H_2C, they determined the threshold for dissociation of ketene on the singlet surface with good precision and characterized energy disposal into CO.[173] Other systems studied recently include glyoxal, acetone, acetaldehyde, carbon suboxide (C_3O_2), and cyclobutanone.[77,172]

Early advances in the dissociation dynamics of NO-containing molecules have been reviewed.[171] Organic nitroso-compounds have a weakly bound $N{=}O$ chromophore, absorb light in readily accessible regions in the visible and/or near-UV region through an (n, π^*) electronic transition, but show very contrasting photophysics and dynamical behavior. O-, N-, and S-nitroso compounds rapidly dissociate from repulsive states or potentials with very shallow wells by direct dissociation or rapid predissociation on the subpicosecond time scale, whereas the C–NO compounds form much longer-lived excited-state complexes with bound excited states often sustaining fluorescence and spectra with well-developed rovibronic structure.

S_1 excitation of C-nitroso compounds close to their dissociation threshold leads to the formation of long-lived complexes. For systems with moderate-level densities, fluorescence is sometimes seen with decay times dominated by nonradiative processes, and the dissociation rates are controlled by the rates of radiationless transitions out of the initially excited S_1 levels. In these cases the complex fluorescence decays observed indicate that the T_1 state participates in the relaxation and subsequent dissociation of levels close to the threshold for dissociation. The NO product-state distributions give little indication of nonstatistical behavior except for levels well above the top of the torsional barriers where intersystem crossing becomes the dominant nonradiative route followed either by rapid dissociation from T_1 or by intramolecular energy randomization (IVR) on that surface.[13,97,108–110,171] Recent results from the group led by Reisler and Wittig indicate that larger perfluoroalkyl nitroso-compounds with high-level densities in the S_0 and T_1 states no longer fluoresce, and the rates of formation of NO measured by delayed LIF are very low, with statistical product state distributions prevailing.[174]

LIF studies of the dissociation dynamics of alkyl nitrites and N-nitroso dimethylamine (NDMA) have utilized 300-K vapor samples and either one- or two-photon excitation of nascent NO. Energy disposal, alignment effects, Λ-doublet preferences, and recoil anisotropies have been measured for CH_3ONO, CH_3CH_2ONO, tertiary butyl nitrite (t-BuONO), and NDMA, as reviewed elsewhere.[13,171] Recent work by Docker et al. on methyl nitrite photodissociation at 300 K has indicated the power but also the limitations of careful Doppler profile analysis for deducing recoil anisotropies and other vector correlations when broad fragment velocity distributions and significant parent velocities need to be taken into account.[21]

The photodissociation of the unstable alkyl thionitrites CH_3SNO and *t*-BuSNO in vapor cells and supersonic jets have recently been studied in our laboratory using one- and two-photon LIF of NO.[12,118,175–177] The absorption spectra reveal weakly structured $S_1 \leftarrow S_0$ transitions near 550 nm and strong unstructured $S_2 \leftarrow S_0$ continua, peaking near 350 nm, with a tail extending well into the visible. Nascent, rotationally hot NO is formed on excitation in the visible and in the UV band. NO formed by 550-nm photolysis of CH_3SNO is strongly aligned ($A_0^{(2)} = +0.75$) and recoils perpendicularly to the transition dipole vector, confirming the symmetry of the S_1-state as A'' and the electronic character as (n, π^*). Figure 4.18 shows part of the LIF spectrum of nascent NO from the photolysis of jet-cooled CH_3SNO near 550 nm. LIF spectra of rotationally relaxed CH_3S have also been recorded, and energy balances have been deduced. The dissociation proceeds directly from unbound surfaces with moderate and very strong repulsive forces developed on S_1 and S_2, respectively. While the Λ-doublets of NO are very unequally populated, there is much less selectivity for population of the spin–orbit states. Similar studies on the photodissociation of jet-cooled N-nitroso methylcyanamide (NMCA) in the UV utilized two-

Figure 4.18 Long wavelength section of the $A\ ^2\Sigma^+$ ($v' = 0$) $\leftarrow X\ ^2\Pi_{3/2}$ ($v'' = 0$) LIF transition of nascent NO from the 560-nm pulsed dye laser dissociation of CH_3SNO jet-cooled with 500 torr He.

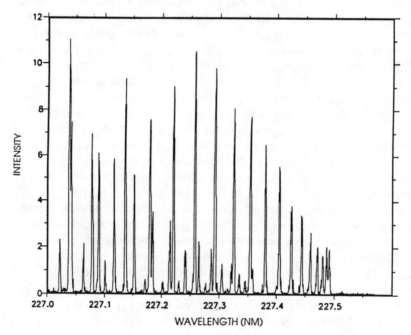

photon LIF to characterize nascent NO and one-photon LIF to probe nascent CN fragments. NO is formed in the primary dissociation process, while CN formation follows the absorption of another photon by the counterradical CH_3N—CN. The NO product state distributions are non-statistical and Gaussian shaped. That is, they are typical for an impact-dominated, direct dissociation, in contrast to the statistical CN distributions, which are more in accord with predissociation or dissociation via a linear transition state.[176] Clearly, dissociation–LIF probe techniques can also assist in revealing the mechanisms involved in multiphoton photochemistry of larger molecules.

4.6.4. Photofragment Yield Spectroscopy

Nanosecond pulsed dissociation–LIF probe experiments lend themselves to spectroscopic studies of predissociating systems that are not amenable to direct LIF spectroscopy because of lack of fluorescence of the molecular parent in the region of interest. In such cases LIF of the fragment can be exploited to obtain spectra of the parent. The dissociation laser wavelength is scanned with the probe laser wavelength fixed to a strong resonance of the appropriate fragment such as NO, CO, OH, SH, CN, or CH_2. Provided the dissociation step is not saturated and operates in the linear regime, the LIF signal maps the product of the parent absorption cross section and the quantum yield for production of fragments in the chosen quantum state. This photofragment yield spectroscopy is sometimes known by the acronym PHOFEX and was first applied by Robra et al. to study the threshold region for photodissociation of jet-cooled NO_2 near 398 nm using LIF of NO.[178]

Accurate dissociation thresholds can sometimes be measured for jet-cooled predissociating molecules by this technique. For NCNO, for example, tuned photodissociation in the visible can be used.[164,165,171] NO-yield spectra of CF_3NO and other C-nitroso compounds provide spectra of the parent molecules in regions where the fluorescence yields are too small to record standard LIF excitation spectra.[97,108,110,164] Bitto et al. monitored production of $^1CH_2$ by scanning the dissociation laser in jet-cooled ketene (H_2C=C=O) and succeeded in resolving individual thresholds for each successive CO(v = O, J) channel.[173] The technique has also provided new information on the overtone spectroscopy of H_2O_2 and similar compounds in the visible, above the dissociation barrier where OH can be monitored by LIF. This illustrates a major advantage of the technique: Low absorption strength of the parent molecule can be overcome by combining high dissociation irradiance with very sensitive fragment detection.

CO photofragment yield spectra of the formyl (HCO/DCO) radical have recently been obtained in the 500–570-nm region using VUV–LIF to monitor the production of CO. The linewidths of the diffuse predissociated transitions strongly increase with K' (the projection of the total angular momentum onto the *a*-axis) and provide predissociation lifetimes as a function of vibronic state in the range of 70–700 fs. Predissociation has been shown to proceed by curve crossing under the influence of the Renner–Teller interaction.[77]

4.7. Reaction Dynamics

The study of reaction dynamics has been regarded traditionally as one of the first but also the most challenging applications of LIF in chemistry and molecular physics. Here the diagnostic power of LIF can be exploited most fully, but the requirements for sensitivity in reactive scattering studies[1,2,8] are often extreme, particularly if atom or radical reactions with high exothermicity that produce the products in a wide range of internal quantum states are to be studied. Dagdigian has reviewed LIF as the principal optical technique in reactive scattering studies and has provided an excellent and detailed summary of the theory and underlying principles.[8] A common problem encountered in such studies is the transformation of number densities, the quantity measured by LIF, to cross sections or relative rate constants, which refer to the flux of molecules in given final states. This density-to-flux transformation has been discussed elsewhere in some detail.[6,8,71]

Gas phase chemical dynamics now represents a relatively mature field of research to which LIF techniques have added new perspectives, such as an increasing insight into fine-structure and spin–orbit effects in open-shell reactive and inelastic scattering systems.[162,163,170] One of the most interesting systems is the $H + CO_2 \rightarrow OH + CO$ reaction, where both OH and CO products are amenable to probing by LIF and where complementary studies under a wide range of conditions as well as calculated potential energy surfaces are available.[181] Thus, the van der Waals complexes of HBr and HI with CO_2 have been prepared in nozzle expansions, and photolyzed to produce H atoms with a restricted range of impact parameters. The OH distributions obtained from such "precursor limited geometries" and full reactive scattering experiments are now well known. Real-time measurements of OH appearance rates from the photolysis of the $HI–CO_2$ complex have been measured.[150] The CO rotational and vibrational distributions recently obtained by VUV–LIF appear rather less statistical than the OH distributions and peak near $J = 20$ for CO in $v = 1$.[181]

Zhang et al. recently reported a state-to-state experiment on the reaction $O(^3P) + HCl \rightarrow OH + Cl$ using LIF of the OH product. This reaction is nearly thermoneutral, and its rate is dramatically enhanced by reagent vibration. HCl ($v = 2$) was prepared in different J levels by excitation with a pulsed tunable IR laser source, and energy disposal into OH was measured as a function of reagent rotation. The observed energy partitioning does not change with HCl rotation, contrary to expectation.[182]

The reaction $H + NO_2 \rightarrow OH + NO$ with 123.6 kJ mol^{-1} energy release has become a prototype for radical–radical reactions that proceed over attractive potentials and a deep well. Two groups have independently studied energy disposal into NO and fine structure branching ratios, both using LIF, thereby complementing earlier and very recent similar studies of the OH product.[162,163] The lower F_1 ($^2\Pi_{1/2}$) spin–orbit component and the symmetric Λ-doublets $\Pi(A')$ of the NO product are preferred, and a mild similar preference is found for OH. The HONO complex does not survive sufficiently long for complete energy randomization: OH is more excited and NO is less than expected on statistical grounds. The observed dynamics is interpreted as being controlled by the ground state HONO surface and the ground state HONO ($\tilde{X}\,^1A'$) intermediate complex.

4.8. Surface Dynamics and Photochemistry

The high sensitivity and the opportunity to probe internal quantum states have offered new perspectives for laser probe techniques in surface scattering, desorption dynamics, and surface photochemistry. This is an area of growing importance with considerable relevance in catalysis, semiconductor and microelectronic technology, and surface analysis and laser desorption mass spectrometry. Recent reviews that cover this emerging new discipline should be consulted where a broader, comprehensive account is needed.[13,183–185] Both LIF and resonantly enhanced multiphoton ionization (REMPI) are suitable for state-resolved probing of scattered or desorbed species above a surface, and these two laser methods have definitely joined the bewildering array of surface diagnostic techniques. Both probe techniques use resonant excitation to an electronic excited state. While LIF utilizes the photons spontaneously emitted by this intermediate state, REMPI relies on a further transition with the same or another suitably chosen frequency to an autoionizing continuum, with detection of the ions so produced. With both LIF and REMPI the laser normally interacts with only one or very few internal quantum states and under certain conditions with only a finite velocity subgroup. Both LIF and REMPI signals are directly proportional to the number density of molecules in the selected set of quantum states in the probe volume. Both techniques have advantages and disadvantages, although their sensitivities are similar at about 10^{12} m^{-3} and quantum state. For applications in the clean UHV environment the REMPI technique is generally preferred and may be somewhat more sensitive.

4.8.1. Molecule-Surface Scattering

NO scattering from Pt(111) and Ag(111) has been researched particularly well for practical reasons: NO is probed readily by LIF or REMPI with pulsed tunable lasers of narrow bandwidth. Both techniques use resonant excitation to the $A\,^2\Sigma^+$ electronically excited state of NO, and the NO-surface interactions in these systems appear to be understood in some detail.[183] Direct-inelastic scattering and trapping-desorption channels have been seen and differentiated through the rotational and angular distributions of the scattered NO. At high surface temperatures, T_{rot} of the scattered NO tends to be significantly lower and substantial rotational cooling can occur in systems dominated by a trapping-desorption mechanism. Jacobs et al. used REMPI to characterize these channels in further detail and succeeded in showing opposite preferences, termed *helicopter* and *cartwheel motion,* for rotational alignment in these two channels.[186] This is achieved by rotating the plane of polarization of the probe laser with a photoelastic modulator and recording changes in the probe signal with the angle between the polarization vector and the surface normal.

4.8.2. Desorption Dynamics

Temperature-programmed desorption (TPD) can be combined with state-specific laser diagnostics, as shown for the first time by Cavanagh and King, who used LIF and TPD to characterize the interaction of NO with several metal surfaces.[185] The rotational distributions deduced were Boltzmann-like, and the rotational temperature of the desorbed NO was essentially equal to the surface temperature. The desorption flux under their experimental conditions was so low that gas-phase collisions could not occur between desorption and probing. Adsorbate-covered surfaces also desorb NO with Boltzmann rotational distributions and a rotational temperature that is within 90% of that of the surface. The fine-structure states of desorbed NO are equilibrated with the rotation, and no significant alignment is observed.

Hsu et al. characterized OH and OD radicals desorbing from Pt(111) by LIF and observed nearly full rotational energy accommodation with equilibrated spin–orbit and Λ-doublet components. The catalytic oxidation of NH_3 and H_2 by O_2 on Pt(111) has also been studied by LIF of desorbing OH and NO.[187,188] The NO rotational temperature remains constant over the surface temperature range 800–1300 K, indicating that dynamical effects control rotational cooling.

Laser-induced thermal desorption (LITD) is currently developing into a promising new technique in surface science that is well suited to combination with state-resolved laser probing of suitable desorbates.[185] A nanosecond pulsed laser generates a transient surface temperature jump, and a delayed synchronized tunable laser probes the desorbate at a suitable distance from the surface. By scanning the probe laser with a fixed delay between the heating and probe pulses, excitation spectra and hence internal state distributions of the desorbate can be obtained. Kinetic energy distributions of the desorbing species in selected quantum states are obtained by optical time-of-flight measurements (i.e., by incrementing the delay between the two laser pulses while monitoring a resonantly excited transition of the desorbate.[185,189] Care must be taken in these measurements to avoid using too high desorption fluxes in order to ensure the absence of collisions in the gas phase prior to probing. LIF of NO has been used to deduce the dynamics of thermal and nonthermal (resonant) desorption of NO from several single-crystal metal surfaces or foils using pulsed laser light ranging from the IR to the short UV. Inverted spin–orbit and non-Boltzmann rotational state distributions have been seen and interpreted together with the wavelength dependence of the kinetic energy distributions.[189–191]

4.8.3. Surface Photochemistry Probed by LIF

Resonant laser desorption and laser-induced surface photochemistry are newly emerging areas that bring together expensive surface science instrumentation with advanced laser techniques.[13,190,192] Recent work by Ertl et al. on NO_2 adsorbed on Pd(111) has set the scene as far as surface photodissociation dynamics of metal adsorbates and state-resolved probing of products is concerned.[192] The dissociative desorption of NO from well-characterized surfaces of NO_2 adsorbed on NO-saturated Pd(111) has been studied at several wavelengths, ranging from 193 nm to 351 nm. At 100 K the NO_2 is dimerized to N_2O_4. The desorbed NO fragments were probed by LIF on the $A\ ^2\Sigma^+ \leftarrow X\ ^2\Pi_i$ transition, that is, the $\gamma(0,0)$ band. Nearly all the NO desorbed is in the $v = 0$ vibrational state, with a few percent in $v = 1$, indicating a vibrational temperature in excess of 1000 K. State-resolved rotational distributions for NO desorbing in $v = 0$ and optical time-of-flight measurements yield the detailed energy disposal into different degrees of freedom. Two desorption channels reflecting accommodation to the surface and fast nonthermal dynamics have been identified.[192] The polarization dependence of the desorption yield indicates that the photodissociation is initiated by excitation of metal electrons rather than by direct absorption by the adsorbate. Electron attachment to N_2O_4 or to NO^+ in the ionized $NO^+NO_3^-$ (i.e., nitrosonium nitrate) form of N_2O_4 may explain the observed photodesorption dynamics. The work vividly illustrates that LIF can be employed as a powerful new tool offering many advantages in surface photochemistry.

4.9. Analytical and Diagnostic Applications of LIF

The high sensitivity and specificity of LIF offers advantages in diverse areas, and the analytical and diagnostic applications of LIF techniques are as numerous as those of nonlaser fluorescence. The replacement of conventional sources in fluorescence spectroscopy by lasers with high intensity and monochromaticity has often opened up entirely new areas of application in medical and clinical diagnostics and genetic engineering (via fluorescence labeling and DNA sequencing). Fluorescence microscopy, remote surveys of water pollution or algal growth, or the visualization of fingerprints with fluorescent markers are just some of the many impressive examples. LIF is particularly suited to studies of biological molecules and the binding of peptides and proteins to DNA, since the aromatic amino acids exhibit intense fluorescence in the UV. Femto- and attomole sensitivity have been achieved in LIF-based chromatographic detection schemes, and single-molecule detection in the condensed phase has been achieved by several groups.[193] LIF analysis in the condensed phase is, however, typically limited to a minimum concentration of 10^{-12}–10^{-13} mol dm^{-3} because of solvent background from Raman scattering and fluorescence. Condensed-phase applications of LIF are excluded from the following paragraphs.

4.9.1. Atomic Laser-Induced Fluorescence

Despite the obvious advantages and the success of narrow-bandwidth tunable lasers in spectroscopy, they have yet to replace well-developed classical spectroscopic tools in analytical spectroscopy. Niemax has recently presented an authoritative account of analytical laser atomic spectroscopy and compared sampling, excitation, and detection methods as well as detection limits of the best-established laser techniques with the classical advanced spectrochemical methods, such as atomic absorption, optical emission, and mass spectrometry.[38] In real analytical applications, sample preparation, optical excitation, and detection are of equal importance, and performance limits are dictated by the weakest part. For atomic LIF the electrothermally heated graphite furnace appears to offer considerable advantages as a sampling method.[38]

Impressive results can be achieved under optimal conditions with CW lasers by employing conditions of optical saturation with hundreds of excitation–fluorescence cycles and coincidence detection techniques to reduce background noise. Consider a Na atom excited on the D line with optical saturation. The lifetime for spontaneous fluorescence is 16 ns. If the atom moves with a mean thermal velocity of 300 m s^{-1} through the 3-mm-diameter beam, it will experience 300 excitation–fluorescence cycles if the probability of being reexcited after spontaneous decay is 0.5. Fairbank et al. detected 10–100 atoms of Na cm^{-3}, and Balykin et al. succeeded in detecting 10 Na atoms per second in an atomic beam.[194,195] In practical atomic LIF the limits of detection are much higher because of the sampling techniques used for atomization, which produce high background radiation and strong fluctuations in the analyte atom number density. Furthermore, pulsed lasers generally have to be used to take advantage of frequency-doubling and other nonlinear wavelength extension techniques. Since the pulse length is typically 10 ns and the repetition rate only 10–100 s^{-1} the duty cycle is small, only a small fraction of the atoms in the excitation volume can be excited, and each atom excited cannot be repeatedly excited. Detection limits are worst if LIF is detected at the same wavelength used for excitation, because of scattered laser light.

Fortunately, this problem can often be avoided by monitoring fluorescence from the pumped level to an intermediate state, which results in a red shift and permits better discrimination of fluorescence and scattered light. Typical detection limits using analytical flames or inductively coupled plasmas (ICP) for atomization are 1–10 ng ml^{-1} for most elements. ICP atomization appears to offer some advantages for LIF, since upper atomic levels are populated significantly because of the high temperature of the plasma. For elements with low ionization limits such as the alkali and alkaline earth metals, the yield of ions can be large. This permits a wider choice of lines for excitation of atoms or their ions in the more convenient longer-wavelength range. Very low detection limits have recently been achieved with graphite tube furnaces using electrothermal atomization of samples. Bolshov et al.[196] claimed to have reached a detection limit of 1.5 fg ml^{-1} for Pb, and Dougherty et al.[197] reported 8 fg ml^{-1} for Ag. Very high sensitivity for lead (30 fg ml^{-1}) by atomic LIF with electrothermal vaporization has recently been confirmed by Omenetto.[198] Laser-produced plasma atomization has also been used by Measures and Kwong in conjunction with time-delayed LIF probing for elemental analysis.[199]

LIF methods for the detection of group IV atoms have been included in a recent review by Hack.[37] Brewer used two-photon LIF and (2 + 1) REMPI near 430 nm to characterize Si atoms formed by multiphoton dissociation of tetramethylsilane (TMS).[200] Progress in the detection of light main group atomic elements has accelerated rapidly in recent years because of the development of tunable pulsed short UV and VUV generation techniques. Cl, Br, I, N, and O atoms can be detected by two-photon LIF with sensitivities of the order of 10^{16} m^{-3}. The detection of H atoms by VUV LIF at the Lyman-α line at 121.56 nm is now a well-established technique that has been applied to studies of photodissociation processes and of H-atom production in chemical reactions.[70,201,202]

4.9.2. Flame, Plasma, and Gas-Flow Diagnostics

These are interesting applications of LIF that have considerable relevance in combustion research, semiconductor technology, and plasma chemistry as well as wind tunnel or other fast-gas-flow diagnostics. The main advantage of LIF in these applications arise from the nonintrusive, optical nature of the laser excitation and fluorescence steps. Thus, the local chemistry and physics are not disturbed, and scalar properties such as number densities, population distributions, and temperatures as well as velocities can be obtained that truly reflect the complex local processes and environment. The spatial resolution of LIF can be made better than 1 mm^3 by careful attention to detail concerning imaging detection and by the use of focused beams. The use of nanosecond pulsed lasers also permits high temporal resolution. These attributes are important in atmospheric pressure flames, where the concentration and temperature gradients are high and in the case of turbulent flames, change very rapidly.

Crosley and Smith have reviewed LIF as a technique for combustion diagnostics.[101] The following combustion intermediates and products have all been observed in flames: H, O, OH, CH, C_2, CN, CO, NH, NO, NO_2, NCO, S_2, SH, SO, and SO_2. OH can be detected in a volume of 1 mm^3 at levels less than 1 ppb. Atomic LIF or LIF of metal oxides or halides can be used through seeding suitable metal compounds into the flame. Collisional electronic quenching and rotational–vibrational relaxation of species in the electronically excited state must be taken into account for quantitative determinations of concentrations and temperatures. Wolfrum and co-workers recently showed, however, that OH fluorescence lifetimes in an atmospheric pressure flame were effectively constant, indicating that collisional quenching rates can be constant, under certain conditions.[203]

One of the most exciting developments of the LIF technique has been in the two- and three-dimensional concentration imaging of transient species. The method utilizes LIF excitation by a planar sheet of pulsed laser light. Provided the collisional quenching rate is constant within the excitation volume, the two-dimensional fluorescence image captured by an image-intensified camera mirrors the species concentrations directly. Powerful lasers are required, since the fluorescence yields can be very small when quenching is significant. Two-dimensional LIF concentration mapping of OH radicals in flames has been used since 1982 by Hanson, Crosley et al. and has been extended to the visualization of velocity distributions using seeding of flames with fluorescent dopants.[204–207] Wolfrum recently described the use of 308-nm XeCl excimer laser light and acetaldehyde as a fluorescent dopant to image flame fronts in internal combustion engines.[203]

In real combustion systems quenching rates can vary greatly because of large temperature and concentration gradients. Fluorescence yields cannot be easily converted then to concentrations, a disadvantage of LIF that has led to coherent anti-Stokes Raman (CARS) techniques being preferred for a long time for spatially resolved but nonimaging flame thermometry of majority species. Quenching effects can be reduced if not eliminated by the technique of laser-induced predissociation fluorescence (LIPF) spectroscopy. LIPF measurements were first realized in the Schumann–Runge bands of O_2.[206,207] The principle of LIPF is to avoid quenching effects by exciting very short-lived levels, as explained in Section 4.2.4. The drawback of small fluorescence yields must be overcome by the use of high intensities and spectral brightness for excitation and requires powerful tunable line-narrowed ArF and KrF lasers, for example.[28,29] The method can be applied to a variety of species, such as OH, NO, O_2, and H_2O, and has been applied successfully to fuel combustion in internal combustion engines by Andresen and coworkers.[28,29,206,207] Two-photon LIF has been used to detect NO,

OH, and CO via their $A^2\Sigma^+ \leftarrow X^2\Pi_i$ and $B^1\Sigma^+ \leftarrow X^1\Sigma^+$ transitions in flames using 450-nm, 615-nm, and 230-nm laser light, but the relatively poor sensitivity as compared to one-photon techniques has thus far prevented wider use of these methods.[208-211]

Optical emission spectroscopy is used extensively for the diagnostics of weakly ionized plasmas of the kind used in materials processing.[212-214] However, emission spectra provide direct information only on species in electronically excited states. Since most plasma species occupy ground electronic states, it is more useful to monitor the ground-state species. Laser excitation techniques are well suited for measurements of ground-state species in discharge plasmas used for deposition or etching.[212-214] Here LIF typically provides information on the identity, number density, spatial distribution, velocity, and internal energy distributions and chemical reactivities of the various species present. Examples of species that have been detected in plasmas by LIF techniques are H, $O(^3P)$, $N(^4S)$, $N(^2D)$, F, Cl, SiH, and SiH_2,[212-216] and many examples of other species are likely to be found in plasmas where LIF spectra are known or have recently been characterized.

Excimer laser ablation, also called laser sputtering or etching, has been studied by pulsed time-delayed LIF, and a variety of species, such as C_2, CN, CuCl, and AlO, have been monitored. LIF spectra of C_2 produced by photoablation of graphite, for example, indicate that the C_2 vibrational and rotational degrees of freedom are in equilibrium with the surface temperature of 3,500 K. Elemental analysis by LIF in laser-produced plasmas has also been demonstrated. The light of a pulsed laser is focused onto the surface, and a microsample is ablated and atomized in the laser-produced plasma. After an optimized delay, a tunable pulsed dye laser excites the atomic LIF resonance in the desorption plume, which is allowed to expand into a low-pressure noble gas.[213,214]

4.9.3. Atmospheric Trace Detection and Remote Sensing

Laser techniques have rapidly developed into important diagnostic tools for the analysis of air pollutants using stationary point sampling as well as remote sensing. In remote sensing by the light detection and ranging (LIDAR) technique, Rayleigh, Mie, and Raman scattering as well as atomic resonance and relaxed molecular fluorescence are used in contrast with long path absorption with retroreflectors.[217-219] While the cross sections for Rayleigh and Raman scattering are very small, those for Mie scattering and fluorescence are larger by many orders of magnitude. The quenching of electronic fluorescence by atmospheric oxygen can reduce fluorescence yields by three to five orders of magnitude. Absorption cross sections in the IR and UV-visible are also substantial. Thus, the highest sensitivity combined with good spatial resolution can be achieved by

combining differential absorption and scattering. In this approach the
back-scattered laser light is compared when the laser is tuned to a resonance line of the species of interest and when it is detuned to the wing of
the line. The Mie scattering from atmospheric aerosols and dust is the same
at both wavelengths, and the ratio of the two signals becomes sensitive to
the number density of the absorbing species. This approach based on the
absorption of the back-scattered light by the analyte has been termed *differential absorption lidar* (DIAL) and offers the greatest sensitivity for long-
range monitoring of specific pollutants or atmospheric constituents. Its
main use is in the monitoring of O_3, NO, NO_2, SO_2, and similar pollutants,
but it has been applied successfully to LIF measurements of alkali atoms
and temperatures in the upper atmosphere.

Figure 4.19 shows a typical LIDAR setup that can be used as an LIF
sensor or as a DIAL instrument for range-resolved measurements of remote atmospheric species in the UV, visible, or near-IR. An intense short
laser pulse is directed through a beam-shaping telescope via a beamsplitter
reference detector combination toward the intended target. The reference
detector provides a start pulse. The back-scattered light is collected by the
receiver optics and imaged onto the entrance slit of a spectrometer with a
suitable gated multichannel detector and time-resolved spectral analysis
system. The microprocessor initiates the firing of the pulsed laser and controls data acquisition and manipulation. In some cases a simple narrow-
pass interference filter, photomultiplier, and transient recorder linked with
a small computer are adequate and may replace the more sophisticated
multichannel detection system.

Figure 4.19 Schematic layout for a fluorescence LIDAR sensor. L: laser system; RD:
reference detector; OO: output beam-shaping optics; RO: receiver optics; SMCD:
spectrum multichannel detector; SAS: spectral analysis system; μ-PC: microprocessor.

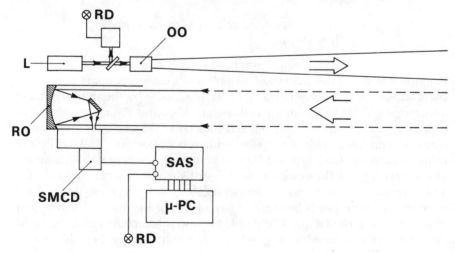

Resonance fluorescence by alkali atoms in the upper atmosphere can be used for LIDAR measurements of vertical temperature profiles. Using a line-narrowed single-axial mode excimer-pumped dye laser, von Zahn et al. succeeded in measuring the Doppler profiles of the Na–D doublet as a function of altitude. Since the Doppler profiles, after suitable deconvolution of the overlap of the two doublet components, are related to the velocities and hence the temperature of the atoms, temperature–altitude profiles can be measured up to a height of 100 km.[220] The use of fluorosensors for aerial surveying of marine oil pollution and for mapping of oceanic phytoplankton by the red fluorescence of chlorophyll are other examples of fluorescence LIDAR.[221,222]

4.9.4. *Ultratrace Analysis of Atmospheric Species*

LIF has considerable potential for the ultratrace analysis of certain key atmospheric constituents, such as OH, NO, NO_2, HONO, and $HONO_2$, in the laboratory or by means of airborne, van-based, or stationary laser equipment. The basic capability for higher sensitivity and higher species selectivity of a laser-based technique has to be weighed against the disadvantages, particularly cost and complexity. For trace analysis of the preceding species in the range above 1 ppb, other techniques, such as differential optical absorption spectroscopy (DOAS), DIAL, and nonspectroscopic methods (e.g., chemiluminescence) are available. These techniques become unreliable for analysis in the part per trillion volume mixing range, and ultratrace methods are needed for intercomparison and calibration of field instruments that are to be used for routine measurements of these species close to their detection limits.

Measurement of the OH radical at number densities of the order of 10^{12} m^{-3} in ambient air remains a challenge. Rodgers et al. have described an LIF technique capable of measuring 10^{13} OH radicals m^{-3} using a frequency-doubled Nd:YAG pumped dye laser to excite the well-known $A\,^2\Sigma^+$ $\leftarrow X\,^2\Pi_i$ fluorescence of OH via the 1–0 transition. The main limitation of their approach is interference from OH produced by $O(^1D)$ via the photolysis of ozone.[223] HONO could be analyzed in the low ppb range by 355-nm pulsed photolysis with the third harmonic of the Nd:YAG laser and delayed pulsed LIF analysis of the OH ($v = 0$) so formed.[224]

We have developed an LIF technique for ultratrace analysis of NO using supersonic jet-cooling and LIF with tunable 225-nm excitation of NO via the $\gamma(0,0)$ bands. Jet-cooling offers the advantage of concentrating the NO population to one or two of the lowest rotational states. It decreases the Doppler width of the NO transitions from 0.15 to less than 0.05 cm^{-1}. The improved match of laser bandwidth and transition linewidth in combination with a substantial increase in the fluorescence yield improves the sensitivity and selectivity of the technique and permits analysis of NO down to a concentration range of 1 ppt[225,226] without interference from other common NO-containing air pollutants.

Near-threshold photolysis of jet-cooled NO_2 in combination with LIF analysis of $O(^3P)$ and NO fragments has also been studied in our laboratory using the combination of jet-cooling with pulsed laser pump–LIF probe techniques. Photofragment LIF via detection of the $NO(^2\Pi_{3/2})$ fragment was found to provide better sensitivity than detection of $O(^3P)$ by two-photon LIF. Figure 4.20 shows an LIF excitation spectrum of NO from the photolysis of a jet-cooled sample of 50 ppm of NO_2 in Ar near 395 nm. Two dye lasers pumped by a single excimer laser were used, with one providing tunable light for photolysis in the near-UV and the other producing frequency-doubled laser light near 225 nm for one-photon probing of the nascent NO.[226] The discrimination of jet-cooled NO and NO formed from NO_2 is accomplished readily while photolysis of HNO_2 at 395 nm produces NO with such highly dispersed rotational and vibrational population distributions that its interference is quite minor. The sensitivity achieved for NO_2 is inferior compared to NO but is still in the interesting range near 1 ppb with little interference from other NO-containing compounds. Recent advances in the analysis of NO_2 by automated diode laser absorption spectroscopy using multiple-pass long-path cells permit measurements in the range 0.5 ppb for NO_2 and 3 ppb for NO. These examples show that LIF techniques are interesting for measurements of atmospheric trace gases in

Figure 4.20 $A\ (v' = 0) \leftarrow X\ (v'' = 0)$, that is, $\gamma(0,0)$ LIF excitation spectrum of nascent NO from the pulsed dye laser photolysis of jet-cooled NO at 395.65 nm, close to the dissociation threshold for $NO_2 \rightarrow NO\ X^2\Pi_{3/2} + O^3P_2$. The features near 226.8 nm are due to the $F_2\ (^2\Pi_{3/2})$ spin–orbit state of NO.

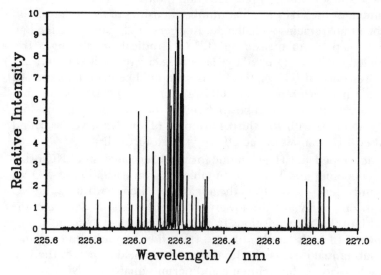

the laboratory and with mobile units of the type recently used for DIAL measurements of NO and NO_2. The DIAL unit can also employ two dye lasers pumped by a single excimer laser.[227]

4.10. Conclusions and Prospects

Laser-induced fluorescence spectroscopy is rapidly becoming one of the most widely practiced laser spectroscopic techniques. Among the main reasons it enjoys increasing popularity are the ease with which spontaneous photon emission can be detected with high sensitivity and selectivity, and the small investment required to implement LIF in practice if a tunable laser system is available. While most important gas-phase applications have been covered in this review, some had to be neglected or could only be treated cursorily, such as energy transfer, gas kinetics, and time-resolved fluorescence. There are also impressive condensed-phase applications of LIF in low-temperature matrices and biological or medical diagnostics that are beyond the scope of this chapter but that have considerable potential for the future.

Acknowledgments

I wish to thank my former and present enthusiastic collaborators, especially Martin McCoustra, Victor Young, Donna Wetzel, Julie Dyet, Michael Hippler, Sandy Yates, and Alec Simpson. I am also grateful for the financial support of the UK Science and Engineering Research Council and the Commission of the European Communities.

REFERENCES

1. J. L. Kinsey, *Ann. Rev. Phys. Chem.* **28,** 349 (1977).

2. R. N. Zare and P. J. Dagdigian, *Science* **185,** 739 (1974).

3. W. Demtröder, *Laser Spectroscopy* (Springer, Berlin, 1982), pp. 417, 434, 468.

4. S. R. Leone, *Adv. Chem. Phys.* **50,** 255 (1982).

5. M. A. A. Clyne and I. S. McDermid, *Adv. Chem. Phys.* **50,** 1 (1982).

6. U. Hefter and K. Bergmann, in *Atomic and Molecular Beam Methods,* Vol. 1, G. Scoles, ed. (Oxford University Press, New York, 1988), p. 193.

7. C. H. Greene and R. N. Zare, *J. Chem. Phys.* **78,** 6741 (1983).

8. P. J. Dagdigian, in Ref. 6, p. 596.

9. R. Altkorn and R. N. Zare, *Ann. Rev. Phys. Chem.* **35,** 265 (1984).

10. G. Herzberg, *Molecular Spectra and Molecular Structure, Vol. I. Spectra of Diatomic Molecules* (Van Nostrand, New York, 1950), p. 208.

11. G. Breit, *Rev. Mod. Phys.* **5,** 91 (1933).

12. D. M. Wetzel, V. M. Young, B. Stewart, and J. Pfab, in preparation.

13. J. Pfab, *Ann. Prog. Phys. Chem. Sect. C,* **84,** 132 (1987).

14. W. Demtröder, *Laser Spectroscopy* (Springer, Berlin, 1982).

15. R. N. Dixon, *J. Chem. Phys.* **85,** 1866 (1986).

16. P. L. Houston, *J. Phys. Chem.* **91,** 5388 (1987).

17. P. L. Houston, *Ann. Rev. Phys. Chem.* **40,** 375 (1989).

18. G. E. Hall, N. Sivakumar, G. Chawla, P. L. Houston, and L. Burak, *J. Chem. Phys.* **88,** 3682 (1988).

19. M. P. Docker, A. Hodgson, and J. P. Simons, *Molecular Photodissociation Dynamics,* M. N. R. Ashfold and J. E. Baggott, eds. (Royal Society of Chemistry, London, 1987), p. 115.

20. T. J. Butenhoff, K. L. Carleton, M.-C. Chuang, and C. B. Moore, *J. Chem. Soc. Faraday Trans. II,* **85,** 1155 (1989).

21. M. P. Docker, A. Ticktin, U. Brühlmann, and J. R. Huber, *J. Chem. Soc. Faraday Trans. II,* **85,** 1169 (1989).

22. D. M. Lubman, C. T. Rettner, and R. N. Zare, *J. Phys. Chem.* **86,** 1129 (1982).

23. J. T. Yardley, *Introduction to Molecular Energy Transfer* (Academic, New York, 1980).

24. R. Clark and A. J. McCaffery, *Mol. Phys.* **35,** 617 (1978).

25. Z. T. Alwahabi, C. G. Harkin, A. J. McCaffery, and B. J. Whitaker, *J. Chem. Soc. Faraday Trans.* II, **85,** 1003 (1989).

26. B. J. Orr and I. W. M. Smith, *J. Phys. Chem.* **91,** 6106 (1987).

27. A. J. McCaffery, M. J. Proctor, and B. J. Whitaker, *Ann. Rev. Phys. Chem.* **37,** 223 (1986).

28. A. M. Wodtke, L. Hüwel, H. Schluter, H. Voges, G. Meijer, and P. Andresen, *J. Chem. Phys.* **89,** 1929 (1988).

29. P. Andresen, A. Bath, W. Gröger, H. W. Lülf, G. Meijer, and J. J. ter Meulen, *Appl. Opt.* **27,** 365 (1988).

30. W. A. Majewsky and W. L. Meerts, *J. Mol. Spectr.* **104,** 271 (1984).

31. W. A. Majewsky, D. F. Plusquellic, and D. W. Pratt, *J. Chem. Phys.* **90,** 1362 (1989).

32. E. Riedle, Th. Weber, U. Schubert, H. J. Neusser, and E. W. Schlag, *J. Chem. Phys.* **93,** 967 (1990).

33. *Lambda Physik Highlights,* No. 15/16, April 1989, p. 6.

34. E. F. Cromwell, T. Trickl, Y. T. Lee, and A. H. Kung, *Rev. Sci. Instrum.* **60,** 1888 (1989).

35. E. F. Cromwell, T. Trickl, Y. T. Lee, and A. H. Kung, *J. Chem. Phys.* **92,** 3230 (1990).

36. T. Imasaka and N. Ishibashi, *Anal. Chem.* **62,** 363A (1990).

37. W. Hack, *Int. Rev. Phys. Chem.* **4,**165 (1985).

38. K. Niemax, *Analytical Aspects of Atomic Laser Spectrochemistry* (Harwood, London, 1989).

39. C. R. Vidal, *Topics Appl. Phys.* **59,** 19 (1986).

40. G. Scoles, ed., *Atomic and Molecular Beam Methods* (Oxford University Press, Oxford 1988).

41. A. Amirav, U. Even, and J. Jortner, *Chem. Phys.* **51**, 31 (1980).

42. V. E. Bondybey, in *Chemistry and Physics of Matrix Isolated Species*, L. Andrews and M. Moskovits, eds. (North-Holland, Amsterdam, 1989).

43. D. H. Levy, *Ann. Rev. Phys. Chem.* **31**, 197 (1980).

44. R. Campargue, *J. Phys. Chem.* **88**, 4466 (1984).

45. J. B. Cross and J. J. Valentini, *Rev. Sci. Instrum.* **53**, 38 (1982).

46. D. Proch and T. Trickl, *Rev. Sci. Instrum.* **60**, 713 (1989).

47. F. J. Grieman, S. H. Ashworth, J. M. Brown, and I. R. Beattie, *J. Chem. Phys.* **92**, 6365 (1990).

48. H.-G. Löhmannsröben, D. Bahatt, and U. Even, *J. Phys. Chem.* **94**, 4025 (1990).

49. T. Imasaka, T. Okamura, and N. Ishibashi, *Anal. Chem.* **58**, 2152 (1986).

50. T. R. Rizzo, Y. D. Park, L. A. Peteanu, and D. H. Levy, *J. Chem. Phys.* **84**, 2534 (1986).

51. J. R. Cable, M. J. Tubergen, and D. H. Levy, *J. Am. Chem. Soc.* **110**, 7349 (1988).

52. J. R. Cable, M. J. Tubergen, and D. H. Levy, *J. Am. Chem. Soc.* **111**, 9032 (1989).

53. T. G. Dietz, M. A. Duncan, D. E. Powers, and R. E. Smalley, *J. Chem. Phys.* **74**, 6511 (1981).

54. V. E. Bondybey and J. H. English, *J. Chem. Phys.* **74**, 6978 (1981).

55. V. E. Bondybey, *Science* **227**, 125 (1985).

56. T. P. Martin, *Angew. Chem.* **98**, 197 (1986).

57. M. D. Morse, *Chem. Rev.* **86**, 1049 (1986).

58. C. J. Whitham, B. Soep, J. P. Visticot, and A. Keller, *J. Chem. Phys.* **93**, 991 (1990).

59. A. E. Ellis, E. S. J. Robles, and T. A. Miller, *J. Chem. Phys.* **94**, 1752 (1991).

60. A. Weber, ed., *Structure and Dynamics of Weakly Bound Molecular Complexes*, NATO ASI Series C, Vol. 212 (Reidel, Boston, 1986).

61. R. R. Bennett, J. G. McCaffery, and W. H. Breckenridge, *J. Chem. Phys.* **92**, 2740 (1990).

62. C. Dedonder-Lardeux, C. Jouvet, M. Richard-Viard, and D. Solgadi, *J. Chem. Phys.* **92**, 2828 (1990).

63. C. Jouvet, C. Lardeux-Dedonder, and D. Solgadi, *Chem. Phys. Lett.* **156**, 569 (1989).

64. W. M. Fawzy and M. C. Heaven, *J. Chem. Phys.* **92**, 909 (1990).

65. J. C. Miller, *J. Chem. Phys.* **86**, 3166 (1987).

66. Th. Weber, A. von Bargen, E. Riedle, and H. J. Neusser, *J. Chem. Phys.* **92**, 90 (1990).

67. M. J. Tubergen and D. H. Levy, *J. Phys. Chem.* **95**, 2175 (1991).

68. J. Sipior and M. Sulkes, *J. Chem. Phys.* **88**, 6146 (1988).

69. R. Bombach, E. Honegger and S. Leutwyler, *Chem. Phys. Lett.* **118**, 449 (1985).
70. D. W. Setzer, ed., *Reactive Intermediates in the Gas Phase* (Academic, New York, 1979).
71. K. Liu, R. G. Macdonald, and A. F. Wagner, *Int. Rev. Phys. Chem.* **9**, 187 (1990).
72. R. G. Macdonald and K. Liu, *J. Chem. Phys.* **93**, 2443 (1990).
73. P. J. Dagdigian, *J. Chem. Phys.* **90**, 6110 (1989).
74. P. Andresen, D. Häusler, and H. W. Lülf, *J. Chem. Phys.* **81**, 57 (1984).
75. L. J. van de Burgt an M. Heaven, *J. Chem. Phys.* **87**, 4235 (1987).
76. D. L. Monts, T. G. Dietz, M. A. Duncan, and R. E. Smalley, *Chem. Phys.* **45**, 133 (1980).
77. S. H. Kable, J.-C. Loison, D. W. Neyer, P. L. Houston, I. Burak, and R. N. Dixon, *J. Phys. Chem.* **21**, 8013 (1991).
78. Q. Lu, Y. Chen, D. Wang, Y. Zhang, S. Yu, C. Chen, M. Koshi, H. Matsui, S. Koda, and X. Ma, *Chem. Phys. Lett.* **178**, 517 (1991).
79. T. A. Miller, *Science* **223**, 545 (1984).
80. D. J. Clouthier and J. Karolczak, *J. Phys. Chem.* **93**, 7542 (1989).
81. J. R. Dunlop, J. Karolzak, D. J. Clouthier, and S. C. Ross, *J. Phys. Chem.* **95**, 3045 (1991).
82. K. Obi, Y. Matsumi, Y. Takeda, S. Mayama, H. Watanabe, and S. Tsuchiya, *Chem. Phys. Lett.* **95**, 520 (1983).
83. J. W. Farthing, I. W. Fletcher, and J. C. Whitehead, *J. Phys. Chem.* **87**, 1663 (1983).
84. M. Fukushima and K. Obi, *J. Chem. Phys.* **93**, 8488 (1990).
85. T. Ishida, H. Abe, A. Nakajima, and K. Kaya, *Chem. Phys. Lett.* **170**, 425 (1990).
86. M. N. R. Ashfold and G. Hancock, in *Gas Kinetics and Energy Transfer, Specialist Periodical Reports*, Vol. 4 (Royal Society of Chemistry, London, 1981), p. 73.
87. D. W. Lupo and M. Quack, *Chem. Rev.* **87**, 181 (1987).
88. S. C. Foster and T. A. Miller, *J. Phys. Chem.* **93**, 5986 (1989).
89. R. Schlachta, G. Lask, and V. E. Bondybey, *Chem. Phys. Lett.* **180**, 275 (1991).
90. S. Sharpe and P. M. Johnson, *Chem. Phys. Lett.* **107**, 35 (1984).
91. S. Sharpe and P. M. Johnson, *J. Chem. Phys.* **81**, 4176 (1984).
92. C. S. Feigerle and J. C. Miller, *J. Chem. Phys.* **90**, 2900 (1989).
93. J. G. Pruett and R. N. Zare, *J. Chem. Phys.* **64**, 1774 (1976).
94. J. E. Butler, *Appl. Opt.* **21**, 3617 (1982).
95. R. E. Smalley, D. A. Auerbach, P. S. H. Fitch, D. H. Levy, and L. Wharton, *J. Chem. Phys.* **66**, 3778 (1977).
96. M. R. S. McCoustra and J. Pfab, *J. Chem. Soc. Faraday Trans. II*, **84**, 655 (1988).
97. J. A. Dyet, M. R. S. McCoustra, and J. Pfab, *J. Chem. Soc. Faraday Trans. II*, **84**, 463 (1988).
98. S. J. Davies and K. W. Holtzclaw, *J. Chem. Phys.* **92**, 1661 (1990).
99. R. N. Dixon, in *High Resolution Spectroscopy*, C. P. Alves, J. M. Brown, and M. J. Hollas, eds. (Plenum, New York, 1988), p. 281.

100. B. Cagnac, in *High Resolution Spectroscopy*, C. P. Alves, J. M. Brown and M. J. Hollas, eds. (Plenum, New York 1988), p. 153.

101. D. R. Crosley and G. P. Smith, Opt. Eng. **22**, 545 (1983).

102. T. Hayashi, T. Imasaka, and N. Ishibashi, *Chem. Phys.* **109**, 145 (1986).

103. R. A. Beaudet, K. G. Weyer, and H. Walther, *Chem. Phys. Lett.* **60**, 486 (1979).

104. M. Asscher and Y. Haas, *J. Chem. Phys.* **76**, 2115 (1982).

105. A. B. Callear and M. J. Pilling, *Trans. Faraday Soc.* **66**, 1618 (1970).

106. A. B. Callear and M. J. Pilling, *Trans. Faraday Soc.* **66**, 1886 (1970).

107. M. J. Tubergen, J. R. Cable, and D. H. Levy, *J. Chem. Phys.* **92**, 51 (1990).

108. M. R. S. McCoustra, J. A. Dyet, and J. Pfab, *Laser Chem.* **9**, 289 (1988).

109. M. R. S. McCoustra, J. A. Dyet, and J. Pfab, *Chem. Phys. Lett.* **136**, 231 (1987).

110. M. R. S. McCoustra and J. Pfab, *Spectrochim. Acta* **46A**, 937 (1990).

111. S.-Y. Chiang and Y.-P. Lee, *J. Chem. Phys.* **95**, 66 (1991).

112. T. A. Stephenson, P. L. Radloff, and S. A. Rice, *J. Chem. Phys.* **81**, 1060 (1984).

113. R. D. Gordon, J. M. Hollas, P. J. A. Ribeiro-Claro, and J. C. Teixeira-Dias, *Chem. Phys. Lett.* **182**, 649 (1991).

114. K. W. Butz, D. J. Krajnovich, and C. S. Parmenter, *J. Chem. Phys.* **93**, 1557 (1990).

115. Y.-D. Shin, H. Saigusa, M. Z. Zgiersky, F. Zerbetto, and E. C. Lim, *J. Chem. Phys.* **94**, 3511 (1991).

116. E. C. Apel and E. K. C. Lee, *J. Chem. Phys.* **85**, 1261 (1986).

117. R. J. Donovan, A. J. Holmes, P. R. R. Langridge-Smith, and T. Ridley, *J. Chem. Soc. Faraday Trans. II*, **84**, 541 (1988).

118. M. R. S. McCoustra and J. Pfab, *Chem. Phys. Lett.* **137**, 355 (1987).

119. M. Heaven, T. A. Miller, R. R. Freeman, J. C. White, and J. Bokor, *Chem. Phys. Lett.* **86**, 458 (1982).

120. L. F. Di Mauro, R. A. Gottscho, and T. A. Miller, *J. Appl. Phys.* **56**, 2007 (1984).

121. H. Nagai, K. Aoki, T. Kusumoto, K. Shibuya, and K. Obi, *J. Phys. Chem.* **95**, 2718 (1991).

122. M. A. Johnson, C. R. Webster, and R. N. Zare, *J. Chem. Phys.* **75**, 5575 (1981).

123. D. E. Reisner, R. W. Field, J. L. Kinsey, and H. L. Dai, *J. Chem. Phys.* **80**, 5968 (1984).

124. E. Abramson, R. W. Field, D. Imre, K. K. Innes, and J. L. Kinsey, *J. Chem. Phys.* **80**, 2298 (1984).

125. C. E. Hamilton, J. L. Kinsey, and R. W. Field, *Ann. Rev. Phys. Chem.* **37**, 493 (1986).

126. Y. S. Choi and C. B. Moore, *J. Chem. Phys.* **90**, 3875 (1989).

127. D. Frye, P. Arias, and H. L. Dai, *J. Chem. Phys.* **88**, 7240 (1988).

128. G. Duxbury, *Chem. Soc. Rev.* **12**, 453 (1983).

129. U. Schubert, E. Riedle, and H. J. Neusser, *J. Chem. Phys.* **90**, 5994 (1989).

130. H. Sieber, E. Riedle, and H. J. Neusser, *Chem. Phys. Lett.* **169**, 191 (1990).

131. E. Riedle, Th. Knittel, Th. Weber, and H. J. Neusser, *J. Chem. Phys.*, **91**, 4555 (1989).

132. A. Levinger and Y. Prior, *J. Chem. Phys.* **94**, 1664 (1991).

133. P. T. Greenland, *Contemp. Phys.* **31**, 405 (1990).

134. M. T. Macpherson and R. F. Barrow, *Ann. Rep. Prog. Chem. Sect. C* **78**, 221 (1981).

135. T. Ishida, H. Abe, A. Nakajima, and K. Kaya, *Chem. Phys. Lett.* **170**, 425 (1990).

136. M. T. Berry, M. R. Brustein, and M. I. Lester, *J. Chem. Phys.* **90**, 5878 (1989).

137. K. Sato, Y. Achiba, H. Nakamura, and K. Kimura, *J. Chem. Phys.* **85**, 1418 (1986).

138. H. Joswig, P. Andresen, and R. Schinke, *J. Chem. Phys.* **85**, 1904 (1986).

139. T. T. Suchizawa, K. Yamanuchi, and S. Tsuchiya, *J. Chem. Phys.* **92**, 1560 (1990).

140. W. J. Orville-Thomas, ed., *Internal Rotation in Molecules* (Wiley, London, 1974).

141. M. Baba, I. Hanazaki, and U. Nagashima, *J. Chem. Phys.* **82**, 3938 (1985).

142. M. Baba and I. Hanazaki, *J. Chem. Phys.* **81**, 5426 (1984).

143. C. M. Cheatham and J. Laane, *J. Chem. Phys.* **94**, 7734 (1991).

144. M. Ito, *J. Phys. Chem.* **91**, 517 (1987).

145. H. Sekiya, K. Sasaki, Y. Nishimura, A. Mori, and H. Takeshita, *Chem. Phys. Lett.* **174**, 133 (1990).

146. K. W. Butz, D. J. Krajnovich, and C. S. Parmenter, *J. Chem. Phys.* **93**, 1557 (1990).

147. R. S. Ruoff, T. D. Klotz, T. Emilsson, and H. S. Gutowsky, *J. Chem. Phys.* **93**, 3142 (1990).

148. X.-Q. Tan, W. A. Majewsky, D. F. Plusquellic, and D. W. Pratt, *J. Chem. Phys.* **94**, 7721 (1991).

149. N. F. Scherer, C. Sipes, R. B. Bernstein, and A. H. Zewail, *J. Chem. Phys.* **92**, 5239 (1990).

150. A. H. Zewail, *Science* **242**, 1645 (1988).

151. L. R. Khundkar and A. H. Zewail, *Ann. Rev. Phys. Chem.* **41**, 15 (1990).

152. J. P. Simons, *J. Phys. Chem.* **88**, 1287 (1984).

153. C. X. W. Qian, A. Ogai, J. Brandon, Y. Y. Bai, and H. Reisler, *J. Phys. Chem.* **95**, 6763 (1991).

154. H. Zacharias, M. Geilhaupt, K. Meier, and K. H. Welge, *J. Chem. Phys.* **74**, 218 (1981).

155. T. Axon, Ph.D. thesis, Heriot-Watt University, Edinburgh, 1991.

156. A. J. Yates, Ph.D. thesis, Heriot-Watt University, Edinburgh, 1991.

157. J. Miyawaki, K. Yamanouchi, and S. Tsuchiya, *Chem. Phys. Lett.* **180**, 287 (1991).

158. P. Andresen, *Ber. Bunsenges. Phys. Chem.* **89**, 245 (1985).

159. R. Vasudev, R. N. Zare, and R. N. Dixon, *J. Chem. Phys.* **80**, 4863 (1984).

160. R. N. Dixon and H. Rieley, *J. Chem. Phys.* **91**, 2308 (1991).

161. J. Pfab, J. Häger, and W. Krieger, *J. Chem. Phys.* **78**, 266 (1983).

162. D. G. Sauder and P. J. Dagdigian, *J. Chem. Phys.* **92**, 2389 (1990).

163. A. M. L. Irvine, I. W. M. Smith, and R. P. Tuckett, *J. Chem. Phys.* **93**, 3187 (1990).

164. I. Nadler, J. Pfab, H. Reisler, and C. Wittig, *J. Chem. Phys.* **81,** 653 (1984).

165. C. X. W. Qian, A. Ogai, H. Reisler, and C. Wittig, *J. Chem. Phys.* **90,** 209 (1989).

166. L. R. Khundkar, J. L. Knee, and A. H. Zewail, *J. Chem. Phys.* **87,** 77 (1987).

167. J. Pfab, *Chem. Phys. Lett.* **99,** 465 (1983).

168. M. Noble, I. Nadler, H. Reisler, and C. Wittig, *J. Chem. Phys.* **81,** 4333 (1984).

169. T. J. Butenhoff, K. L. Carleton, and C. B. Moore, *J. Chem. Phys.* **92,** 377 (1990).

170. P. J. Dagdigian, M. H. Alexander, and K. Liu, *J. Chem. Phys.* **91,** 839 (1989).

171. H. Reisler, M. Noble, and C. Wittig, in Ref. 19, p. 139.

172. K. A. Trentelman, D. B. Moss, S. H. Kable, and P. H. Houston, *J. Phys. Chem.* **94,** 3031 (1990).

173. H. Bitto, D. R. Guyer, W. F. Polik, and C. B. Moore, *Faraday Discus. Chem. Soc.* **82,** 149 (1986).

174. E. Kolodny, P. S. Powers, L. Hodgson, H. Reisler, and C. Wittig, *J. Chem. Phys.* **94,** 2330 (1991).

175. P. G. Giovanacci, Ph.D. thesis, Heriot-Watt University, Edinburgh, 1990.

176. P. G. Giovanacci, M. R. S. McCoustra, and J. Pfab, in preparation.

177. J. Pfab, D. M. Wetzel, and V. M. Young, *Ber. Bunsenges. Phys. Chem.* **94,** 1322 (1990).

178. U. Robra, H. Zacharias, and K. H. Welge, *Z. Phys. D* **16,** 175 (1990).

179. J. A. Dyet, M. R. S. McCoustra, and J. Pfab, *Faraday Discus. Chem. Soc.* **82,** 206 (1986).

180. J. A. Dyet, M. R. S. McCoustra, and J. Pfab, *Chem. Phys. Lett.* **136,** 231 (1987).

181. J. K. Rice and A. P. Baronavski, *J. Chem. Phys.* **94,** 1006 (1991).

182. R. Zhang, W. J. van der Zande, M. J. Bronikowski, and R. N. Zare, *J. Chem. Phys.* **94,** 2704 (1991).

183. M. C. Lin and G. Ertl, *Ann. Rev. Phys. Chem.* **37,** 587 (1986).

184. D. S. King and R. R. Cavanagh, in *Chemistry and Structure at Interfaces: New Laser and Optical Techniques,* R. B. Hall and A. B. Ellis, eds. (VCH, Weinheim, 1986), p. 25.

185. D. S. King and R. R. Cavanagh, *Adv. Chem. Phys.* **76,** 45 (1989).

186. D. C. Jacobs, K. W. Kolasinski, S. F. Shane, and R. N. Zare, *J. Chem. Phys.* **91,** 3182 (1989).

187. D. S. Y. Hsu and M. C. Lin, *J. Chem. Phys.* **88,** 432 (1988).

188. L. V. Novakovski and D. S. Y. Hsu, *J. Chem. Phys.* **92,** 1999 (1990).

189. D. Weide, P. Andresen, and H.-J. Freund, *Chem. Phys. Lett.* **136,** 106 (1987).

190. F. Budde, A. V. Hamza, P. M. Ferm, and G. Ertl, *Phys. Rev. Lett.* **60,** 1518 (1988).

191. S. A. Buntin, L. J. Richter, D. S. King, and R. R. Cavanagh, *J. Chem. Phys.* **91,** 6429 (1989).

192. E. Hasselbrink, S. Jakubith, S. Nettesheim, M. Wolf, A. Cassuto, and G. Ertl, *J. Chem. Phys.* **92,** 3154 (1990).

193. W. B. Whitten, J. M. Ramsay, S. Arnold, and B. V. Bronk, *Anal. Chem.* **63,** 1027 (1991).

194. W. M. Fairbank, T. W. Hänsch, A. L. Schawlow, *J. Opt. Soc. Am.* **65,** 199 (1975).

195. V. I. Balykin, V. S. Lethokov, V. I. Mishin, V. A. Semchishen, *JETP Lett.* **26,** 357 (1977).

196. M. A. Bolshov, A. V. Zybrin, and I. I. Smirenkina, *Spectrochim. Acta* **36B,** 1143 (1981).

197. J. P. Dougherty, F. R. Prely, and R. G. Michel, *J. Anal. Atom. Spectrom.* **2,** 429 (1987).

198. N. Omenetto, *Appl. Phys.* **B46,** 209 (1988).

199. H. S. Kwong and R. M. Measures, *Anal. Chem.* **51,** 428 (1979).

200. P. D. Brewer, *Chem. Phys. Lett.* **136,** 557 (1987).

201. U. Gerlach-Meyer, E. Linnebach, K. Kleinermanns, and J. Wolfrum, *Chem. Phys. Lett.* **133,** 113 (1987).

202. K. Tsukiyama and R. Bersohn, *J. Chem. Phys.* **86,** 745 (1987).

203. J. Wolfrum, *Appl. Phys.* **B46,** 221 (1988).

204. G. Kychakoff, R. Howe, R. K. Hanson, and J. C. McDaniel, *Appl. Opt.* **21,** 3225 (1982).

205. B. Hiller and R. K. Hanson, *Opt. Lett.* **10,** 206 (1985).

206. *Lambda Physik Highlights,* No. 8, December 1987.

207. *Lambda Physik Highlights,* No. 14, December 1988.

208. D. R. Crosley and G. P. Smith, *J. Chem. Phys.* **79,** 4764 (1983).

209. K. P. Gross and R. L. McKenzie, *J. Chem. Phys.* **76,** 5260 (1982).

210. M. Alden, S. Wallin, and W. Wendt, *Appl. Phys.* **B33,** 205 (1984).

211. P. J. H. Tjossem and K. C. Smith, *J. Chem. Phys.* **91,** 2041 (1989).

212. R. A. Gottscho and T. A. Miller, *Pure Appl. Chem.* **56,** 189 (1984).

213. R. W. Dreyfus, J. M. Jasinski, G. S. Selwyn, and R. E. Walkup, *Laser Focus* **22**(12), 62 (1986).

214. R. M. Measures and H. S. Kwong, *Appl. Opt.* **18,** 281 (1979).

215. R. A. Gottscho, G. P. Davis, and R. H. Burton, *Plasma Chem. Plasma Process* **3,** 193 (1983).

216. R. E. Walkup, K. L. Saenger, and G. E. Selwyn, *J. Chem. Phys.* **84,** 2668 (1986).

217. B. J. Finlayson-Pitts and J. N. Pitts, Jr., *Atmospheric Chemistry* (Wiley, New York, 1986).

218. R. M. Measures, ed., *Laser Remote Chemical Analysis* (Wiley, New York, 1988).

219. B. L. Sharp, *Chem. Brit.* **18,** 342 (1982).

220. U. von Zahn and R. Neuber, *Atmos. Phys.* **60,** 294 (1987).

221. *Lambda Physik Highlights,* No. 17, June 1989.

222. *Lambda Physik Highlights,* No. 30/31, October 1991.

223. M. O. Rodgers, J. Bradshaw, S. T. Sandhold, S. KeSheng, and D. D. Davis, *J. Geophys. Res. C: Oceans Atmos.* **90,** 12819 (1985).

224. M. O. Rodgers and D. D. Davis, *Environ. Sci. Technol.* **23,** 1106 (1989).

225. M. Hippler, A. J. Yates, and J. Pfab, in *Optogalvanic Spectroscopy,* R. S. Stewart and J. E. Lawler, eds. (IOP, Bristol, 1990), p. 303.

226. A. J. Yates and J. Pfab, in preparation.

227. H. J. Kölsch, P. Rairoux, J. P. Wolf, and L. Wöste, *Appl. Opt.* **28,** 2052 (1989).

5

High-Resolution Infrared Spectroscopy

Brian J. Howard and John M. Brown

5.1. Introduction

Infrared spectroscopy (in the wavenumber range 200–4,000 cm^{-1}) has been a valuable source of information on the vibrational motions of molecules for much of this century.[1] It has been used to study and characterize an enormous number of molecules. Most applications are analytical ones in which the use of group frequencies has aided the identification of molecules and the elucidation of structures.

In this chapter we limit our discussions to high-resolution spectroscopy. This is taken to mean rotationally resolved spectroscopy, which almost exclusively involves molecules in the gas phase. It should, however, be noted that occasionally high-resolution spectra are obtained in the condensed phases when the interaction of the molecule with the solvent, lattice, or cavities is sufficiently weak for almost free rotation still to occur.

Until the mid-1970s the highest-resolution spectra were recorded with large-grating spectrographs using a continuous broadband source of radiation. Although the spectrometers are relatively easy to use, the linewidths are limited to about 0.01 cm^{-1} (or 300 MHz) and at best can only approach Doppler resolution (a limiting feature arising from the random motion of molecules along the direction of propagation of the radiation). We shall return to this point later. In addition, because of the low intrinsic spectral brightness of the radiation sources used, these spectrometers suffer from a lack of sensitivity. It is difficult to detect absorptions of much less than

1%. Improved sensitivity in these experiments has been achieved by the use of multipass cells of considerable length. However, attempts to detect the spectra of transient molecules and other species in low concentration have met with little success.

These problems have largely been overcome with the advent of the infrared laser. However, its use in high-resolution spectroscopy has lagged behind the application of lasers in other spectral regions. Coherent sources of radiation in the microwave region ($0.1–10 \text{ cm}^{-1}$), namely klystrons, magnetrons, and backward-wave oscillators, have been much used for high-resolution rotational spectroscopy since the early 1940s.[2] The use of visible and ultraviolet lasers in high-resolution spectroscopy was well advanced by 1970.[3] Most developments involving infrared lasers date from after 1975.

Before presenting a detailed account of infrared laser sources and their application, we should note that, in recent years, the laser has had to compete with an alternative high-resolution, high-sensitivity instrument: the Fourier-transform spectrometer.[4–6] In this, a continuum source of radiation is dispersed with a Michelson-type interferometer, the spectrum being scanned by changing the path length in one of the optical arms. The resulting interferogram can be transformed to provide the spectrum. By moving the reflecting mirror over a considerable distance, it is possible to obtain very high resolution indeed. A modern state-of-the-art instrument with a travel of 500 cm can yield a resolution of $1/(2 \times 500) = 0.001 \text{ cm}^{-1}$, close to the Doppler linewidth. Although they operate mainly at Doppler resolution like high-quality diffraction grating spectrographs, Fourier-transform instruments show a considerable gain in sensitivity known as the *multiplex advantage*. This results from the fact that at all times the whole of the spectrum is being sampled as the spectrometer is being scanned. High-resolution instruments are now commercially available, thanks in part to developments in instrument making and in part to the availability of powerful microcomputers to store the interferograms and to perform the Fourier transformation. As a result of their high resolution, good accuracy, and reasonable sensitivity, they are now probably the instruments of choice for the study of most stable gas-phase molecules.

Fourier-transform spectrometers nonetheless still use broadband radiation sources of low intrinsic spectral brightness, and by and large they lack the overall sensitivity and versatility of laser-based spectrometers. Lasers in the infrared are very fine spectroscopic sources, producing highly collimated beams of high spectral brightness and very narrow linewidths (typically a few megahertz, which is much less than the Doppler linewidth). As a result, IR spectroscopy at the Doppler linewidth has become commonplace. Indeed, several studies at sub-Doppler resolution have used saturation effects or other specialized techniques. Far more important is the great advance in sensitivity that has become possible since the advent of the laser. It is now possible to detect IR spectra of many unstable or otherwise transient species such as free radicals, molecular ions, and van der Waals complexes.

In this chapter we present some of the landmarks in the field. A brief account of the infrared laser systems available is given, highlighting those features that make them most suitable for high-resolution spectroscopy. However, we do not limit ourselves to the wavenumber range 200–4,000 cm^{-1} mentioned earlier and termed the mid-infrared. Instead we include the far infrared (10–200 cm^{-1}), where many important advances are being made. The next section contains a discussion of some of the important techniques (many of them very recent) that have been developed for producing molecular species in large concentrations and studying their infrared spectra. The final section is devoted to specific examples of the molecular spectra, in particular, of transient species. These have been chosen to illustrate the various experimental methods used and the wide variety of information that is now available.

5.2. Infrared Laser Systems

It is not our intention in this chapter to give a detailed description of the many infrared laser systems available. An introduction has been given in Chapter 2 and detailed accounts are given elsewhere.[7–9] However, some lasers are much used and are intrinsically associated with particular applications. In those cases, a knowledge of the mode of operation of the laser and its particular properties can be useful when assessing its value for given problems. Some of the main features of the laser systems described here are summarized in Table 5.1.

Before discussing individual lasers, some general comments can be made. There are two main types of systems, *tunable* lasers, whose frequency can be varied continuously over a range, and *fixed-frequency* lasers, which can only oscillate at certain fixed frequencies (corresponding to spectroscopic transitions in the lasing molecules). Tunable lasers require a near continuum of energy levels and have a condensed-phase active medium. In contrast, fixed-frequency lasers use a gaseous gain medium. It is evident from Table 5.1 that considerably more power is available from the gas lasers, and it is this property that makes them competitive with the less powerful, tunable lasers.

Different lasers operate in different regions of the infrared. Since a chosen molecule often has transitions in only a limited range of frequencies, this frequently dictates which laser system can be used. Also, for tunable lasers, it is important to appreciate how far each device can be tuned in a single scan. Individual scans are usually quite narrow (~1 cm^{-1}), but with the use of a computer for control of the laser and collection of spectral data it is now possible to achieve far broader scans by stitching individual scans together.

Table 5.1 Characteristics of Infrared Lasers

Laser	Active Medium	Wavenumber Range, cm^{-1}	Typical Output Power (CW)	Range of Tunability (Single Scan)	Linewidth	Disadvantages
CO_2	CO_2/N_2	850–1,100	30 W	Line tunable	10^{-5} cm^{-1}	Not tunable, 10 μm only
CO	$CO/N_2/He$	1200–2,050 (2450–3,500)[a]	2 W	Line tunable	3×10^{-5} cm^{-1}	Not tunable
Diode	p–n junction diode	380–3,500	100 μW	100 cm^{-1} (1–5 cm^{-1})	10^{-3} cm^{-1}	Multimode lasing, patchy coverage
F center	$F_{A(II)}$ center	2,900–4,000	10 mW	500 cm^{-1} (50 cm^{-1})	10^{-4} cm^{-1}	No long λ coverage
Difference	Dye laser + Ar^+ + $LiNbO_3$	2,400–4,500	10 μW	100 cm^{-1} (1 cm^{-1})	2×10^{-4} cm^{-1}	No long λ coverage
Optically pumped FIR	Low-pressure gas	5–250	1 mW to 1 W	Line tunable	10^{-5} cm^{-1}	Not tunable
FIR sideband	FIR + microwave	10–30	1 μW	Up to 1 cm^{-1}	$<10^{-5}$ cm^{-1}	Low power
Difference FIR	$CO_2 + CO_2$	20–200	0.1 μW	180 MHz	10^{-6} cm	Limited tunability

[a]$\Delta v = 2$ transitions.

5.2.1. *Tunable Mid-Infrared Lasers*

For high-resolution molecular spectroscopy a tunable source of radiation is very desirable. Several tunable infrared lasers have been developed. Among these are the spin-flip Raman laser and the optical parametric oscillator. Dye lasers, so common in the visible, can also be made to operate far into the infrared, especially by Raman shifting. However, the principal lasers for infrared spectroscopy have been the diode laser, the color-center laser, and the difference-frequency laser. These all combine high-quality radiation, at moderate powers, with significant ease of operation. Commercial versions are available. The main features of these lasers are as follows.

5.2.1.1. *The Diode Laser*

The power emitted from a diode laser results from the recombination of electron and holes at the junction of a p-type and an n-type semiconductor that has been forward biased.[10,11] The principle for emission is the same as that for a light-emitting diode (LED). On crossing the heterojunction, electrons from the conduction band of the n-type semiconductor fall back to the valence band of the p-type semiconductor with emission of radiation (Figure 5.1). If the junction is placed in a resonant cavity (usually formed by the polished faces of the semiconductor crystal) and if the radiation field is sufficiently intense, stimulated emission exceeds spontaneous processes and laser action results.

Diode lasers in the mid-infrared are of the lead salt type. They consist of nonstoichiometric compounds of lead (or cadmium and tin) with sulfur, selenium, and tellurium. The frequency of the radiation is largely determined by the band gap and is affected by the chemical composition of the diode. Laser action can occur over several cm^{-1}. Thus, a typical diode operates simultaneously on several modes and these must be separated using a monochromator. Each mode has a potentially very narrow linewidth; values as low as 100 kHz have been measured. However, mechanical, electrical, and temperature instabilities result in effective linewidths of several megahertz.

The laser modes can be tuned by a variety of means such as applying a mechanical pressure or a magnetic field. More usually, temperature changes provide coarse tuning with current changes providing fine tuning of the frequency. A typical diode laser operates over a range of about 100 cm^{-1} and a typical mode can be scanned about 1 cm^{-1} without retuning. Early lead salt lasers required very low temperatures (10–50 K), achieved by using closed-cycle refrigerators. New diodes, however, can operate above liquid nitrogen temperatures, permitting the use of mechanically quiet temperature-controlled nitrogen cryostats.

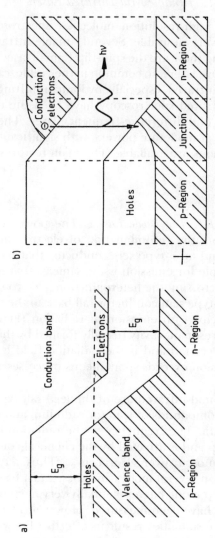

Figure 5.1 Energy level diagram of the p-n junction of a diode laser: (a) without bias, (b) with a forward bias applied.

5.2.1.2. The Color-Center Laser A convenient source of moderately high-power, high-resolution, tunable radiation for the higher-frequency region of the infrared is the color-center (or F-center) laser.[12] It can be viewed as a form of solid-state dye laser. The active medium is typically an $F_A(II)$ or F_2^+ center in alkali halide crystals, formed by irradiation of the crystal with high-energy radiation. The resulting color centers have strong absorption profiles in the visible region of the spectrum. As a consequence the medium is conveniently excited by radiation from powerful visible light sources such as ion lasers, the red lines of the Kr^+ laser being the most frequently used.

After excitation, rapid "vibrational" relaxation of the color center occurs and the subsequent fluorescence is moved to lower frequency, into the infrared (see Figure 5.2). The quantum efficiency of the luminescence decreases with increasing temperature. Consequently, lasers are operated with the crystal mounted on a liquid nitrogen–cooled cold finger. The emission is broadband but single wavelengths can be selected by means of a diffraction grating or similar frequency-selective components. When the crystal is placed in a suitable resonant cavity, there is usually sufficient gain for laser action to occur.

The laser linewidth is narrowed and the laser forced to oscillate on a single mode with the introduction of a thin and a thick etalon. Typical las-

Figure 5.2 Schematic diagram of the operation of an F-center laser. The excited electronic state ($s \rightarrow p$ transition) possesses a different relaxed structure from the ground state. Excitation by the pump laser gives an unrelaxed structure for the p state. Rapid relaxation of the lattice around the color center results in emission to an unrelaxed configuration of the ground electronic state.

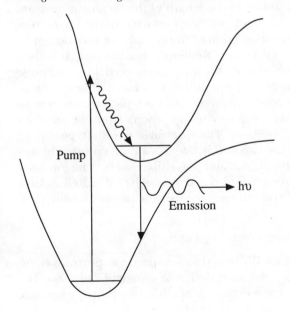

ers have a power of a few tens of milliwatts in a bandwidth of ~1 MHz. Single-mode scans are significantly less than 1 cm^{-1}. However, several computer-controlled scan systems[13,14] have been developed in which adjacent scans are joined together to give a wide apparent tuning range (up to ~100 cm^{-1}).

5.2.1.3. The Difference-Frequency Laser   The difference-frequency laser works by producing the difference of two visible laser frequencies in a nonlinear crystal[15-17]—in other words one whose response does not vary linearly with the power of the incident radiation. The resulting radiation is typically in the mid-infrared. It is not a conventional laser in that a population difference is not created between molecular energy levels. Instead, coherent emission results from mixing the output of two coherent laser sources.

A typical difference-frequency laser system consists of a fixed-frequency argon ion laser, Lamb-dip stabilized on a single emission line, and a stabilized single-mode dye laser. These two frequencies are then mixed in a nonlinear crystal that is typically LiNbO$_3$. The difference radiation appears in a direction determined by conservation of momentum of the interacting photons. The requirement that the mixing crystal be transparent at the wavelengths of each of the visible sources and at the wavelength of the infrared difference radiation places a serious constraint on the mixing crystals that can be used. The efficient LiNbO$_3$ crystal cannot be used below 2,400 cm^{-1} because of the onset of photon absorption. The radiation can be extended to longer wavelengths by using the less-efficient LiIO$_3$ crystal.

Since coherent emission occurs, the potential power of the difference-frequency laser varies as the square of the length of the nonlinear medium excited. This assumes that emission from different parts of the crystal is *in phase*. However, because of the dispersion of the crystal (the variation of its refractive index with wavelength) this is frequently not the case. It is then necessary to change the orientation of the anisotropic crystal or to change its refractive index, by varying the temperature, to achieve *phase-matching*.

The power from the difference-frequency laser is typically a few tens to a few hundreds of microwatts using 1-W pump powers from the ion laser and about 100-mW from the dye laser. The spectral linewidth (typically ~1 MHz) reflects the quality of the two visible lasers. Calibration often makes use of the substantial technology available for visible lasers. The ion laser is of accurately known frequency and the dye laser is calibrated against accurately known standards and interpolated using accurate etalons.

5.2.2. Fixed-Frequency Lasers

It is undoubtedly true that for spectroscopic transitions, a continuously tunable laser is the most desirable source of radiation. Despite this, a fixed-frequency laser may be a much better source when one considers its operating frequency, intensity, spectral purity, and reliability. In these circumstances, it may still be possible to record spectra by tuning the molecular transitions into coincidence with the laser frequency in some manner. The ways in which this can be accomplished are discussed later. The gain medium of a fixed-frequency laser is invariably gaseous (and usually at low pressure) because only then is the linewidth sufficiently narrow. The lines of such lasers correspond to the individual transitions in particular molecules.

5.2.2.1. The Carbon Dioxide Laser The carbon dioxide laser was one of the earliest infrared lasers. The gain medium consists of a low-pressure mixture of CO_2, N_2, and He (in the approximate proportions 12:20:68) excited in a water-cooled DC electric discharge. The upper level of the main lasing transitions, corresponding to excitation of the CO_2 antisymmetric stretch and designated the (001) level, is populated directly by resonant energy transfer from vibrationally excited nitrogen. Lasing occurs from this level to lower unpopulated levels. A total population inversion is achieved. The lower levels of these transitions are the Fermi-perturbed dyad, derived from the levels with one quantum of the CO_2 symmetric stretch and two quanta of the bend, designated (100) and (02^00). This scheme is shown in Figure 5.3. Lasing thus occurs for a wide range of J values in the P and R branches of these bands centered at 9.4 and 10.4

Figure 5.3 Vibrational levels of CO_2 involved in the operation of the CO_2 laser.

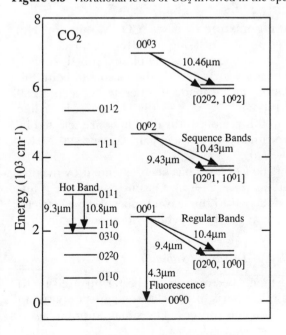

μm. It is also possible to make the laser oscillate on higher sequence bands, starting from the more excited levels (002) and (003).

Individual lines are selected by a diffraction grating that forms one end of the laser cavity. Each line can be tuned by a small amount over its Doppler-limited gain curve ($\sim$60 MHz). Lines occur roughly 1.5 cm^{-1} apart between 900 and 1,100 cm^{-1}. Single-mode, CW powers of over 100 W have been achieved. The laser can be operated either with a slowly flowing stream of gas or in the sealed-off mode. In the latter case, it is possible to use rarer isotopomers of CO_2 and hence obtain additional laser frequencies.

5.2.2.2. The Carbon Monoxide Laser

Although the CO laser is superficially similar to the CO_2 laser (a suitably chosen mixture of gases is excited in an electric discharge to create a population difference between vibration–rotational levels), the principle on which its operation is based is very different. The objective is to obtain a pronounced disequilibrium between the vibrational and rotational degrees of freedom so that $T_{vib} \gg T_{rot} \approx T_{trans}$. In the discharge, higher vibrational levels of CO are populated very effectively by vibrational (VV) transfer, but the more efficient rotational relaxation keeps the rotational temperature close to that of the walls of the discharge tube. Thus, the vibrational temperature is very high with a typical population ratio, N_{v+1}/N_v, of 0.95. Since the rotational temperature is low, it is possible to create a small population inversion between the levels ($v + 1, J - 1$) and (v,J). The laser therefore oscillates on P lines ($J - 1 \rightarrow J$) only. An excellent review of the CO laser has been written by Urban.[9]

Most CO lasers use a flowing mixture of gases (CO, N_2, air, and He) whose composition is adjusted to obtain lasing at different wavelengths. The efficiency of the gain medium increases as the plasma tube is cooled. It is now common to cool the jacket surrounding the plasma to liquid nitrogen temperatures. The latter can be made to operate between 1,250 cm^{-1} ($v = 37 \rightarrow 36$) and 2,160 cm^{-1} ($v = 1 \rightarrow 0$). There is a laser line approximately every 2 cm^{-1} in this region. Individual lines are selected by a diffraction grating and are tunable over their gain curve ($\sim$90 MHz). Typical output power is a few watts.

Very recently, the range of the CO laser has been extended by making it oscillate on $\Delta v = 2$ transitions ($v + 2 \rightarrow v$). Output powers are a few hundred milliwatts. This gives access to the wavenumber region between 2,500 and 3,500 cm^{-1} where many X—H stretching vibrations occur.

5.2.3. The Far-Infrared (Submillimeter) Laser

For a long time, spectroscopy in the far-infrared (FIR) region lagged behind that in other parts of the electromagnetic spectrum because of a lack of suitable radiation sources. The situation nowadays is very different, with some 5,000 laser lines having been identified for use in this region. The earliest lasers were based on electrical excitation of the lasing molecule (either H_2O, D_2O, or HCN) in a discharge. Because the gain of the medium was low, very long discharge tubes were required. However, only a few lines were available in this way, providing rather poor coverage of the spectral region. Despite this, the electrically excited laser lines are still very useful because they provide relatively high-power output (e.g., the 3.37-μm line of HCN gives 600 mW).

Optically pumped FIR lasers were first developed in 1970 and provide a large number of laser lines. New transitions are still being discovered today. The principle of operation is very simple. Radiation from an infrared laser (usually a CO_2 laser because of its ease of operation and power) is used to pump molecules to a particular rotational level of an excited vibrational state. In this way a population inversion is created with respect to lower rotational levels of the same vibrational state (see Figure 5.4). It is also possible to create a partial population inversion between rotational levels of the ground vibrational state. In this situation, lasing action at FIR frequencies can be stimulated by placing the sample in a laser cavity of the appropriate length. The lasing gas is at low pressure (~0.1 torr). Output powers vary from 1 mW up to 1 W. Lasing has been achieved for about 5,000 lines between 40 μm (250 cm^{-1}) and 2 mm (5 cm^{-1}). The actual

Figure 5.4 Schematic diagram of the molecular energy levels involved in an optically pumped FIR laser.

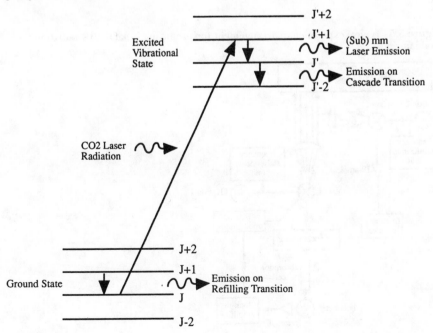

frequency is that of the transition involved in the lasing molecule. These frequencies have been tabulated, together with the lasing gas and the pump line, in a recent book by Douglas.[18] For example, the 118.8-μm line of CH_3OH is pumped by the P(36) line of $^{12}C^{16}O_2$ and has a measured frequency of 2,522,781.6 MHz. This is the most powerful FIR line so far discovered and can be made to deliver more than 1 W!

5.2.4. Tunable Far-Infrared Radiation

So far, it has not proved possible to develop a tunable laser at FIR wavelengths. Tunable coherent radiation is nevertheless available in this region, generated as a difference in frequency of two other sources. Two techniques have been developed. In the first, the radiation is generated as a microwave sideband on a fixed-frequency FIR laser. In the second, the difference of two CO_2 laser frequencies is generated in a diode.

5.2.4.1. Microwave Sideband Source   The basic arrangement as described by van den Heuvel, Meerts, and Dymanus[19] is shown in Figure 5.5. Radiation from a fixed-frequency FIR laser and that of a millimeter-wave klystron are incident on a diode. In the nonlinear medium of the diode, mixing occurs and two beams of FIR radiation are generated with frequencies equal to the sum and difference of the fundamental frequencies of the laser and klystron. The diode is of the Schottky-barrier diode type and is mounted in a mixer with a semi-open structure (i.e., the

Figure 5.5 Typical experimental arrangement for producing tunable FIR radiation (at diode mixer 1).

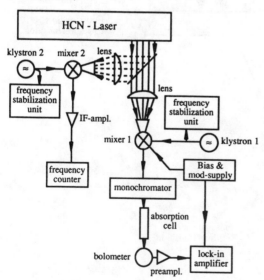

microwave radiation is transmitted to the diode via a normal closed rectangular waveguide penetrated by a diode stud, whereas the FIR radiation is transported to and from the diode by a whisker antenna mounted in free space). A Michelson-type interferometer is used to separate the sidebands from the fundamental laser radiation and a simple monochromator selects the sideband beam that corresponds to either the sum or the difference frequency. Sideband power is between 0.1 and 1 μW for input powers of 50 and 150 mW from the laser and klystron, respectively. The FIR laser is free-running and there is some uncertainty (~2 MHz) in its frequency. This uncertainty can be eliminated with the help of a second mixer, which monitors the laser frequency during measurements.

The main limitation of this approach to the generation of coherent FIR radiation is the requirement for a strong FIR laser line at the appropriate frequency. The suitable lines are almost all near the long-wavelength end of this region of the spectrum and in practice this imposes an upper-frequency limit of about 1 THz. Its advantage is that it is possible to scan over a wide frequency range, particularly if a solid-state synthesizer is used as the microwave source.

5.2.4.2. Difference-Frequency Generation   In this approach, the FIR radiation is generated as the difference between a fixed-frequency CO_2 laser and a tunable waveguide CO_2 laser. The waveguide laser has about ± 120 MHz of tunability. The method was devised by Evenson, Jennings, and Petersen.[20] The experimental arrangement used to record molecular spectra is shown in Figure 5.6. The common isotopomer of CO_2 is used in the waveguide laser and one of four different isotopic forms is used in the fixed-frequency laser. In this mode of operation, 80% of all

Figure 5.6 Experimental arrangement of a tunable FIR laser based on the mixing of two CO_2 laser frequencies at a MIM diode.

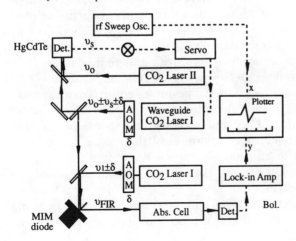

frequencies from 0.3 to 4.5 THz can be synthesized. Operation can be extended to frequencies in excess of 6 THz, but the output then drops to a few percent. Two 90-MHz optoacoustic modulators are used to isolate the output beams of the CO_2 lasers from the mixing diode and also to increase the tunability by 180 MHz. Two fixed-frequency CO_2 lasers are used, stabilized to saturated fluorescence signals in CO_2; one is focused onto the diode and the other is used as a frequency reference for the waveguide lasers. The mixing diode is of the metal–insulator–metal (MIM) type, using a tungsten tip on a metal base; the oxide layer provides the insulation. Typical FIR powers of a few tenths of a microwatt are obtained from 200 mW of CO_2 power. Because the CO_2 laser frequencies have been measured to high accuracy, the FIR frequency is also known to about 25 kHz.

The scanning range of this method can be increased by forming the FIR radiation from two fixed-frequency CO_2 lasers with the addition of microwave sidebands. Again, a MIM diode can be used to mix the frequencies, but the third-order generation is much more efficient if the base is made of cobalt rather than nickel. In this case, the two CO_2 lasers are stabilized to CO_2 sub-Doppler saturated fluorescence dips using separate low-pressure cells. Their radiation is focused onto the diode along with that of a frequency-synthesized microwave source (2–20 GHz) used to add the sidebands. The great advantage of this difference-frequency generation technique compared with the sideband method is its ability to cover essentially the whole of the FIR region (0.5–6.5 THz).

5.3. Techniques

The features of a laser that make it ideal for detecting very weak spectra, and in particular molecules present in low concentration, are its high spectral brightness, its monochromaticity, and the ease with which it can be collimated. In addition, efficient detectors of infrared radiation have been developed. Most of these are photoconductive or photovoltaic and depend upon excitation of electrons in a semiconductor by the incoming photons. They are essentially quantum devices, with one electron excited per photon. To improve sensitivity and reduce noise from thermal fluctuations, most detectors are operated at liquid nitrogen (77 K) or liquid helium ($\sim$4 K) temperatures. The most sensitive detectors are indium antimonide (InSb), for wavelengths shorter than 5.5 μm, and mercury cadmium telluride (HgCdTe). The wavelength response of HgCdTe depends upon the stoichiometric composition of the crystal, but it has a satisfactory response out to 25 μm (400 cm^{-1}). For longer wavelengths, quantum devices become less efficient and it is normal to use liquid helium–cooled bolometers in the FIR.

The signal from a weakly absorbing sample can be increased by using very long absorption pathlengths. This can be implemented in the remote

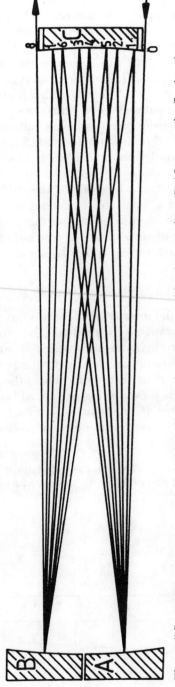

Figure 5.7 The optical arrangement of a White cell. Radiation enters at position 0, near mirror C. After several reflections it emerges on the other side of mirror C (at position 8).

monitoring of, say, pollutants in the upper atmosphere. In the laboratory it is best achieved using multiple-reflection cells. The classic design is that due to White.[21,22] It consists of a confocal arrangement of two spherical mirrors, separated by the sum of their focal lengths. One of the mirrors is split into two, each part of which can be separately aligned. The optical arrangement is shown in Figure 5.7. Radiation enters close to mirror C and is reflected alternately on sections A and B. With a proper alignment of A and B, the spots on C move sideways, odd reflections in one direction and even reflections in the opposite direction, eventually missing the mirror. This radiation can be focused onto a suitable detector. Since the multipass cell is in a near-confocal configuration, it is stable to slight misalignments of the laser beam.

The infrared laser is particularly good for the study of unstable and transient molecules present in small concentrations. In the rest of this section we discuss some of the specific techniques that have been developed to improve the production of such species as free radicals, ions, or molecular dimers and to help extract weak signals from noise.

5.3.1. Discharge–Flow Methods

One of the most convenient ways to generate transient molecules involves the discharge–flow method. Two streams of gases are mixed in a reaction zone to generate the species of interest. One of the streams is usually passed through an electric discharge to generate atoms or radicals that can readily react with molecules in the second stream. A typical production scheme is shown in Figure 5.8. Although the discharge can be operated DC (or at low AC frequencies), it is preferable to use either an rf or microwave discharge as they do not require electrodes within the

Figure 5.8 A schematic diagram of a discharge–flow apparatus. The reactive products of a discharge through a flowing gas A are mixed with a secondary flowing gas B to generate the desired short-lived molecules.

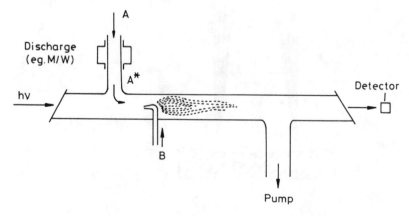

gaseous volume. Electrodes tend to provide efficient sites for the removal of transient molecules and are often attacked by the reactive plasma. Powers typically of the order of 100 W are dissipated within the discharge.

An example of this is the formation of fluorine atoms in a discharge of CF_4 or F_2. These atoms can be used for subsequent hydrogen atom abstraction as in the reaction with formaldehyde to produce HCO[23]:

$$H_2CO + F \rightarrow HCO + HF$$

Many free radicals have been produced by this method.

5.3.2. Electric Discharges

Much of the progress in the spectroscopy of transient molecules during the past two decades has depended upon the use of electric discharges to generate the molecules concerned. This is particularly true for ionic species. One major advantage of an electric discharge is that it provides a method of generating molecules continuously over quite a large volume. This makes it easier to achieve long absorption paths (using a White cell mirror arrangement) and hence very high sensitivity. Many of the alternative techniques (which will be subsequently discussed) tend to produce species in a far more confined volume.

Gaseous electric discharges have been used for a variety of purposes for a long time. However, the details of their operation are still poorly understood. Recent years have seen some attempt at characterizing them from the point of view of the formation of both charged and neutral species. The main regions of a low-frequency (or DC) electric discharge are shown in Figure 5.9. The voltage drop between the anode and cathode is not uniform; in fact, most of it occurs in a region close to the cathode known as the cathode dark space. The concentration of positive ions is greatest in the

Figure 5.9 Regions of an electric discharge. The potential drop is not uniform through the discharge, the largest fall occurring in the cathode dark space. The largest concentration of positive ions is found in the negative glow region.

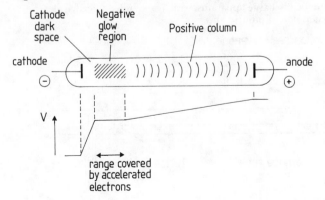

negative glow region, which proves to be short under normal operating conditions. Its length is related to the distance traveled by the high-energy electrons that have been accelerated across the cathode dark space and is proportional to the electron energy. At typical pressures of 1 torr, the negative glow region has a length of a few centimeters.

The mechanism for producing ions in this region is electron bombardment:

$$M + e^- \rightarrow M^+ + 2e^-$$

The characteristics of these glow discharges are determined principally by phenomena occurring in the cathode region. In a normal discharge the voltage across the "cathode drop" region remains essentially constant with change in current ($R_{\text{eff}} \sim 0$), hence the need for a ballast resistor in series with the applied voltage to limit the current that flows. In conjunction with this, the cathode volume covered by the discharge plasma increases. In so-called anomalous discharge, the plasma covers the entire cathode region and the voltage across the "cathode drop" rises roughly in proportion to the current. As a result, the energy of the electrons and the distance they travel both increase with current. A number of alternative designs have been developed. Two major improvements of the discharge design have been used to increase the production of ions.

5.3.2.1. Hollow-Cathode Discharge Cell   In this design the cathode is wrapped around as much of the discharge volume as possible. This increases the extent of the negative glow region and thereby the number of ions formed in the discharge. Such a design based on one published by van den Heuvel and Dymanus[24] is shown in Figure 5.10. The discharge is struck between the hollow cathode and an anode positioned at the end of a glass insert into the main tube. Stable operation is possible over a wide range of currents and pressures. The positive column is confined to the glass insert between the anode and cathode, and the negative glow together

Figure 5.10 Hollow-cathode discharge apparatus, with an extended negative glow region to increase the generation of positive ions.

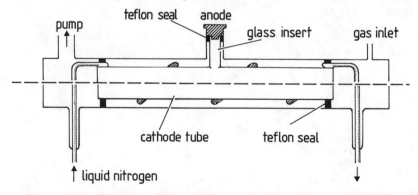

with the positive ions is concentrated along the axis of the cathode tube. At about 10 cm from the cathode surface, the glow changes abruptly into the cathode dark space. The length of the glow is directly related to the voltage applied to the anode. The cathode is efficiently cooled by a slow flow of liquid nitrogen through a helical copper tube soldered around the cathode tube.

5.3.2.2. Magnetic Field Enhancement of the Negative Glow Region

In the conventional design of a discharge tube (see Figure 5.9), the electrons accelerated through the cathode gap tend to spread out and are subsequently lost through collisions with the wall of the tube. The distance they travel down the tube is consequently less than it might otherwise be. It was observed many years ago that the length of the negative glow could be extended by applying a longitudinal magnetic field, the electrons spiraling around the magnetic lines of force. De Lucia and his group[25] have exploited this observation to obtain dramatic increases in the signals arising from ions produced in the discharge. By passing a current through a solenoid wound around the outside of the discharge tube, magnetic field strengths of several tens of millitesla were produced. Fields of this magnitude are sufficient to confine electrons with several hundred volts of transverse energy to a cyclotron radius of the order of 1 cm, roughly the transverse dimension of the discharge probed by the laser beam. By restricting the size of the electrodes, it is possible to force the discharge to run in the anomalous mode. The increase in positive ion concentration achieved in this way is about a hundredfold.

5.3.3. Supersonic Expansions

Most high-resolution spectroscopic studies of molecules involve gaseous samples that are either in a static cell or flowing slowly through a tube. Such species are essentially at thermal equilibrium with a characteristic temperature (often the temperature of the sample container). The population of the molecular energy levels and the Doppler linewidth of spectral lines are both determined by this temperature.

To improve the resolution of spectra one can try to reduce the temperature, but this is often of limited applicability because of the onset of condensation. Lower temperatures also alter the population of quantum states involved in spectroscopic transitions. There is a reduction in the overall molecular partition function, yielding a significant increase in the population of the lower rotational levels. This results in greater sensitivity for detecting these states and frequently results in reduced spectral congestion.

A radically different approach and one that has revolutionized high-resolution spectroscopy in the past 20 years is the development of the supersonic nozzle source.[26–29] At its most basic it consists of a small orifice through which gas is expanded from a moderately high pressure (possibly a few atmospheres) to a low-pressure (often near collision-free) region. If the conditions of the expansion are such that the mean free path λ of the expanding gas is significantly larger than the dimensions d of the nozzle orifice (assumed circular for the present), there is a negligible number of molecular collisions and the gas is said to be undergoing effusion. The distribution of molecular speeds $N(v)$ reflects the Maxwell–Boltzmann distribution in the nozzle source at a temperature T_0:

$$N(v) \propto v^3 \exp\left(-\frac{mv^2}{2kT_0}\right) \tag{5.1}$$

where m is the molecular mass. The extra power of v in the preexponential term arises from faster molecules having a greater probability of passing through the orifice.

If, on the one hand, $\lambda \ll d$, each molecule undergoes many collisions during the expansion and much of the enthalpy of the gas is converted into forward translational motion. The relative motion of the molecules is considerably reduced and, in the continuum region of the expansion (where the rate of molecular collisions is sufficient to maintain near-thermal equilibrium locally), is described by a substantially reduced temperature. The resulting distribution of speeds is

$$N(v) \propto \exp\left(-\frac{m(v - \bar{v})^2}{2kT}\right) \tag{5.2}$$

where $\bar{v}$ is the mean flow velocity and T is the new temperature. A comparison of the distribution of molecular speeds for an effusion source and a high-pressure supersonic nozzle is given in Figure 5.11. The properties of the expanding gas are largely determined by the heat capacity ratio, $\gamma = C_p/C_v$. For example,

$$\frac{T}{T_0} = \left(\frac{P}{P_0}\right)^{\gamma-1} \tag{5.3}$$

gives the variation of the temperature of the gas with density. Such an expression assumes an isentropic expansion with negligible viscosity effects.

A useful quantity for characterizing the expansion is the Mach number M, which is defined as the ratio of the bulk flow speed of the gas to the local speed of sound a.

$$M = \frac{\bar{v}}{a} \tag{5.4}$$

where

$$a = \left(\frac{\gamma kT}{m}\right)^{1/2} \tag{5.5}$$

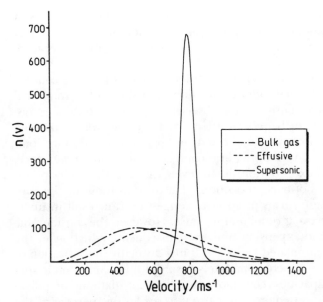

Figure 5.11 Relative velocity distributions for a bulk gas, an effusion source, and a supersonic nozzle expansion.

For a circular orifice the variation of Mach number with distance down-stream of the nozzle is given by

$$M = \left(\frac{x}{d}\right)^{\gamma - 1} \left(C_1 + C_2\left(\frac{d}{x}\right) + C_3\left(\frac{d}{x}\right)^2 + C_4\left(\frac{d}{x}\right)^3 + ...\right) \quad (5.6)$$

where d is the diameter of the nozzle and for $\gamma = 5/3$, $C_1 = 3.232$, $C_2 = -0.7563$, $C_3 = 0.3937$, and $C_4 = -0.0729$. Similar expressions have been derived for diatomic molecules ($\gamma = 7/5$) and for planar ($2 - D$) nozzle expansions.[30,31] The state functions of the gas are related to the Mach number by the following expressions:

$$\left(\frac{T}{T_0}\right) = \left(1 + \frac{\gamma - 1}{2} M^2\right)^{-1} \quad (5.7)$$

$$\left(\frac{P}{P_0}\right) = \left(1 + \frac{\gamma - 1}{2} M^2\right)^{-\gamma/(\gamma - 1)} \quad (5.8)$$

$$\left(\frac{\rho}{\rho_0}\right) = \left(1 + \frac{\gamma - 1}{2} M^2\right)^{-1/(\gamma - 1)} \quad (5.9)$$

It is easy to get confused by all these mathematical expressions. An important result to appreciate is that in these expansions a translational temperature of the order of 1 K is readily obtainable. As the rotation is virtually in equilibrium with the translation, molecules are cooled to their lowest rotational levels. It takes many collisions to relax the vibrational degrees of freedom, so the vibrational temperature lags considerably behind that of translation and may be no lower than 50–200 K for a room-temperature expansion.

The huge reductions in molecular partition functions and the concentration of the bulk of the molecular population in the lowest few rotational levels result in increased sensitivity for observing molecules in these levels. The cooling has a further beneficial effect for molecules with congested spectra. Asymmetric tops have complicated energy levels and produce complex spectra with many overlapping branches. Also, at room temperature, many molecules have excited vibrational levels that are thermally populated. These can give hot-band transitions that overlap those of the fundamental. Both of these forms of spectral congestion can result in unresolved vibrational spectra at room temperature. However, the considerable cooling in a supersonic expansion and the consequent reduction in the number of thermally populated levels result in a dramatic simplification of the spectrum with (it is hoped) completely resolved lines. Monatomic species, with low heat capacities, cool more rapidly than diatomic or polyatomic molecules. As a consequence it is usual to seed a few percent of the molecule of interest into a vast excess of a rare gas such as argon or helium. During the expansion the seed molecules achieve the velocity and temperature of the carrier gas.

It must be emphasized that supersonic expansions are dominated by kinetic considerations and should not be regarded as at equilibrium. For example, at true temperatures in the region of 1 K all gases should be converted to solids. There is, however, some condensation in these expansions. Species such as molecular dimers (known as van der Waals molecules) are formed. Depending upon the initial conditions (pressure, temperature, nozzle diameter, and gas composition) a wide variety of species from dimers, through polymers, to macroscopic clusters and liquid droplets can be formed. The spectroscopy of these van der Waals molecules is discussed in Section 5.4.3.

5.3.4. Molecular Beams

When molecules are expanded through a small hole into a near vacuum, whether the source be a supersonic nozzle or an effusion orifice, the mean free path downstream quickly becomes sufficiently large for intermolecular collisions to become negligible. Under these conditions it is easy to place collimating slits (or a skimmer) in the expansion to create a well-defined beam of molecules. Such molecular beams have a small angular divergence. By interrogating the beam with perpendicular radiation, far greater spectral resolution is possible.[32] The Doppler width of spectral lines is greatly reduced and the possible effects of collision- (or pressure-) broadening are almost eliminated. This type of experiment provides the ultimate in resolution. Although a few experiments based on this technique have been performed and will be described in this chapter, the full potential of this approach has yet to be realized.

5.3.5. High-Temperature Sources

At first sight, the use of high-temperature furnaces to generate species for infrared spectroscopic studies may not appear very promising, especially as a furnace is a good source of infrared radiation. However, laser beams are easy to collimate. Consequently, it is quite easy to separate the signal beam spatially from the background radiation arising from the furnace.[33,34] Several groups have recorded infrared spectra of transient molecules formed in a furnace. Jones and co-workers have exploited the technique to the full in studying a wide range of diatomic hydrides.[35] The sample cell is made of an alumina ceramic and has two cylindrical water-cooled stainless steel electrodes fitted coaxially to the ends of the tube. The cell is heated in a furnace to temperatures up to 1,300°C. Powdered metal M is placed in a ceramic boat in the center of the cell and the cell is filled with a low-pressure mixture of hydrogen and helium. When a discharge is passed through the gas, H atoms are formed, which combine with the hot metal to produce MH.

5.3.6. Photolysis and Ablation

The production of free radicals by the irradiation of samples with UV lamps has been known for a long time. Since the 1960s this has provided many electronic spectra using the technique of flash photolysis. However, the concentration of transient species produced is usually insufficient to permit the detection of IR spectra. More recently, the availability of high-power excimer lasers has dramatically changed the situation and most molecules can now be dissociated into two or more fragments using one or more photons (the latter known as multiphoton dissociation). For example, the methyl radical is formed in large concentrations by the photolysis of CH_3I[36]:

$$CH_3I + h\nu \rightarrow CH_3^{\cdot} + I^{\cdot}$$

Both radicals and ions have also been generated by photolysis or photoionization of a stable precursor seeded in a rare-gas free-jet expansion.[37] As explained in Section 5.3.4, the molecules formed in this way are cold and the spectrum is considerably simplified.

5.3.7. Spectroscopy with Fixed-Frequency Lasers

As mentioned earlier, fixed-frequency lasers have advantages over their tunable counterparts, chiefly those of higher power and greater spectral purity. However, the question arises as to how one can record a frequency spectrum using such a laser. The answer is to tune the molecular transitions into coincidence with the laser rather than the other way around. Molecular tuning can be achieved in one of three ways: Stark tuning, Zeeman tuning, or Doppler tuning.

Stark tuning is achieved by applying a variable electric field. All polar molecules exhibit a Stark effect.[2] The electric field **E** interacts with the electric dipole moment μ, causing a shift in the energy of the molecule. Each rotational level J is split into its M_J components, the actual shift being given by

$$\Delta E = -<\mu \cdot \mathbf{E}> \tag{5.10}$$

Most molecules show a second-order Stark effect, that is, the energy change is proportional to E.[2] A more rapid shift is given in symmetric top states with $K \neq 0$; here a first-order Stark effect is observed. As a result of these effects, the frequency of a transition between a given pair of M_J states varies as a function of the electric field. With a sufficiently close laser transition, it is often possible to find a field at which the transition frequency equals that of the laser and absorption is observed.

Zeeman tuning is the analogous experiment performed with a magnetic field. In this case the M_J states are shifted through the interaction of the magnetic dipole **m** with the applied magnetic flux density **B**.[38]

$$\Delta E = -<\mathbf{m} \cdot \mathbf{B}> \tag{5.11}$$

Such an interaction can be quite large in open-shell molecules (free radicals), which possess an unpaired electron orbital or spin angular momentum. The interaction is always nearly proportional to the applied field and to the magnetic quantum number M_J. As in Stark tuning, application of the magnetic field can bring the frequency of a transition between a given pair of M_J states into resonance with a nearby laser line. This technique is known as laser magnetic resonance (LMR).

Doppler tuning is a technique that can be used to study the spectra of molecular ions using a fixed-frequency laser. The Doppler effect results from the fact that a molecule moving toward a radiation source of frequency ν_0 with a velocity v perceives a shifted frequency

$$\nu = \nu_0 \left(\frac{1 + v/c}{1 - v/c} \right)^{1/2} \tag{5.12}$$

Thus, by changing the speed at which the molecule moves relative to the source, it is possible to change the apparent frequency in the molecule's frame of reference according to the preceding expression. The velocity of charged particles can readily be controlled by accelerating them through a known voltage:

$$\frac{1}{2} mv^2 = Ze\mathrm{V} \tag{5.13}$$

for molecules of mass m and charge Ze. This method was first applied to molecular ions by Wing and co-workers.[39] There is also an additional advantage from a spectroscopic point of view. When ions are accelerated through a fixed potential, the initial velocity spread is considerably compressed. This is because energy must be conserved in the acceleration process. The spectroscopic linewidth, which depends on this velocity spread, is consequently reduced, the more so the faster the ions are moving.

5.3.8. Saturation Spectroscopy

An ideal situation in high-resolution spectroscopy would be to have infinitely narrow spectral lines. This, of course, is never the case. There are several forms of *homogeneous* broadening (all molecules contributing to the overall linewidth) such as lifetime broadening (due to the finite lifetime of the excited state of a molecular transition) and pressure broadening resulting from collision-induced lifetimes. These contributions to the linewidth are usually not very important in the infrared; the latter, if significant, can usually be reduced by performing the experiments at reduced pressure.

As mentioned earlier, the Doppler effect is the dominant source of broadening. It results from the perceived shift in frequency of the radiation in an axis system moving with the molecules. Absorption of radiation occurs over a range of frequencies due to the Maxwell–Boltzmann distribution of molecular velocities. This is a *heterogeneous* form of broadening as each molecule contributes to a different part of the absorption line. As mentioned in Section 5.3.4, this contribution to the linewidth can be substantially reduced by using a well-collimated beam of molecules and an orthogonal laser beam. An alternative approach is to monitor saturation effects in a gaseous sample. Consider an intense beam of radiation passing through a gas at a frequency slightly different from the center of a molecular transition. This beam only interacts with those molecules with a suitable component of velocity along its direction of propagation to be tuned into resonance. If the radiation is sufficiently intense, as from a laser source, saturation occurs; the absorption is nonlinear in power because a significant fraction of molecules with the required velocity component are transferred to the excited level.

Now consider a counterpropagating laser beam (Figure 5.12 shows a typical arrangement). This beam interacts with molecules with a velocity component in the opposite direction. If the laser is now tuned to the center of the transition, both beams interact with the same molecules, those moving essentially perpendicular to the two laser beams. Under conditions of saturation the magnitude of the absorption signal for either beam de-

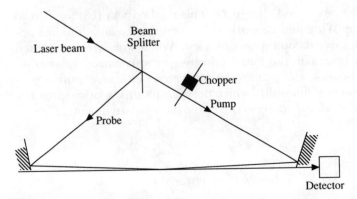

Figure 5.12 Basic arrangement of a saturation spectrometer. The sample is placed between the two mirrors and interacts simultaneously with the two counterpropagating laser beams.

creases. Such a reduction is known as a Lamb dip. The Doppler effect is essentially eliminated and the residual linewidth of the Lamb dip is dominated by the other sources of broadening: power, natural, or collisional.

5.3.9. High Resolution by Computer Enhancement

It is not always possible to obtain spectra with as high resolution as one would like. With poorly resolved spectra, it is common to attempt to resolve the underlying structure by simulating the spectrum and obtaining a best fit with the observed spectrum. However, this method is often unsatisfactory and prone to errors.

A much better approach is to reduce the observed spectral linewidth artificially and hence resolve the underlying structure. For a spectral line recorded at Doppler-limited resolution, the experimental line function $L(v)$ is a convolution of the Doppler profile $D(\Delta v)$ with the molecular lineshape $M(v)$, which is the resultant of the natural, collisional, and instrumental functions:

$$L(v) = \int_{-\infty}^{\infty} M(v')D(v - v') \, dv' \tag{5.14}$$

Since the Doppler function is precisely known for a molecule of molecular mass m at temperature T, it is possible to convolute the experimental lineshape to remove the Doppler broadening. Pliva and co-workers have demonstrated that a Doppler-limited difference-frequency laser spectrum can have its resolution enhanced by a factor of about 3, provided the signal-to-noise ratio is high enough ($\sim 10^3:1$).[40] Such an enhancement in resolution is shown in Figure 5.13.

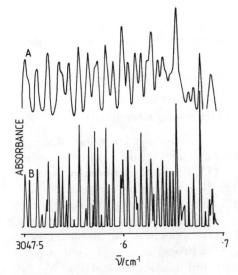

Figure 5.13 Example of the improvement of the resolution of a spectrum (top) by computer-enhanced deconvolution (bottom).

5.4. Applications

The remainder of this chapter is devoted to descriptions of individual molecules that have been studied by high-resolution infrared laser spectroscopy. All species are transient molecules. They have been chosen because they are in some way important and because they demonstrate the range and power of the various methods described earlier. We shall discuss molecules under three headings: free radicals, molecular ions, and van der Waals molecules.

5.4.1. Free Radicals

Free radicals are the short-lived, uncharged intermediates of many chemical reactions. Because of their reactivity, free radicals are usually present in very small concentrations. Many of these species have open shells and possess unpaired electron spins or nonzero orbital angular momentum. Consequently, interaction with a magnetic field will form a large part of many of the studies that follow.

5.4.1.1. The Hydroperoxyl Radical, HO_2 HO_2 is a short-lived free radical that plays an important part in combustion processes and in atmospheric chemistry. It has been studied spectroscopically in many ways and its major structural properties are well characterized. Its ground electronic state is antisymmetric with respect to reflection in the plane of the molecule ($\tilde{X}^2A''$) and its geometry is well determined: $r_{OH} = 0.0997$ nm, $r_{OO} = 0.1334$ nm, $<HOO = 104°$.

HO_2 has three vibrational modes, all of a' symmetry and all of which have been studied in the infrared by laser spectroscopy. The wavenumbers of the vibrations are:

$$\tilde{\nu}_1 = 3{,}436.20 \text{ cm}^{-1} \quad \text{(H—O stretching vibration)[41]}$$
$$\tilde{\nu}_2 = 1{,}391.75 \text{ cm}^{-1} \quad \text{(bending vibration)[42]}$$
$$\tilde{\nu}_3 = 1{,}097.63 \text{ cm}^{-1} \quad \text{(O—O stretching vibration)[43]}$$

Note that the O—O stretching vibration has a lower frequency than the bending vibration, an unusual occurrence that is consistent with the anomalous length (and weakness) of the O—O band. In the laboratory the molecule is usually formed in a discharge flow system by the reaction of F atoms with H_2O_2[44] or of O atoms with allyl alcohol.[43] Part of the ν_2 band recorded by diode laser spectroscopy is shown in Figure 5.14. All the observed transitions[42] are induced by the a component of the dipole moment and hence obey the selection rule $\Delta K_a = 0$. From the inertial point of view, the molecule is a near-prolate symmetric top. Each rotational level N_{KaKc} is split into a doublet by spin–rotation coupling $\mathbf{J} = \mathbf{N} + \mathbf{S}$. The two Q branches $(\Delta J = 0)$ that arise in this way for $K_a = 3$ are shown in Figure 5.14. The spin splitting is seen to be quite large, showing that the spin–rotation interaction changes with vibrational excitation. Each line in the spectrum appears as a second derivative of an absorption lineshape. This is because the signals were recorded with Zeeman modulation and detected at twice the modulation frequency.

Figure 5.14 Part of the ν_2 band of HO_2 recorded using a diode laser. Second derivative lineshapes are shown. The two Q branches shown arise from spin doubling.

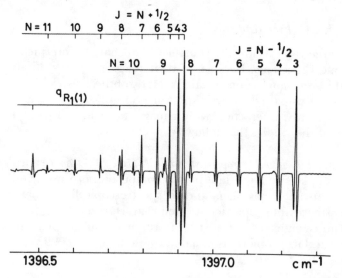

5.4.1.2. The Formyl Radical, HCO   The HCO radical plays an important part in combustion processes and, like HO_2, has been studied in almost every conceivable spectroscopic experiment. It is nonlinear in its ground $^2A'$ state with the geometry: $r_{CH} = 0.1124$ nm, $r_{CO} = 0.1175$ nm, $<HCO = 125°$. The three vibrational modes (all of a' symmetry) are well established:

$$\bar{\nu}_1 = 2{,}434.48 \text{ cm}^{-1} \quad \text{C—H stretching vibration}^{[45]}$$
$$\bar{\nu}_2 = 1{,}080.76 \text{ cm}^{-1} \quad \text{bending vibration}^{[46,47]}$$
$$\bar{\nu}_3 = 1{,}868.17 \text{ cm}^{-1} \quad \text{C—O stretching vibration.}^{[48–50]}$$

The C—H bond is much longer than expected. The weakness of this bond is further demonstrated by the low C—H stretching frequency.

As mentioned earlier the most common method for generating HCO is the reaction of F atoms with H_2CO in a discharge flow system.[23] The radical can also be formed quite efficiently by the photolysis of CH_3CHO with an excimer laser.[45] HCO is a good candidate for study by fixed-frequency laser spectroscopy because the ν_2 and ν_3 vibrations fall in the 10-μm (CO_2) and 5-μm (CO) regions, respectively. The radical has a sizable magnetic moment from the spin of the unpaired electron and it has been studied extensively by LMR spectroscopy. However, it is almost the only transient molecule to have been studied by laser Stark spectroscopy.[46] This technique requires the sample to be located between two closely spaced metal plates across which the electric field is generated. In practice, it is very difficult to form reactive molecules in such an environment because they tend to recombine on the metal surface. Largely for this reason, laser Stark spectroscopy has been used to study stable molecules.

Because laser Stark spectroscopy depends on the interaction between the electric dipole moment and the applied electric field, it also provides a measurement of the dipole moment. In the case of HCO, the dipole moments are determined to be $\mu'_{a2} = 1.3474(48)D$, $\mu''_a = 1.3626(39)D$. The electric dipole moment does not lie along the a inertial axis but only its a component can be determined.

5.4.1.3. The Methylene Radical, CH_2   The methylene radical is a fundamental hydrocarbon fragment that has proved to be a very elusive spectroscopic target. It took a long time before its electronic spectrum was firmly identified.[51] More recently, it has succumbed to attack in the far infrared and mid-infrared, and as a result most of its major properties are now characterized. The ground state has an open-shell electronic structure ($\tilde{X}^3B_1$) with a very obtuse bond angle[52]: $<HCH = 135.5°$. There is also a low-lying singlet state ($\tilde{a}^1A_1$), only $3{,}165$ cm^{-1} above the triplet state in which the bond angle is more nearly acute at $104°$.[53] Of its three vibrational modes, only the bending vibration $\bar{\nu}_2$ has been measured[54]: $\bar{\nu}_2 = 963.0995$ cm^{-1}. The two C—H stretching vibrations are thought to be only

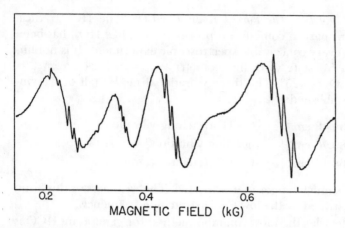

Figure 5.15 Part of the LMR spectrum of CH_2 observed using the $P(34)$ laser line of $^{12}C^{16}O_2$. The spectrum shows saturation dips, which give resolved proton hyperfine structure.

weakly infrared active and have not yet been detected. Their calculated wavenumbers are 2,992 cm^{-1} ($\bar{\nu}_1$) and 3,212 cm^{-1} ($\bar{\nu}_3$).

The bending vibration has been widely studied by LMR (the transitions fall conveniently in the range of the CO_2 laser) and also by diode laser spectroscopy.[55] The LMR spectrum of the $\bar{\nu}_2$ fundamental shows characteristic Zeeman patterns that provide the key to analysis of both this spectrum and the FIR spectrum obtained previously.[52,56] Figure 5.15 shows part of the mid-IR LMR spectrum. The first-derivative lineshapes are caused by the use of Zeeman modulation. Notice also the saturation dips on each M_J component. The triplet structure arises from the proton hyperfine interaction and confirms that ortho levels are involved in this transition.

      5.4.1.4. The Methyl Radical, CH_3   The methyl radical is one of the most important intermediates in chemical reactions. Infrared laser spectroscopy has provided the means both to determine its structure and to monitor its concentration in situ in gas-phase reactions. Other spectroscopic methods of structural determination fail because the molecule is planar (so that it has no conventional rotational spectrum) and because higher electronic states are predissociated so that rotational structure can barely be observed in the electronic spectrum. Yamada and co-workers have recorded the vibration–rotation spectrum of CH_3 in the region 600–750 cm^{-1} with a diode laser.[57] The radical was generated in high concentration ($\sim 2 \times 10^{13}$ molecule cm^{-3}) by either heating or running an electric discharge through di-*tert*-butyl peroxide at about 1 torr pressure. The transition observed involves the out-of-plane bending coordinate $\bar{\nu}_2$. The fundamental, first, and second hot bands were all observed. An example of

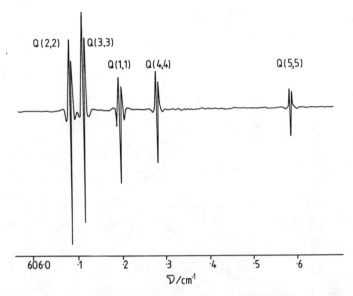

Figure 5.16 Part of the diode laser spectrum of CH_3. Zeeman modulation is used to help discriminate against nonparamagnetic species.

the Q-branch region in the fundamental band is shown in Figure 5.16. The spectrum is recorded with Zeeman modulation, which has the advantage of suppressing the lines from diamagnetic impurities; lines are displayed as second derivatives through phase-sensitive detection at twice the modulation frequency.

Analysis of the CH_3 spectrum has provided definitive evidence that the radical is planar in its ground state, a point that had been conjectural beforehand. The planar structure follows from the observation that levels with odd N and $K = 0$ are missing and from the regularity of the spacing between the vibrational levels $v_2 = 0$, 1, 2, and 3. The spectrum also gives accurate values for the rotational constant B from which the C—H bond length can be determined.

5.4.1.5. The N_3 Radical Several polyatomic molecules have been studied by mid-IR LMR. One of the latest examples is N_3, a linear symmetrical molecule with a $^2\Pi_g$ ground state. The electronic spectrum of this molecule was first studied some time ago[58] but gave no information as to the ground-state vibrational internals. Very recently, vibration–rotational transitions involving the antisymmetric stretching vibration $\tilde{\nu}_3$ have been detected, first by FTIR[59] and then by LMR with a CO laser.[60] The vibrational wavenumber is determined from these studies to be 1,644.679 cm^{-1}. The LMR experiment suffers from being a very poor search technique but once transitions are detected, it offers much greater

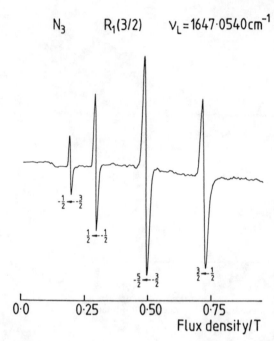

Figure 5.17 LMR spectrum associated wtih the $R(3/2)$ transition in the v_3 fundamental band of N_3.

sensitivity. It has therefore proved possible to detect several hot bands and also to resolve ^{14}N hyperfine structures with Lamb dips. The N_3 radical is generated in a discharge flow system by the reaction of F atoms with hydrazoic acid, HN_3. Figure 5.17 shows the LMR spectrum associated with the $R(3/2)$ transition of the fundamental ($\bar{v}_3$) band of N_3.

5.4.2. Molecular Ions

Most of the ions that have been studied are positive ions. These are most readily formed in discharges or by direct ionization using multiphoton or electron bombardment techniques. Because of the effect of external fields on the motions of ions, it is difficult to use Zeeman or Stark tuning of the energy levels. However, the electric charge makes Doppler tuning readily applicable. Examples of the various approaches follow.

5.4.2.1. The Hydrogen Molecular Ion, HD$^+$ H_2^+ is a molecule of fundamental importance. It is the simplest possible molecular ion, with only one electron. It has been the subject of many very accurate ab initio calculations but, despite its importance, few spectroscopic studies. One problem is that H_2^+ is a symmetrical, nonpolar molecule and therefore has no vibration–rotation spectrum. The isotopic modification, HD^+, is polar and has been intensively studied instead. The origin of the dipole moment is the noncoincidence of the center of mass and the center of charge. As the ion vibrates, there is also an oscillating dipole that gives rise to an infrared spectrum.

HD^+ (in its ground $^2\Sigma^+$ state) has been most studied using the technique of Doppler-tuning ion beams. Because HD^+ is a one-electron system, its bond is comparatively weak and vibrations appear within the compass of the CO laser (around 1,915 cm^{-1}). This work was started by Wing et al.[39] The layout of the apparatus used is shown in Figure 5.18. A beam of HD^+ is formed by electron bombardment of HD. The ions are accelerated to high energy by a voltage of several kilovolts applied to a region of constant electrostatic potential. The ion beam is almost coincident with a CO laser beam. As the accelerating voltage is changed, the vibrational transitions of the ion are brought into resonance with a nearby laser line through the Doppler effect. The effect of the spectroscopic transition is monitored by passing the ions through a low-pressure target gas, where partial charge transfer takes place, before being collected on a Faraday cup. The efficiency of the charge transfer changes slightly between vibrational levels so that when HD^+ is pumped from one vibrational level to another, the intensity of the ion beam at the detector changes. In this way several vibration–rotation transitions can be detected. Furthermore, the compression of the velocity distribution on acceleration of the ions gives sufficiently narrow linewidths tht proton hyperfine structure can be resolved.

This work has been extended by Carrington and his co-workers to high-lying vibrational levels of HD^+ close to the dissociation limit.[61-63] Doppler

Figure 5.18 Typical apparatus for observing the vibration–rotation transitions of HD^+ by Doppler tuning.

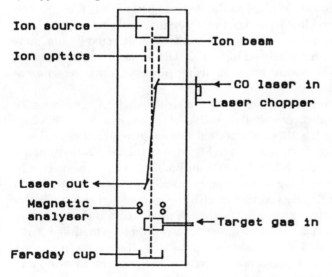

tuning is used to bring transitions into resonance with a CO_2 laser. The transitions are detected via the absorption of a second IR photon, which leads to photodissociation. For example,

$$HD^+ (v'', J'') \xrightarrow{hv} HD^{+2}(v', J') \xrightarrow{hv} H + D^+$$

The appearance of D^+ follows the absorption of photons in the initial step. In this way levels within two infrared quanta of dissociation can be studied. Very recently, Carrington et al. have shown that levels of HD^+ very close to dissociation ($v = 21$) can be dissociated simply by applying an electric field.[64] This is a more efficient method than photodissociation and has led to a dramatic improvement in sensitivity.

5.4.2.2. H_2F^+ H_2F^+ has been chosen as a representative example of the work of Saykally and his group, who have studied a large number of positive and negative ions. They have also developed the velocity modulation method, which has greatly increased the sensitivity for detection of molecular ions in a discharge.[65] Although quite high concentrations of positive ions can be produced in a discharge plasma, they are always much less (by about three orders of magnitude) than the concentrations of parent neutral molecules. Because the latter often cause infrared absorptions in the same wavenumber region as the ions, it is frequently difficult to extract the weak ion signals.

In the positive column region of a discharge plasma, the local electric field is typically 10 V cm^{-1}. This accelerates the ions; however, because of collisions they experience a net drift velocity (v_D) of about 500 m s^{-1}. Within the moving frame of the ions, the radiation frequency v_{IR} is Doppler shifted by $\Delta v = (v_D/c) v_{IR}$. This shift is about the same size as the normal Doppler linewidth. In the velocity modulation technique, a bipolar AC voltage (at a few kilohertz) is applied to the discharge. This has the effect of shifting the absorption line in and out of resonance with the tunable laser at the modulation frequency. Neutral species are not affected by the alternation of the discharge polarity. The ion signal can thus readily be separated from that of the parent molecule using phase-sensitive detection at the modulation frequency.

H_2F^+ is isoelectronic with water. It has a closed-shell $^1A_1$ ground state and its geometry is also very similar to that of water: $r_{FH} = 0.0968$ nm, $<$HFH $= 113.9°$. It has three vibrational modes (two stretches and one bend), all of which are infrared active. The symmetric and antisymmetric stretches, $\bar{v}_1$ and $\bar{v}_3$, are at 3,348.71 cm^{-1} and 3,334.67 cm^{-1}, respectively, and have been studied using a color-center laser. H_2F^+ was formed in an AC discharge through HF and the spectrum was detected using velocity modulation; part of the rotational structure of the $\bar{v}_3$ band is shown in Figure 5.19. Like water, H_2F^+ is an asymmetric rotor and the rotational structure is rather irregular. The spectrum also illustrates the effects of nuclear spin statistics. The two H nuclei are equivalent and give rise to ortho and para spin states with weights 3 and 1, respectively.

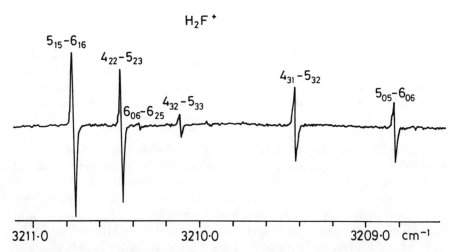

Figure 5.19 Portion of the P branch of the v_3 band of H_2F^+, recorded using a color-center laser. Note the 3:1 relative intensity of equivalent transitions due to nuclear spin statistics.

5.4.2.3. DCl^+ In addition to the large number of neutral free radicals that have been studied in the infrared by LMR, a few positive ions have been detected. DCl^+ in its $X^2\Pi$ state is a good representative of this class of molecules. It was the first ion to be detected by LMR in the far infrared, by Saykally and Evenson.[66] It has also been studied in the mid-IR (in its deuteriated form because the vibrational frequency of HCl^+ lies outside the range of the CO laser) by Urban's group in Bonn using an LMR spectrometer based on the Faraday rotation effect.[67,68] This system offers some advantages of increased sensitivity and is ideally suited to the study of long-discharge plasmas since the magnetic field is provided by a longitudinal, superconducting solenoid.

DCl^+ was formed in an electric discharge through a few millitorr of DCl in 2 torr of He. The discharge was operated in the anomalous mode so that the negative glow extended over the whole absorption region (30 cm in length). The discharge was cooled down to about 150 K with liquid nitrogen, causing an increase in the population of the lower rotational levels. Under these conditions it was easy to resolve the Cl nuclear hyperfine splitting. On the other hand, hot bands up to $v = 7 - 6$ were observed suggesting extensive population of excited vibrational levels. It is postulated that this vibrational distribution may be the result of Penning ionization by metastable He atoms.

5.4.2.4. H_3^+ H_3^+ is of fundamental importance. It is the simplest possible polyatomic molecule; it is a very stable component of ion–molecule reactions and acts as a protonating source in many situations; it is important in both terrestrial and astrophysical environments, largely because of the high abundance of hydrogen in the universe. The molecule has the symmetrical shape of an equilateral triangle in its ground electronic state $(\tilde{X}^1 A'_1)$. However, it has no bound electronic excited states of the same geometry so that it is not possible to detect and study the molecule through its electronic spectrum. The detection of the IR spectrum by Oka in 1980 provided a landmark in the molecular spectroscopy of ions.[69] The H_3^+ was generated by a discharge through H_2 in a long tube, the walls of which were cooled to liquid nitrogen temperatures. It has been possible to improve the signal-to-noise ratio greatly in later experiments, some of which have involved the use of diode lasers.

The H–H bond length of H_3^+ in its ground state is now known to be 0.08763 nm. The molecule has two vibrational modes that can be classified in its point group (D_{3h}). The totally symmetric stretching vibration v_1 (breathing mode) is of a'_1 symmetry and the twofold degenerate ring distortion mode v_2 is of e' symmetry. The former is not infrared active and has still not been measured experimentally. Oka detected transitions in the v_2 band induced by the perpendicular components of the dipole moment and obeying the selection rule $\Delta K = \pm 1$. His observations and subsequent work[70,71] have established the wavenumber for this vibration to be 2,521.31 cm^{-1}.

5.4.3. Van der Waals Molecules

The interaction between a pair of molecules is normally attractive at some orientation. If the resulting intermolecular well is sufficiently deep, it supports bound vibrational levels. Species held together in such a way are termed van der Waals molecules. Such complexes are stable in a collision-free environment. Even in the presence of collisions they have a finite lifetime and can be studied spectroscopically.[72,73]

All gaseous samples contain a finite number of dimers and higher polymers. Their concentrations can be increased by reducing the temperature. However, the concentration of dimer rarely exceeds 0.01% that of the monomer. An alternative approach is to use supersonic nozzles. (See Section 5.3.3.) Here the very low temperatures attained lead to substantial clustering. In many cases (especially in seeded beams) about 10% of the monomer can be conveniently converted to dimer. The following examples illustrate IR absorption studies in both equilibrium cells and in supersonic nozzle expansions.

5.4.3.1. The Argon–Hydrogen Chloride Complex
One of the most studied van der Waals molecules is that formed between argon and hydrogen chloride. Extensive infrared spectra[74] have been obtained with a low-temperature ($T = 127$ K), long-path-length ($L = 72$ m) White cell. Using a mixture of 2 torr HCl and 7 torr Ar, any interfering spectrum due to $(HCl)_2$ can be reduced. Between the $P(1)$ and $R(O)$ lines of HCl appears an extensive spectrum due to the HCl stretch within a linear Ar–HCl (Figure 5.20). The monomer lines are very strong. Spectra can

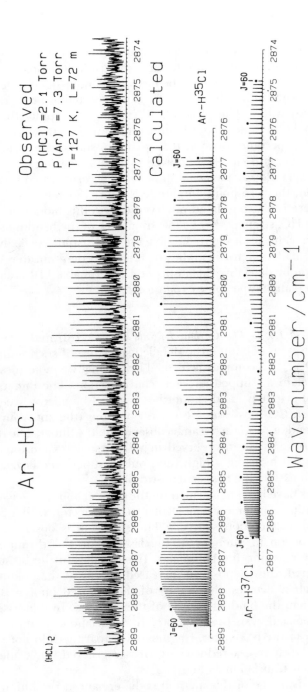

Figure 5.20 High-resolution difference-frequency laser spectrum of Ar–HCl observed in a low-temperature static cell. Beneath is a simulation of the contributions from different chlorine isotopomers; every tenth rotational transition is marked with a dot.

be observed only for complexes formed from such molecules as HCl, which have a very open rotational structure to their spectrum. Rotational transitions in the complex up to $J = 60$ can be measured and assigned. Beyond this level the spectrum suddenly stops. This is interpreted as being due to rotational predissociation. The molecule is rotating so rapidly that the ground vibrational state is no longer bound, and tunneling through a centrifugal barrier has become so rapid that quantum states with J' greater than 60 are no longer sufficiently long-lived to give narrow lines.

Ar—HCl is a very floppy molecule, possessing a very large amplitude bending vibration; the zero-point amplitude is approximately $\pm 45°$. As a consequence, there is a strong band observed due to this motion in combination with the HCl stretch.[74,75] The low frequency of the bending vibration, about 34 cm^{-1}, makes it highly suitable for far-IR study. Saykally and co-workers have made extensive studies of this and other van der Waals modes of vibration by detecting the absorption of a tunable FIR laser due to molecules in a planar supersonic expansion.[76-79]

5.4.3.2. The Carbon Dioxide Dimer   Infrared spectroscopy is an ideal way of obtaining the detailed structure of such nonpolar molecules as $(CO_2)_2$. This complex is formed by placing two molecules parallel to each other in a "slipped-parallel" configuration. The dimer possesses a center of symmetry and cannot be observed by microwave spectroscopy. However, CO_2 possesses IR-active vibrations (the antisymmetric stretch, ν_3, and bend, ν_2) and these can be observed in the dimer. The direct absorption spectrum has been observed in a supersonic expansion. The intensity is increased by using a high-flux, pulsed supersonic nozzle and gated detection to extract the transient absorption.[80]

Further studies have been performed on the $\nu_1 + \nu_3$ and $2\nu_2 + \nu_3$ combination bands using a bolometric molecular beam detector, with a color-center laser used as radiation source.[81] The spectrum is observed, not through the reduction in intensity of the laser beam but through the change in energy reaching the bolometer. If molecules in the beam are vibrationally excited, they deposit an increased energy at the bolometer. For van der Waals molecules, the energy of the infrared photon is frequently greater than the binding energy of the complex. In that case the van der Waals molecule can predissociate. If dissociation occurs during the time of flight to the bolometer, fragments fly apart and do not reach the detector, in which case a reduced signal is registered (molecules no longer deposit their translational energy).

A significant advantage of this method is the enhanced resolution that is possible. A small detection area on the bolometer can define a well-collimated molecular beam. If the laser radiation is applied orthogonally to such a beam, the potential Doppler width is substantially reduced. Spectral linewidths approaching the laser bandwidth of 1 MHz are now possible.

REFERENCES

1. R. Robertson, *Faraday Soc. Trans.* **25,** 899 (1929).
2. C. H. Townes and A. L. Schawlow, *Microwave Spectroscopy* (McGraw-Hill, New York, 1955).
3. K. Shimoda (ed.), *High Resolution Laser Spectroscopy, Topics in Applied Physics,* Vol. 13 (Springer, Berlin, 1975).
4. J. Connes and P. Connes, *J. Opt. Soc. Am.* **56,** 895 (1966).
5. J. Pinard, *Ann. Phys.* (Paris) **4,** 147 (1969).
6. P. R. Griffiths (ed.), *Transform Techniques in Chemistry* (Heyden, London, 1978).
7. J. Hecht, *The Laser Guidebook* (McGraw-Hill, New York, 1987).
8. L. F. Mollenauer and J. C. White (eds.), *Tunable Lasers, Topics in Applied Physics,* Vol. 59 (Springer, Berlin, 1987).
9. W. Urban, in *Frontiers of Laser Spectroscopy of Gases,* ed. A. C. P. Alves, J. M. Brown, and J. M. Hollas (Kluwer Academic, Dordrecht, 1988).
10. R. W. Campbell and F. M. Mims III, *Semiconductor Lasers* (Howard W. Sams, Indianapolis, 1972).
11. E. D. Hinkley, K. W. Nill, and F. A. Blum, *Topics in Applied Physics,* Vol. 3 (Springer, Berlin, 1976), p. 127.
12. L. F. Mollenauer and D. H. Olsen, *J. Appl. Phys.* **46,** 3109 (1975).
13. J. V. V. Kasper, C. R. Pollock, R. F. Curl, Jr., and F. K. Tittel, *Appl. Opt.* **21,** 236 (1982).
14. Z. S. Huang, K. W. Tucks, and R. E. Miller, *J. Chem. Phys.* **85,** 3338 (1986).
15. A. S. Pine, in *Laser Spectroscopy,* Vol. 3 (Springer, Berlin 1977), p. 376.
16. A. S. Pine, *J. Opt. Soc. Am.* **64,** 1683 (1974).
17. A. S. Pine, *J. Opt. Soc. Am.* **66,** 97 (1976).
18. N. G. Douglas, *Millimetre and Submillimetre Wavelength Lasers* (Springer, Berlin, 1990).
19. F. C. van der Heuvel, W. L. Meerts, and A. Dymanus, *Chem. Phys. Lett.* **88,** 59 (1982).
20. K. M. Evenson, D. A. Jennings, and F. R. Peterson, *Appl. Phys. Lett.* **44,** 576 (1984).
21. J. U. White, *J. Opt. Soc. Am.* **32,** 285 (1942).
22. H. J. Bernstein and G. Herzberg, *J. Chem. Phys.* **16,** 30 (1948).
23. I. C. Bowater, J. M. Brown, and A. Carrington, *J. Chem. Phys.* **54,** 4957 (1971).
24. F. C. van der Heuvel and A. Dymanus, *Chem. Phys. Lett.* **92,** 219 (1982).
25. F. C. De Lucia, E. Herbst, G. M. Plummer, and G. A. Blake, *J. Chem. Phys.* **78,** 2312 (1983).
26. J. B. Anderson, in *Molecular Beams and Low Density Gas Dynamics,* P. P. Wegener, ed. (Dekker, New York, 1974).
27. H. Mikami, *Bull. Res. Nuclear Reactors* **7,** 151 (1982).
28. B. Fenn, *Appl. Atom. Collision Phys.* **5,** 349 (1982).
29. R. Campargne, *J. Phys. Chem.* **88,** 4466 (1984).
30. H. Ashkenas and F. S. Sherman, in *Rarefied Gas Dynamics, Proceedings of 4th*

International Symposium, J. H. deLeeuw, ed. (Academic Press, New York, 1965), p. 784.

31. D. R. Miller, in *Atomic and Molecular Beam Methods,* G. Scoles, ed. (Oxford University Press, New York, 1988).

32. G. Scoles (ed.), *Atomic and Molecular Beam Methods,* Vol. 2 (Oxford University Press, New York, 1991).

33. N. N. Haese, D. J. Liu, and R. S. Attman, *J. Chem. Phys.* **81,** 3766 (1984).

34. B. Lemoine, C. Demuynck, J.-L. Destombes, and P. B. Davies, *J. Chem. Phys.* **89,** 673 (1988).

35. U. Magg and H. Jones, *Chem. Phys. Lett.* **146,** 415 (1988).

36. G. E. Hall, T. J. Sears, and J. M. Frye, *J. Chem. Phys.* **90,** 6234 (1989).

37. S. C. Foster, R. A. Kennedy, and T. A. Miller, in *Frontiers of Laser Spectroscopy of Gases,* A. C. P. Alves, J. M. Brown, and J. M. Hollas, ed. (Kluwer Academic, Dordrecht, 1988).

38. A. Carrington, *Microwave Spectroscopy of Free Radicals* (Academic, London, 1974).

39. W. H. Wing, G. A. Ruff, W. E. Lamb, and J. J. Spezeski, *Phys. Rev. Lett.* **36,** 1488 (1976).

40. J. Pliva, A. S. Pine, and P. D. Willson, *Appl. Opt.* **19,** 1833 (1980).

41. C. Yamada, Y. Endo, and E. Hirota, *J. Chem. Phys.* **78,** 4379 (1983).

42. K. Hagai, Y. Endo, and E. Hirota, *J. Mol. Spectrosc.* **89,** 520 (1981).

43. J. W. C. Johns, A. R. W. McKellar, and M. Riggin, *J. Chem. Phys.* **68,** 3957 (1978).

44. H. E. Radford, K. M. Evenson, and C. J. Howard, *J. Chem. Phys.* **60,** 3178 (1974).

45. C. B. Dane, D. R. Lander, R. F. Curl, F. K. Tittel, Y. Guo, M. I. F. Ochsner, and C. B. Moore, *J. Chem. Phys.* **88,** 2121 (1988).

46. B. M. Landsberg, A. J. Merer, and T. Oka, *J. Mol. Spectrosc.* **67,** 459 (1977).

47. W. C. Johns, A. R. W. McKellar, and M. Riggin, *J. Chem. Phys.* **67,** 2427 (1977).

48. J. M. Brown, J. Buttenshaw, A. Carrington, K. Dumper, and C. R. Parent, *J. Mol. Spectrosc.* **79,** 47 (1980).

49. J. M. Brown, K. Dumper, and R. S. Lowe, *J. Mol. Spectros.* **130,** 445 (1983).

50. A. R. W. McKellar, J. B. Burkholder, J. J. Orlando, and C.J. Howard, *J. Mol. Spectrosc.* **130,** 445 (1988).

51. G. Herzberg, *Proc. Roy. Soc. Lond. A,* **262,** 291 (1961).

52. T. J. Sears, P. R. Bunker, A. R. W. McKellar, K. M. Evenson, D. A. Jennings, and J. M. Brown, *J. Chem. Phys.* **77,** 5348 (1982).

53. A. R. W. McKellar, P. R. Bunker, T. J. Sears, K. M. Evenson, R. J. Saykally, and S. R. Langhoff, *J. Chem. Phys.* **79,** 5251 (1983).

54. T. J. Sears, P. R. Bunker, and A. R. W. McKellar, *J. Chem. Phys.* **77,** 5363 (1982).

55. M. D. Marshall and A. R. W. McKellar, *J. Chem. Phys.* **85,** 3716 (1986).

56. J. A. Mucha, K. M. Evenson, D. A. Jennings, and C. J. Howard, *Chem. Phys. Lett.* **66,** 244 (1979).

57. C. Yamada, E. Hirota, and K. Kawaguchi, *J. Chem. Phys.* **75,** 5256 (1981).

58. A. E. Douglas and W. E. Jones, *Can. J. Phys.* **43**, 2216 (1965).

59. C. R. Brazier, P. F. Bernath, J. B. Burkholder, and C. J. Howard, *J. Chem. Phys.* **89**, 1762 (1988).

60. R. Pahnke, S. H. Ashworth, and J. M. Brown, *Chem. Phys. Lett.* **147**, 179 (1988).

61. A. Carrington and J. Buttenshaw, *Mol. Phys.* **44**, 267 (1981).

62. A. Carrington, J. Buttenshaw, and R. A. Kennedy, *Mol. Phys.* **48**, 775 (1983).

63. A. Carrington and R. A. Kennedy, *Mol. Phys.* **56**, 935 (1985).

64. A. Carrington, I. R. McNab, and C. A. Montgomerie, *Chem. Phys. Lett.* **151**, 258 (1988).

65. C. S. Gudeman and R. J. Saykally, *Ann. Rev. Phys. Chem.* **35**, 387 (1984).

66. R. J. Saykally and K. M. Evenson, *Phys. Rev. Lett.* **43**, 515 (1979).

67. A. Hinz, W. Bohle, D. Zeitz, J. Werner, W. Seebass, and W. Urban, *Mol. Phys.* **53**, 1017 (1984).

68. W. Bohle, J. Werner, D. Zeitz, A. Hinz, and W. Urban, *Mol. Phys.* **58**, 85 (1986).

69. T. Oka, *Phys. Rev. Lett.* **45**, 531 (1980).

70. J. K. G. Watson, *J. Mol. Spectrosc.* **103**, 350 (1984).

71. J. K. G. Watson, S. C. Foster, A. R. W. McKellar, P. Bernath, T. Amano, F. S. Pan, M. W. Crofton, R. S. Altmann, and T. Oka, *Can. J. Phys.* **62**, 1875 (1984).

72. *Faraday Discuss. Far. Soc.* **83**, 1982.

73. A. Weber (ed.), *Structure and Dynamics of Weakly Bound Molecular Complexes*, NATO ARW (Reidel, Dordrecht, 1987).

74. B. J. Howard and A. S. Pine, *Chem. Phys. Lett.* **122**, 1 (1985).

75. D. J. Nesbitt and C. M. Lovejoy, *Faraday Discuss. Far. Soc.* **86**, 13 (1988).

76. R. L. Robinson, D.-H. Gwo, D. Ray, and R. J. Saykally, *J. Chem. Phys.* **86**, 5211 (1987).

77. R. L. Robinson, D.-H. Gwo, and R. J. Saykally, *J. Chem. Phys.* **87**, 5149 (1987).

78. R. L. Robinson, D.-H. Gwo, and R. J. Saykally, *J. Chem. Phys.* **87**, 5156 (1987).

79. R. L. Robinson, D.-H. Gwo, and R. J. Saykally, *Mol. Phys.* **63**, 1021 (1988).

80. M. A. Walsh, T. H. England, T. R. Dyke, and B. J. Howard, *Chem. Phys. Lett.* **142**, 265 (1987).

81. K. W. Jucks, Z. S. Huang, R. E. Miller, G. T. Fraser, A. S. Pine, and W. J. Lafferty, *J. Chem. Phys.* **88**, 2185 (1988).

6 Modern Techniques in Raman Spectroscopy

M. D. Morris

6.1. Introduction

A small fraction of the light incident on any molecule is inelastically scattered and appears at shifted frequencies. The frequency shifts correspond to rotational, vibrational, or electronic transitions of the molecule. The Raman-scattering effect is weak. For vibrational transitions the scattering efficiency is about 1 part in 10^7. By contrast, the quantum efficiency for fluorescence can approach 100% and is commonly above 1%. Although Raman scattering is a weak effect, modern lasers and detectors make a Raman spectrum quite easy to acquire, and scientists are increasingly exploiting its very attractive properties.

Raman scattering on rotational transitions and on vibrational–rotational transitions can be observed in the gas phase. These are intrinsically interesting, of course, and gas-phase Raman spectroscopy has applications in combustion chemistry and several other fields. Electronic Raman scattering is observable in crystals of transition metal and rare earth metal salts. The electronic Raman effect has proven useful in physical inorganic chemistry. But by far the greatest interest in Raman scattering is in measurement of vibrational spectra. Here the variety of applications to pure and applied chemistry, to materials science, and to biology is immense. For this reason we will confine our discussion to vibrational Raman spectroscopy.

The broad applicability of Raman spectroscopy may be traced to four factors. First, *every* molecule has a vibrational Raman spectrum. The selection rules are different from those of infrared absorption, but they exclude nothing with two or more atoms. Second, the experimental configurations for observation of scattering are more flexible than those for absorption, particularly for irregular solids or turbid liquids. Third, the effect is observed in the experimentally convenient UV-visible, near-IR region. Finally, quartz, glass, and water are weak Raman scatterers. Silica and glass can be used for optics and sample containers and Raman spectroscopy works well in aqueous solutions.

6.2. Principles of Raman Scattering

6.2.1. Normal Raman Scattering

The theory of Raman scattering has been discussed in detail.[1,2] Ordinarily, the optical frequency of the exciting light is lower than the energy of the first electronic transition of the target molecule. An incident photon can undergo a weak interaction, often described as promotion to a virtual state. The electric field around the molecule is weakly perturbed, and this leads to weak scattering at new frequencies. Figure 6.1 shows this process schematically. Classically, the molecule can be considered to be polarized in the external electric field, **E,** and Raman scattering results from modulation of the induced dipole movement, **P.** The scaling factor is called the polarizability, α, and is a tensor. Equation (6.1) describes this relation.

$$\mathbf{P} = \alpha \, \mathbf{E} \tag{6.1}$$

From this model we can derive the basic Raman selection rule. A vibrational mode is Raman active if there is some component of its normal coordinate, Q, for which the polarizability derivative, $\delta\alpha/\delta Q \neq 0$. That is, a vibrational mode is Raman active if the vibrational motion causes a change in the dipole moment that is induced in an external electric field. The significance of higher-order corrections to equation (6.1) is considered in Chapter 7.

Some of the basic features of Raman selection rules are easily understood from this classical picture. First, any Raman feature associated with the stretching or bending of highly polar bonds, such as O—H, should be weak. The dipole moment is so large that there is little change in an external field. Conversely, lines associated with stretches of electron-rich but nonpolar bonds should be strong, because the electron distribution, and with it the induced dipole moment, changes as a consequence of the motion. Examples include the infrared-silent stretches of H_2, Cl_2, and the symmetric C—H stretch of methane. Raman-scattering intensity is proportional to the square of the polarizability derivative for a given mode. It is beyond the scope of this review to discuss modern theories of Raman in-

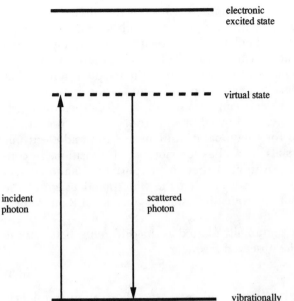

electronic
excited state

virtual state

incident
photon

scattered
photon

vibrationally
excited state

vibrational energy

ground state

Figure 6.1 The normal (Stokes) Raman-scattering process. An incident photon interacts weakly with a molecule. The Raman-scattered photon appears at lower energy than the incident photon.

tensities. The interested reader is directed to the references for both classical and quantum mechanical treatments.

The intensity of Raman scattering can be expressed, in terms of its cross section, σ, defined by

$$\sigma = \frac{P}{I} \tag{6.2}$$

where P is the total radiated Raman power and I is the incident power density, or irradiance. It is more common to use the differential cross section, the cross section per solid angle, $d\sigma/d\Omega$. The units of the differential cross section are $cm^2 sr^{-1}$, and normal vibrational Raman cross sections are about 10^{-30} cm^2 sr^{-1} at 500 nm. The Raman cross section is related to the transition moment for Raman scattering. It is a measurable quantity, although for its determination great care must be taken to establish a known illumination and collection geometry and to calibrate the transmission efficiency of the spectrometer and response of the detector system. Absolute Raman cross sections have been measured for a number of gases and liquids.[2]

The polarizability is a tensor quantity because of the directional properties of a molecular vibration, and this property leads to both strong scattering directionality and useful polarized light relations. In a freely rotating gas or liquid, totally symmetric vibrations will give rise to scattering strongly polarized parallel to the plane of polarization of the incident light. Other vibrations will give depolarized scatter. The relations are more complicated in the solid state, because solids can be anisotropic. Most spectroscopists measure Raman scattering caused by vibrational transitions. However, the Raman effect is observed for rotational transitions in gases and electronic transitions in certain crystals, as well as for vibrational transitions in condensed or gaseous media. Although they are important, rotational and electronic Raman effects have not enjoyed the widespread usage of the vibrational effect. We shall limit our discussion to vibrational Raman spectra.

Raman spectra are conventionally plotted as the difference between the exciting frequency and the observed frequency:

$$\omega_{Raman} = \omega_{ex} - \omega_{obs} \tag{6.3}$$

where ω_{ex} is the exciting frequency and ω_{obs} is the observation frequency. The Raman frequency shift, ω_{Raman}, is independent of the exciting frequency. Equation (6.3) and Figure 6.1 are based on the assumption that the scattering molecule is in the ground electronic state. If this is true, then $\omega_{obs} < \omega_{ex}$. The red-shifted scattered light is called Stokes-shifted. At any finite temperature a small fraction of the molecules is, however, in the excited state of the vibrational transition. In this case, the scattering starts from the vibrationally excited state and ends in the ground state, generating blue-shifted, or anti-Stokes, scattering. The anti-Stokes/Stokes intensity ratio can be used as a measure of temperature, through the Boltzmann equation.

As does any other scattering, Raman-scattering intensity increases with the fourth power of the scattering frequency. Thus, it is theoretically advantageous to use blue or ultraviolet lasers to maximize the Raman signal. However, experimental considerations, such as fluorescence interference, the possibility of photochemical damage to the sample, or wavelength limitations of the available instrument, may dictate the actual preferred excitation wavelength.

6.2.2. Resonance-Enhanced Raman Scattering

If the exciting wavelength is coincident, or almost coincident, with an electronic transition of the molecule, the intensity of certain Raman bands will increase, often by factors of 10^2–10^4. This increase, called resonance enhancement, ultimately arises because the photon–molecule interaction is strong in true absorption.[3,4] The most common mechanism is Franck–Condon enhancement. A totally symmetric vibrational mode is enhanced if some component of its normal coordinate is along the direction in which the molecule expands when it is promoted to the electronic excited state. The enhancement intensity is proportional to the change in the nuclear coordinates between ground and excited states. Symmetric or nontotally symmetric vibrations are enhanced if they couple two electronic transitions.

In the resonance region the scattering intensity does not have a quartic dependence on excitation frequency. Rather, the intensity roughly follows the shape of the absorption spectrum.[3] The Raman intensity versus frequency (or wavelength) plot is called an excitation profile and can be calculated from a knowledge of the ordinary Raman spectrum and the absorption spectrum of the molecule. Conversely, the absorption spectrum can be derived from the resonance Raman excitation profile.

Experimentally, the importance of resonance enhancement is that it makes otherwise unobservable Raman spectra intense and, more important, that it introduces an additional element of selectivity into spectra. Resonance enhancement has been used extensively by physical biochemists to probe the structure and dynamics of specific chromophoric regions of complicated molecules, such as proteins. The classic examples are probes of the conjugated polyenes of the visual system proteins and the porphyrin active site of hemoglobin.

6.2.3. Surface-Enhanced Raman Scattering

Surface-enhanced Raman scattering (SERS) provides strong Raman signals from molecules adsorbed on structural metal surfaces.[5] The effect was originally discovered in attempts to use Raman spectroscopy as a diagnostic for adsorption occurring on metal electrodes. With excitation in the 500-nm region, only silver is a useful SERS substrate, and most SERS has been performed with silver. If red or near-infrared lasers are used, gold and copper become good SERS substrates as well. Although the enhancement mechanism remained controversial through most of the 1970s, it gradually became clear that both electromagnetic effects and formation of charge transfer complexes could contribute to SERS.[6,7] Some controversies remain about relative contributions of these effects and about details of their description. Since the early 1980s attention has shifted from elucidation of the nature of SERS to development of applications, such as biochemical Raman spectroscopy[8,9] and chemical analysis.[10]

The largest enhancement at a metal surface comes from increasing the amplitude of the local electromagnetic field with which a molecule interacts. Interaction of a light electromagnetic field with a smooth metal surface can generate about a sixfold enhancement in Raman-scattering intensity. On a roughened surface, light can excite conduction electrons in the metal. When the optical frequency is coincident with the excitation frequency of an extended surface state, called a surface plasmon, protrusions on the metal become polarized. The electromagnetic field in the interior of the particle becomes much larger than the surface field and falls off with the cube of the distance from the particle. The resonant frequency and the magnitude of the effect can be calculated for idealized geometries, such as spherical or ellipsoidal particles on a flat plane.[11]

There are two major requirements for strong surface enhancement at a metal. First, the refractive index of the metal should be small and have nearly a purely imaginary value. This condition occurs at the frequency for excitation of the surface plasmon. Second, the incident wavelength should match the plasmon resonance frequency of the substrate. This frequency will depend on the size and shape of the substrate particles. Together these requirements imply that only the coinage metals (Ag, Au, Cu) and the alkali metals will be SERS-active in the visible.

The factors governing charge transfer enhancement are diverse. In general, charge transfer complexes between molecules and metal atoms are similar to those between the same molecules and the corresponding metal ions. Thus, chemical enhancement is particularly strong for aromatic molecules, including pyridine and phenols. Charge transfer, or chemical, enhancement is often accompanied by shifts in Raman frequencies of 3–20 cm^{-1}.

SERS provides strong Raman spectra from solutions in the 10^{-3}–10^{-6} M range. Detection limits are usually 10^{-6}–10^{-8} M. Under resonance SERS conditions (i.e., when an electronic transition of the adsorbed molecule matches the surface plasmon frequency of the metal), the sensitivity rivals that of laser-induced fluorescence.[12]

The selection rules for SERS are analogous to the surface selection rules for infrared spectroscopy. In the electromagnetic enhancement model, vibration will be strongly SERS active if there is a component of the induced dipole normal to the metal surface.[6,7] For example, the observation of C—H stretches in pyridine implies that the ring is oriented perpendicular to the electrode. Because of the contributions from different enhancement mechanisms, SERS selection rules are not completely rigorous and might better be called *propensity rules*.[7]

6.3. Specialized Instrumentation

6.3.1. The Basic Raman Experiment

A basic Raman spectrometer is shown in block diagram form as Figure 6.2. Almost every known laser with output in the range 50–1,100 nm has been used as a light source at one time or another. However, the most commonly used laser remains the argon ion laser. For many purposes 100–500 mW is adequate power, although weakly scattering samples can require several watts. Routine liquid and solid samples are analyzed in small glass containers, which typically hold between 10 μl and 1 ml. Gases

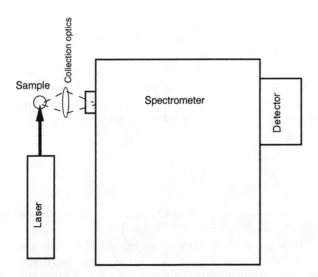

Figure 6.2 Schematic of a typical dispersive Raman spectrometer. The sample is illuminated by a laser and scattered light is collected at 90° to the illumination direction. The light is focused into the spectrometer. Either a scanning monochromator and a photomultiplier detector or a spectrograph and an array detector may be used.

often require multipass cells, similar to the classical White cells of infrared spectroscopy. The discussion of some specialized sampling techniques is deferred to later sections.

Although it is rapidly giving way to multichannel systems, the sequentially scanned double Czerny–Turner monochromator with a photomultiplier operated in photon counting mode is still commonly encountered. Two gratings in additive dispersion, with an intermediate slit, solve the problem of reduction of Rayleigh-scattered light and, in addition, provide high spectral resolution. Alternatively, a multichannel detector can be used to observe a large fraction (usually 500–$1,500$ cm^{-1}) of the spectrum simultaneously. The standard configuration has been a triple spectrograph. The first two stages are operated in subtractive dispersion as a coarse filter to eliminate Rayleigh scatter, while spectral dispersion is provided by the third stage. Such systems are effective, although losses at the 7–10 reflective surfaces reduce throughput.

Alternatively, single-stage spectrographs can be used with sharp-cut filters, such as multilayer dielectric interference filters or diffraction filters. The single-stage instruments are simpler and less expensive than triple spectrographs and have both higher throughput and better collection efficiency (lower f/number). Modern interference filters have steep slopes and high optical densities but have ripples in their passbands. Diffraction filters have equal or better slopes and flat pass bands and somewhat better transmission (75% to 85%). The optical density varies from O.D. 4 to O.D.

8, depending on the wavelength, the thickness of the diffracting layer, and other design considerations. Suspensions of colloidal charged polystyrene microbeads behave as pseudocrystals and make effective prefilters.[13] Recently, holographic diffraction filters have been used as spectrograph prefilters.[14] These are fabricated as interference patterns in dichromated gelatin. Unlike the colloidal arrays, the holographic filters are commercially available for many common laser lines. They allow observation as close as 100 cm^{-1} from the Rayleigh line.

6.3.2. The Charge-Coupled Device Array Detector

Since its introduction to Raman spectroscopy by Murray and Dierker,[15] the charged-coupled device (CCD) has largely supplanted intensified photodiode arrays and even photon counting. The CCD can have an essentially zero dark current and has a low readout noise and better quantum efficiency than the photocathodes of intensifiers or photomultipliers. Except where the time-gating properties of image intensifiers is needed, the CCD is probably the detector of choice for dispersive spectrograph Raman spectroscopy.

The properties and general spectroscopic applications of the CCD have been reviewed.[16,17] The CCD is a two-dimensional metal–oxide–semiconductor device, which operates in a charge storage (integrating) mode. The CCD stores electrons and can typically hold about 250,000. The charge is measured by shifting it systematically from element to element in reverse-biased *p–n* junctions. The charge is ultimately read out in a single specialized low-capacitance element. The low capacitance provides a low readout noise, which justifies use of a 14–16-bit A/D converter. The shift from well to well eliminates the noisy multiplexing step that is required in the conventional photodiode array. If it is operated at low temperature, usually about −110°C, the device has a negligibly small dark current. Bilhorn and co-workers calculate that minimum detectable signal is about an order of magnitude smaller for a typical CCD than for a photomultiplier.[16] The improvement comes from a combination of higher quantum efficiency and lower dark current.

There is only one readout amplifier on a CCD. The pixel charges are clocked row by row to the row containing the readout, and then along that row to the readout edge. Because of this construction, the device can be operated to acquire spectra in a framing mode. The data are taken at the row farthest from the readout and then are clocked to the readout, one row at a time. When the first data are clocked away from the acquisition row, another data set can be acquired there. In this way several hundred consecutive spectra, spaced a few microseconds apart, can be acquired on the CCD. The device can then be read out at the normal rate, which is usually 20–50 μs per pixel (i.e., about 2–5 s per frame).

Charges can be electronically summed on the chip, a process called *binning*. Adjacent rows and column elements can be binned to create any aspect ratio or resolution desired. Binning improves signal-to-noise ratio, because the electronic summing is essentially noiseless. The binned pixels are measured with only a single unit of read-out noise. Careful use of binning allows the spectroscopist to optimize the trade-off between resolution and signal-to-noise ratio.

The CCD does have some disadvantages. First, it is subject to blooming. As the potential well of a pixel fills, charge spills into adjacent wells. Thus, some care is required in measuring weak bands located close to much stronger ones. Second, the CCD is always on. The integration is started and stopped with an electromagnetic shutter or other device that physically prevents light from reaching the CCD. There is no gating mechanism. The CCD must be preceded by a conventional image intensifier if nanosecond gating is required. Ironically, this strategy generally reduces the signal/noise ratio of the detector and is only recommended if nanosecond timing is really needed. Because they are silicon-based detectors, CCDs do not operate to the red of 1.1 μm. Thinned, backside-illuminated devices are sensitive in the UV. Front-illuminated devices can be made UV-sensitive by coating with a phosphor.

Finally, most currently available devices were developed for imaging, not spectroscopy. Typically, they have 1:1 (512 × 512 pixel) or 3:2 (576 × 378 pixel) aspect ratios. The pixels are usually 20–25 μ square, and such detectors cannot image the full 25-cm flat field length of a modern spectrograph. However, manufacturers are now beginning to provide CCDs with spectroscopically useful aspect ratios and 1,000–1,200 pixels on the long side.

6.3.3. Fourier-Transform Raman Spectroscopy

Fluorescence problems arise in Raman scattering because the conventional excitation frequency is relatively high. Since the late 1960s most Raman spectroscopy has employed Ar^+ 488 nm (20,492 cm^{-1}) or 514 nm (19,436 cm^{-1}) excitation. It has been known that fluorescence problems become less severe as the excitation wavelength is moved to the red. Very few organic molecules have electronic transitions below 13,000 cm^{-1} (769 nm), although some transition metal ions may fluoresce at much lower excitation frequencies. Until recently, deep-red or near-infrared excitation has been considered impractical. This opinion was based on the assumption that the ν^4 intensity dependence of Raman signals would make them too weak to be useful, and on the absence of photon detectors with response much beyond 1 μm. The development of CCD detectors has made far-red (750–800 nm) excitation a practical route to elimination of many fluorescence problems. The development of Fourier-transform (FT)

Raman spectroscopy with 1.06-μm excitation has eliminated almost all fluorescence problems from organic materials and most fluorescence interference from transition metals.

With visible excitation, modern spectrographs and CCD array detectors or scanning spectrometers and cooled photomultipliers can provide Raman spectra that are at or nearly at shot noise limits. For this reason, a multiplexed instrument, such as the Michelson interferometer used for FT infrared spectrometry, can provide no further improvement. In fact, shot noise from the very intense Rayleigh line would make a multiplexed instrument perform substantially worse than a dispersive instrument. In the infrared, on the other hand, only very noisy detectors are available, so that a multiplex advantage is possible.

In the gas phase, Rayleigh scattering is weak, particularly at high Raman shifts, so that Fourier-transform Raman spectroscopy can be used to obtain high-resolution Raman spectra.[18] In a landmark paper Hirschfeld and Chase described prefilter configurations that could be used to attenuate the Rayleigh line and so make Fourier-transform Raman spectroscopy generally applicable.[19] They demonstrated that the use of high-average-power near-infrared lasers, such as the widely available Nd:YAG (1.064-μm) laser, allowed acquisition of fluorescence-free Raman spectra from most samples. The multiplex advantage and Jacquinot (throughput) advantage provided by a Michelson interferometer made acquisition times reasonably short, even using a noisy PbS detector with a figure of merit $D* = 6 \times 10^{11}$. ($D*$ is the rms signal-to-noise ratio per unit rms incident power per unit square root of detector area.)

Within two or three years of the Hirschfeld–Chase demonstration, FT–Raman spectroscopy was accepted as the most generally useful technique for obtaining Raman spectra of complex materials, such as industrial polymers and natural products. In addition to providing fluorescence-free spectra, the technique effectively makes a Raman spectroscopy system into an accessory to an FTIR instrument. There are some economies, as well as the advantage of having infrared and Raman spectra automatically in the same file format.

The basic FT–Raman instrumentation remains much as defined by Hirschfeld and Chase. Recent refinements have been summarized by Cutler.[20] Figure 6.3 shows a typical example of a laboratory-constructed FT–Raman system. The light source is a CW Nd–YAG laser, with a Pellin–Broca prism to filter out spontaneous emission. A lens is used to focus or defocus the laser light at the sample, and a second lens recollimates it for presentation to the Michelson interferometer. A dielectric filter is used for Rayleigh line rejection before the monochromator. Quartz beam splitters are most common. However, quartz has an absorption around 1,400 cm^{-1} relative to Nd–YAG, which can degrade signal-to-noise ratios somewhat. The most common detector is cooled InGaAs.

Although early workers often employed high-resolution research-grade FTIR instruments, more compact (and less expensive) benchtop instru-

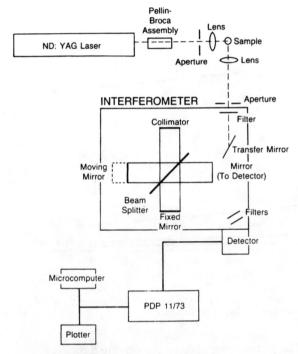

Figure 6.3 Schematic of a typical FT–Raman apparatus. (Reprinted from Ref. 64 with permission.)

ments appear adequate for most purposes.[21] FT–Raman is employed primarily with condensed-phase samples. In condensed phases 2–5 cm^{-1} resolution is usually adequate, and better than 1 cm^{-1} resolution is almost never needed. Good prefilters and high-D^* detectors (or low-noise equivalent power, NEP) are more important than high resolution.

The multiplex advantage relative to a single-channel scanned detector is a strong function of Rayleigh line rejection, as well as the number of resolution elements and detector NEP.[22] Figure 6.4 is a plot of the Fellgett advantage as a function of the ratio of Rayleigh power to benzene 992 cm^{-1} and benzene 1,606 cm^{-1} Raman power as a function of detector NEP. For 3,000 resolution elements the full multiplex advantage of 3,000$^{1/2}$ (i.e., 54.8) is achievable only for a very noisy detector and for no Rayleigh line power incident on the detector. For more realistic conditions, the multiplex advantage for a strong Raman band is 10–20, at most. In fact, Cutler[20] has suggested that it is detector amplifier noise, not detector noise itself, that allows any multiplex advantage at all. Of course, the best signal-to-noise ratio for the Raman spectrum itself is obtained with the quietest detector.

Transmission-type long-pass interference filters are most commonly employed as prefilters. Modern designs are useful down to about 150 cm^{-1}, but suffer from 5% to 10% ripple in the passband. Figure 6.5 shows the performance of some contemporary designs, as well as some early versions.

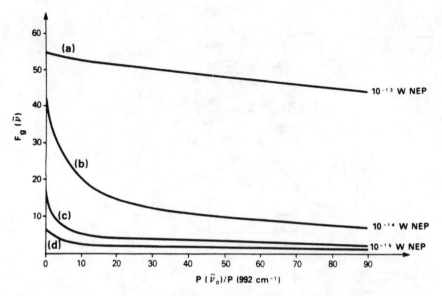

Figure 6.4 Plot of Fellgett advantage F_g versus ratio of Rayleigh line power to benzene 992 cm^{-1} power. (Reprinted from Ref. 22 with permission.)

As with any interference filter, these edge filters are designed to work with collimated light. Filter performance is optimized by angle-tuning, because manufacturing tolerances are not tight enough to guarantee optimal performance on-axis. Reflective interference filters are also satisfactory. The simplest commercial design is a stack of four filters at 45° to each other, sold as a "chevron" filter. Although there have been no published reports of their use yet, diffraction-based filters may well offer excellent performance in the near infrared, as in the visible.

FT–Raman can use any of the many sampling techniques employed with dispersive Raman spectroscopy. For routine work, a backscattering geometry is commonly used. Most workers employ simple glass cuvettes, but Cutler[20] has recommended an externally aluminized 6-mm-diameter spherical cell for liquids. This cell functions as an integrating sphere and provides about a fivefold stronger signal than is obtained with a transparent cell.

Optical fiber beam delivery and Raman signal collection are particularly useful in near-infrared work. Once aligned to the interferometer optics, a fiber bundle stays aligned indefinitely. This is quite helpful, because the usual visual cues to alignment are absent, unless a monitor He–Ne laser is also used.[23] Fiber bundles are also useful in remote sampling[24,25] or with turbid liquids. Most of the standard fiber sampler designs have been tested. Bifurcated fibers and fiber bundles with a single central delivery fiber are the most common versions. Although quartz optical fibers are satisfactory for work in the 1–2μm region, the choice of fiber is important.[25] A low

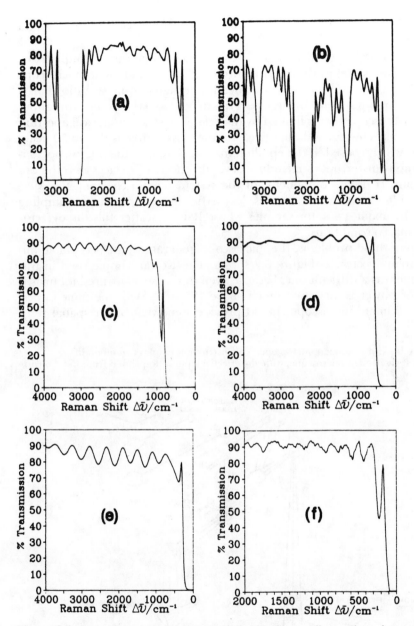

Figure 6.5 Transmission of some Nd:YAG blocking filters. Filters a, b, and c are early designs. Filters d, e, and f are recent prototype and production filters. (Reprinted from Ref. 20 with permission.)

hydroxyl group content is necessary to keep the fiber background low and unstructured.

A standard all-reflective optics infrared microscope can be coupled to a FT–Raman system, as shown in Figure 6.6.[26] The FT–Raman microscope has a lower throughput and provides lower signal-to-noise ratio than a macro sampling system. In large part this arises from the lower throughput of the microscope that is limited by the relatively small numerical aperture (0.65) of its Schwarzchild objective. Another loss source is the 50:50 beam splitter, which is used in the epi-illumination system of the microscope. Finally, the transfer optics introduce several reflective surfaces into the optical train. If other components of the system remain constant, the Messerschmidt–Chase microscope is less efficient than a macro-sampling system by about 4.4× for samples whose Raman scatter fills the detector under macro conditions.

Despite these problems, the microscope generates quite acceptable micro-Raman spectra, as Figure 6.7 shows. The Kevlar sample used in this experiment is a difficult one, because even in the near infrared it burns if the laser power is increased much above 125 mW. With the same optics, the FT–Raman microscope should provide somewhat better spatial reso-

Figure 6.6 An FT–Raman microscope. The instrument employs an unmodified Spectra-Tech IR-plan microscope, with all-reflective optics. (Reprinted from Ref. 26 with permission.)

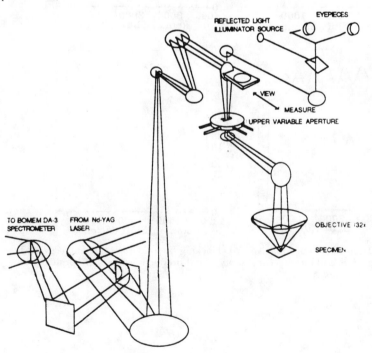

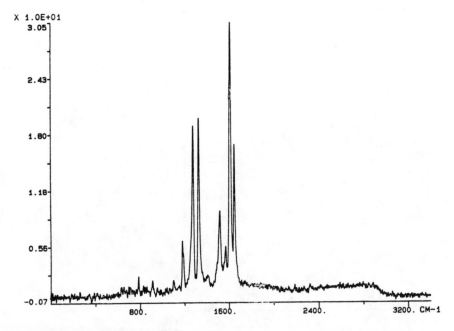

Figure 6.7 The micro-FT–Raman spectrum of a single strand of Kevlar (TM) fiber, 12-mm diameter, 15 min accumulation. The incident laser power is approximately 125 mW. (Reprinted from Ref. 26 with permission.)

lution in the C—H stretching region, and dramatically better resolution in, for example, the aromatic ring mode region.

Microscopes with refractive optics function well in FT–Raman systems.[27,28] The mismatch between microscope numerical aperture and interferometer numerical aperture remains an important source of reduced efficiency. A wider range of magnifications and numerical apertures is available in refractive optics objectives than in reflecting types, and visible light image quality is better. In general, refractive optics are less expensive as well. Commercial designs are intended for use at visible and, in some cases, near UV wavelengths. They are not necessarily achromatic or free of aberrations throughout the 1–1.8-μm range. However, for generation and collection of a Raman spectrum from a small region of a sample, small deviations from optimum imaging are unimportant, as Figure 6.8 shows.

6.3.4. Other Approaches to Red–NIR Raman Spectroscopy

There are alternatives to FT–Raman spectroscopy. Hadamard multiplexing is equally feasible.[29] A prototype Hadamard system employing an electro-optic mask system and a cooled InGaAs detector has been demonstrated. The detector is quite quiet (NEP = 2×10^{-15}

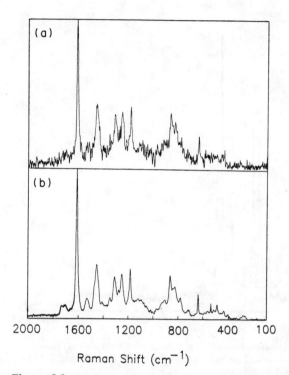

Raman Shift (cm^{-1})

Figure 6.8 FT–Raman spectrum of a polyurethane elastomer. Upper trace, microscope spectrum, incident laser power 550 mW focused to approximately 50 mm, collection time 7.5 mins. Lower trace, the same sample with macro-optics, 37.5 min collection time. (Reprinted from Ref. 28 with permission.)

W Hz$^{-1/2}$), and the expected multiplex advantage would be quite small. The system employs a stationary 127-element electro-optic Hadamard mask system at the exit plane of a grating spectrograph. This forces a trade-off between spectral range and resolution. However, the system is compact, inherently rugged, and potentially inexpensive. It may be ideally suited to process control and other applications where repetitive monitoring of a limited spectral region is desired.

Because the multiplexer is a stationary shutter array, spectral registration is perfect (Connes advantage) and spectral subtraction is easily accomplished. The system uses a single entrance slit, so that the Jacquinot advantage is absent. Since it is difficult to fill even a small detector with Raman-scattered light, this objection is more academic than practical.[30] Indeed, scanning double monochromators fitted with quiet near-infrared detectors generate quite acceptable Nd–YAG excited Raman spectra. Good results have been obtained using 5–30-mW average power from a Q-switched Nd–YAG laser[31] and with 40–600-mW power from a CW Nd–YAG.[32] A multiplexed system is faster, but a scanned system can produce a good spectrum in 15–20 minutes.

Spectrograph–CCD detection systems are feasible with excitation wavelengths in the 800-nm region. Diode lasers, which are available at nominal wavelengths of 783 and 830 nm, are especially attractive excitation sources.[33–35] Figure 6.9 shows the response curve for a commercial CCD and the Raman shift range relative to 784 nm. This excitation wavelength is about the longest for which CCD quantum efficiency is 5% or greater over the complete Stokes shift region, including the C—H stretching region. Fluorescence is present but is often not particularly severe at 780 nm. In Figure 6.10, the spectrum of Nylon, a traditionally difficult or impossible sample with 514.5-nm excitation, shows only moderate fluorescence, with 782 nm. Rhodamine 6G, another favorite test case, is also only weakly fluorescent under 780- or 830-nm excitation. In both cases, software baseline correction can easily remove the fluorescence background.

Commercial Ga–Al–As diode lasers, which operate in the 780–830-nm region, have bandwidths of less than 1 cm^{-1}. The approximate wavelength is determined by the diode composition. The exact wavelength depends upon the laser dimensions—in other words, on the device operating temperature. The greatest long-term wavelength stability is obtained if the laser is actively temperature-stabilized with a feedback-controlled Peltier (thermoelectric) cooler. However, an inexpensive diode laser operating at ambient temperature is adequately stable if it is allowed to come to thermal equilibrium prior to use.[34] It is only necessary to operate the laser at constant power (i.e., approximately constant current) for several minutes prior to using it.

All spectrograph measurements have the problem that the frequency resolution (pixels per wave number) systematically increases with increas-

Figure 6.9 Quantum efficiency curve for a typical CCD (Thompson TH7882). Common laser wavelengths and the Raman shift range for a 784 nm diode laser are shown. (Reprinted from Ref. 33 with permission.)

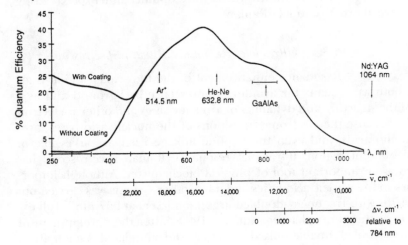

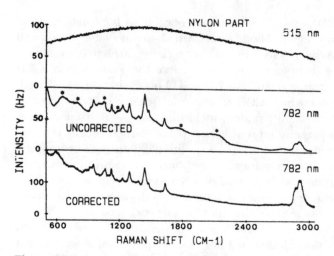

Figure 6.10 Raman spectra of nylon, obtained with 515- and 782-nm excitation. The 515-nm spectrum is recorded with a scanning spectrometer. The 782-nm spectra are composites of six CCD spectra, each obtained with a 0.6-m single-stage spectrograph, 300-g mm^{-1} grating, 75-mm slit, long-pass thin-film prefilter and 20-mW laser power at the sample. The uncorrected (center) spectrum shows spurious bands that result from local oscillations in the passband transmission of the interference filter. The lower spectrum is obtained after ratioing to the system response function. (Reprinted from Ref. 35 with permission.)

ing Stokes shift, because the spectrograph output is linear in wavelength. The variation is small enough to be ignored with green excitation but is quite pronounced with near-infrared excitation. Although multichannel Raman spectroscopy requires lower laser power and can provide higher signal-to-noise ratios than multiplex spectroscopy, it does have complications. Array detectors operating in the 1–2-μm region have recently become available. These devices are still expensive and have low quantum efficiencies. In the near future they may make multichannel spectroscopy an attractive alternative to FT–Raman.

6.3.5. Ultraviolet Resonance Raman Spectroscopy

Resonance enhancement in the ultraviolet region is attractive both as a means for avoiding fluorescence background and as a probe of the structure and dynamics of aromatic rings and other π-electron systems.[4] Because the electronic transitions of the nucleic acid bases, aromatic amino acids, and even the peptide linkage itself are accessible in the 200–300-nm region, ultraviolet resonance Raman spectroscopy has emerged as an important tool of physical biochemistry. Although important work in the nucleic acid bases and a few other systems was carried out in the 1970s with frequency-doubled argon ion lasers at 257 nm,[36] full exploitation of the ultraviolet began in the 1980s, after the development of frequency-doubled tunable pulsed dye lasers and stimulated Raman effect laser frequency shifters.

Conventional dispersive spectrographs and scanning spectrometers are used in the ultraviolet. It is necessary to use finer gratings or longer focal lengths than are typical in the visible, because in the ultraviolet the number of wavenumbers per unit wavelength, $\Delta\tilde{v}/\Delta\lambda$, is 100–200 cm^{-1} nm^{-1}. With the usual 0.5–0.6-m focal length spectrograph, a 2,400 grooves mm^{-1} grating is satisfactory. Reflective (aluminum) foreoptics can be used for completely achromatic focusing. However, quartz or MgF$_2$ lenses also work well.

Dye lasers pumped by the second or third harmonic of a Q-switched Nd–YAG laser are the most common source lasers. Tunable ultraviolet radiation is obtained by frequency-doubling the dye laser output or by summing the dye laser and pump laser. Nd–YAG-based systems have low (10–30 pulses s^1) repetition rates and high peak powers. Even with frequency conversion, the peak powers are high enough to cause saturation or even photodecomposition. These effects may be nuisances, or they may be used to study excited-state or transient species.[37,38] To some extent, saturation and photochemical effects can be reduced by use of high-repetition-rate (100–500 pulses s^{-1}) excimer laser pumped dye lasers, which provide high average power with low peak power.

Intensified diode array detectors and, in recent work, CCD detectors have been preferred for ultraviolet Raman. Single-channel scanned systems require the use of gated integrators to process the burst of photons arising from the 10-ns laser pulse, and spectrum acquisition can be painfully slow.

Sample or matrix fluorescence is not a major problem in the ultraviolet. If excitation occurs into S_2 or a higher excited state of a molecule, any fluorescence appears harmlessly far to the red of the Raman spectrum. Small molecules whose first excited state lies above 35,000–40,000 cm^{-1} are usually not efficient emitters.

The use of high-peak-power pulsed lasers requires special care with samples. Loose focusing is used when possible to prevent decomposition in the intense laser beam. Generally, solutions must also be kept stirred or flowing. Pumping a solution sample through a thin quartz capillary usually works well. Delicate samples can sometimes decompose and deposit on the capillary walls. This problem can be avoided by pumping the sample through a dye laser jet nozzle to generate an open, stable stream of solution. Even so, the quartz tube technique has been used successfully with live bacteria.[39]

6.3.6. Optical Fiber Probes

Multimode optical fibers make efficient and versatile Raman probes. They are small, are usable in remote sampling or hostile environments, and provide stable, reproducible signals from solids, liquids, and gases. A common contemporary design employs fiber bundles, in which a central fiber delivers the laser beam and a surrounding array of fibers collects the Raman scatter.[40,41] The fibers may be parallel or the collection fibers may be angled toward the delivery fiber. Alternatives include a delivery fiber and single collecting fiber oriented at an angle to each other,[42] opposing single fibers,[42,43] and a single fiber operating as both beam delivery and signal collection device.[44]

A bundle of parallel fibers can collect several percent of the Raman-scattered radiation generated in an isotropic medium.[40] The collection efficiency is about an order of magnitude greater than is available from a pair of closely spaced parallel fibers. It is important to keep the fibers closely spaced, or even touching. Hard-clad (fluorocarbon) fibers are commonly used, as are fibers from which the cladding is removed entirely. Collection efficiency can be improved further by orienting the collection fibers at an 8°–10° angle toward the delivery fiber. Figure 6.11a is a diagram of a

Figure 6.11 (a) Multiple optical fiber probe. The central fiber delivers the beam to the sample and the surrounding fibers collect scattered light. The collection fibers are oriented at an angle α to the delivery fiber. (b) Raman-scattering intensity as a function of fiber angle and distance to polymeric solid sample. (Reprinted from M. A. Leugers and R. D. McLachlan, Proc. SPIE **990**, 88 (1988), with permission of the Society of Photo-Optical Instrumentation Engineers.)

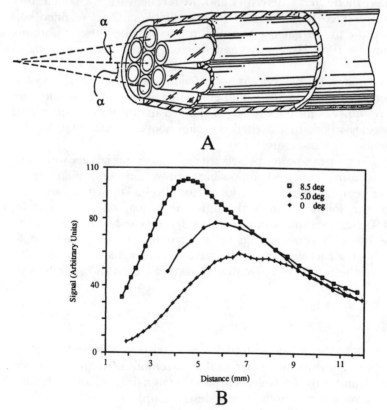

fiber bundle probe with angled fibers. As Figure 6.11b shows, the effect of fiber angling can be as much as a factor of 2 in collection efficiency.

Angled dual fiber configurations can also be efficient, especially if one or both of the fibers is terminated in a ball or graded index refractive index (GRIN) lens.[42] However, these still remain inefficient, compared to single fibers, which automatically provide perfect overlap between the delivery cone and the backscattered light cone. Single fibers, however, require a dichroic mirror to separate the exciting laser beam entering the fiber from the Raman scatter exiting from it. The dichroic mirror limits the low-frequency performance of the device.

The optical fiber itself and, if any is used, the epoxy or other cement that holds the fiber bundle together are sources of Raman scatter and fluorescence. The bands of silica are often observed from the core. Both the core and cladding may show weak bands from various impurities, or for organic claddings, from the polymer itself.[25,42] Fluorescence generally arises from organic claddings, the protective jacket, or cements in the fiber assembly. Fibers from different manufacturers are not necessarily equally suited for probe use, although they might have similar transmissions, numerical apertures, and core and cladding compositions. Most fibers have been developed either for communications applications or for sensing applications with intense fluorescence or transmitted light. In neither case is weak Raman scatter or fluorescence in the material a problem. Because fiber background cannot be eliminated completely, it is advisable to investigate fibers of similar optical properties from different sources and choose one that minimizes spectral interferences.

Alternatively, it is possible to mount interference filters on the fiber ends to attenuate spurious Raman and fluorescence signals generated in the delivery fiber and Rayleigh scatter generated in the sample.[45] The filters themselves must be chosen for low fluorescence. Filters are easily added to angled dual fiber probes or dual opposing fiber probes. An inevitable disadvantage is the added bulk of the filters themselves, and the added mechanical complexity of the probe structure.

For remote sensing applications it is necessary to choose an exciting laser wavelength that maximizes the Raman signal. The choice depends on the interplay between attenuation in the fiber, which is not monotonic with wavelength, and the inverse fourth power of wavelength dependence of Raman-scattering intensity.[46] For short fibers (<0.1 km) the dominant effect is the efficiency of Raman scattering, and short-wavelength excitation gives stronger signals. At longer lengths (>1 km) the fiber attenuation is important, and the wavelength for maximum signal shifts to the transmission maximum of the fiber, which is usually in the red end of the visible spectrum.

The major attraction of optical fiber probes is their ease of use. They can be placed over solid samples or inserted into liquid samples. If the sample container is a long capillary, internal reflection can be used[47] to

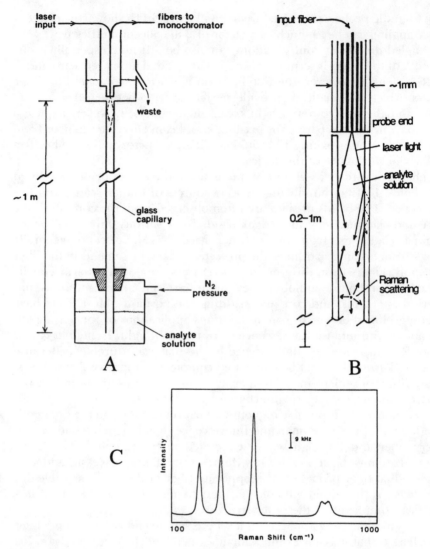

Figure 6.12 Capillary sample cell, using internal reflection to obtain intense Raman signals. (a) Overall view. (b) Detail of cell. (c) Raman spectrum of neat CCl4 in long pass cell with 130 mW of 441.6-nm light entering the sample. Bandpass = 9 cm^{-1}. (Reprinted from Ref. 47 with permission.)

generate strong Raman signals from low-power lasers, as shown in Figure 6-12. Unfortunately, optical fiber probes are also readily susceptible to contamination. In time, microscopic cracks develop between the cement and the fibers at the distal end of the probe. Contaminants can accumulate in these cracks and generate Raman scatter or luminescence. Although the problem can be avoided by using coupling lenses at the fiber ends, a simpler solution is to wrap the probe in transparent plastic food wrap.[48] Commercial food wraps are made from polyethylene or polyvinylidene and are transparent and resistant to most laboratory solvents, as well as extremes of pH. The films are usually about 10–15 μm thick and are invisible with most probes, because the collection fiber field of view does not overlap with the delivery fiber laser radiation field close to the probe surface.

6.3.7. Raman Microscopy

Since development of the first practical systems in the early 1970s, the Raman microprobe has emerged as a powerful sampling system, with applications to many different fields. Virtually all designs use the epi-illumination system developed by Delhaye and Dhamelincourt.[49]

The bulk of the applications have been to Raman microspectroscopy (i.e., to the use of the device for obtaining a spectrum from a small, well-defined region of a material). Raman imaging, or mapping, after a brief flurry of early demonstrations to strong Raman scatterers, was dormant for some years but has recently been enjoying a revival. An excellent discussion of the optics of Raman microscopy, as well as a review of early applications, can be found in Ref. 50.

A schematic diagram of a typical Raman microprobe is shown as Figure 6.13. The microscope is usually a modified research-grade fluorescence or metallographic microscope. These instruments have provisions for illuminating a sample through the microscope objective. This mode of illumination is called epi-illumination or vertical illumination in the microscopy literature. Typically, there is also provision for illuminating the sample with incoherent light, parallel to the incident laser beam, for visual inspection. Almost all commercial instruments (and many home-made ones) include a video camera and monitor for observation of the sample and positioning of the laser beam. In addition to the very real quantitative and convenience advantages of video microscopy,[51] substitution of a camera for the traditional binocular eyepieces is an important safety feature.

Well-corrected apochromatic objectives are usually used on microprobes. Depending on the objective, the magnification will be 10–100× and the numerical aperture (N.A.) will be 0.5–0.9. The diameter, d, of the focused laser spot is approximated by

$$d = 1.22\frac{\lambda}{\text{N.A.}} \tag{6.4}$$

For green radiation ($\lambda = 0.5$ μm), the focused beam diameter can approach 1 μm.

Although the microprobe has excellent spatial sensitivity, it also generates high local power densities. If the incident laser beam of irradiance I_0 (W m^{-2}) is absorbed by a particle of radius r_p (m) in contact with a substrate

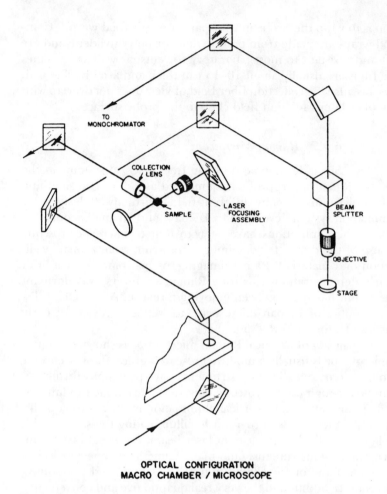

**OPTICAL CONFIGURATION
MACRO CHAMBER / MICROSCOPE**

Figure 6.13 Optical schematic for a Raman microprobe. The beam splitter, objective, and stage are contained in the body of the microscope. (Reprinted from Ref. 52 with permission.)

or host of thermal conductivity K (Wcm^{-1} K) with efficiency Q_a, then the steady-state temperature rise, ΔT_s, is given by[50]

$$\Delta T_s = \frac{Q_a I_o r_p}{4K} \tag{6.5}$$

For a typical transparent material, Q_a is approximately 1×10^{-4}. If the sample is in contact with air ($K = 2.6 \times 10^{-2}$ W m^{-1} K) the temperature rise can be quite high (>200°C). However, for a particle in contact with a glass substrate, or other reasonably good thermal conductors ($K = 10$ W m^{-1} K), the temperature rise will be only 10°–20°C, if the laser power is held to a few milliwatts.

Resonance Raman microspectroscopy is possible but usually requires some form of external cooling. Cold stages, which are common microscope accessories, are a suitable approach. Alternatively, one can increase the thermal conductivity of a water-compatible sample by immersing it in water. This may be done by using a water-immersion objective, which will also provide a higher numerical aperture than a conventional objective. In some cases resonance enhancement is great enough to allow reduction of sample power to microwatts, thus avoiding the heating problem.

Fluorescence interference is less troublesome with the Raman microprobe than with macro sample illumination systems.[52] Three factors contribute to this effect. First, with visible excitation, most fluorescence is caused by impurities, not by the sought component(s) of the sample. Impurities are typically heterogeneously distributed, so that a weakly fluorescent or nonfluorescent region can often be found. Second, the high laser flux from a tightly focused laser beam will burn out fluorescence more quickly than in a macro system. Third, because fluorescent lifetimes are nanoseconds long, excited-state migration often causes emission to occur several micrometers away from the illumination point. Spatial filtering at any of several intermediate image planes in the optical system can block much of this radiation.

The possibility of thermal damage necessitates the use of relatively low laser power densities. In turn, the intensity of the Raman scatter is much lower than is normally encountered with macro sampling systems. Moreover, it can be shown that the total signal is proportional to $(N.A.)^2/M^2$, where N.A. is the numerical aperture and M is the magnification of the microscope objective. For a given objective optical design numerical aperture increases, but less than proportionately, with magnification. Consequently, high magnification, and with it high resolution, will be obtained at the cost of a weaker signal.

A 50/50 beam splitter is commonly used in the Raman microprobe. The device is reasonably achromatic. But it is inefficient because half of the incident radiation and half of the collected Raman-scattered light are lost. Some workers prefer to use a 90% T, 10% R beam splitter. The system laser usually provides much more power than can be used safely, so attenuation of the incoming radiation by the beam splitter is less important than efficient transmission of the scattered light. A dichroic mirror can be used in place of the beam splitter, allowing 90% reflection of the incident laser and 90% transmission of the Raman scatter. However, a separate dichroic mirror is needed for each laser line, and the transmission efficiency falls off rapidly below about 800 cm^{-1}.

Polarization ratio measurements under a microscope require special precautions.[53,54] The illumination and gathering angles through microscope objectives are both large. For example, a numerical aperture of 0.85 is associated with a subtended angle of 39.1°. The standard formulas, which are based on the assumptions of illumination with collimated light and collection through a small angle, do not apply. Neglecting the polarizing effects of the microscope optics themselves, for isotropic samples, the polarization ratio, ρ, is given by

$$\rho = \frac{(A + B)\beta^2}{15A\alpha^2 + \left(\frac{4}{3}A + B\right)\beta^2} \tag{6.6}$$

where α and β are the mean polarizability and anisotropy of the polarizability tensor, as conventionally defined.[1] A and B are defined by equations (6.7) and (6.8), where Θ_m is the gathering half-angle of the objective:

$$A = \pi^2\left(\frac{4}{3} - \cos \Theta_m - \frac{1}{3}\cos^3 \Theta_m\right) \tag{6.7}$$

$$B = \left(\frac{2}{3} - \cos \Theta_m + \frac{1}{3}\cos^3 \Theta_m\right) \tag{6.8}$$

For $B << A$, equation (6.6) becomes the familiar depolarization equation

$$\rho = \frac{3\beta^2}{45\alpha^2 + 4\beta^2} \tag{6.9}$$

The apparent depolarization ratio is a function, then, of the illumination angle. For totally polarized bands, equation (6.6) may not be adequate, because it neglects the depolarizing effects of the microscope objective itself. Similar effects are observed in single-crystal polarization measurements.

Although Raman scattering is intrinsically quite weak, imaging Raman microscopy is possible. Most of the practical techniques were understood (and many were demonstrated) early in the development of Raman microscopy.[49] Most of these have proven practical only for very strong scatterers and are not widely used. We review them briefly here.

The sample can be illuminated globally, and the spectrograph operated as a filter.[49] In addition to requiring a highly stigmatic spectrograph, this approach requires wide entrance slits, causing loss of resolution and susceptibility to stray light. Consequently, it is satisfactory only for use with isolated lines from strong scatterers having little fluorescent background. In principle, the spectrograph can be replaced by any narrowband filter, such as an interference filter, holographic filter, or interferometer, in exact analogy to fluorescence microscopy. If the filter is not tunable, then a complete spectrum could be obtained with a tunable laser and fixed filters. This approach was attempted by Delhaye and Dhamelincourt but failed for lack of adequate filters.[49] Even now, existing filter systems cannot adequately reject laser light to make this approach practical for most real samples.

Recently, global illumination has been combined with one-dimensional spatial Hadamard multiplexing for Raman imaging.[55,56] The image is multiplexed in the spatial direction parallel to grating dispersion and is imaged directly parallel to the spectrograph entrance slit. An image is recovered by computer processing after a series of exposures with different multiplexing masks. With a two-dimensional detector, multispectral imaging is possible. This approach allows high incident powers and correspondingly strong Raman signals, so that image acquisition times are short.

The laser can be focused into a line with cylindrical lenses. Raman scatter from the line is imaged on an array detector, oriented parallel to the spectrograph entrance slit. The image is built up by moving the sample under the line or by scanning the line across the sample. With a linear detector, a Raman image at a single band is obtained. With a two-dimensional (video) detector, multiband (multispectral) imaging is possible. Versions of this technique are in current use.[57,58]

The focused laser beam can be raster scanned across the sample, or the sample can be moved systematically under the laser beam. With an array detector multispectral imaging is possible. This technique is slow and is most useful for coarse mapping. Raster scanning of the sample stage is readily automated, however.[59]

6.3.8. Special Techniques for Surface-Enhanced Raman Scattering

There are several convenient ways to produce a structured metal surface. First, the metal can be used as an electrode. A silver surface can be roughened by oxidizing and re-reducing it in the presence of chloride. About 250–500 C m^{-2} is optimum for production of surface features with plasmon resonances near 500 nm. Gold and copper electrodes can be prepared similarly, although several oxidation–reduction cycles are required for good results. Colloidal metals can also be used. Colloids are prepared by rapid chemical reduction of metal salts from solution. The preparations generally give heterogeneous distributions, and some improvement in sensitivity can be obtained by size separation and selection of the fraction with optimum particle size.[60] Metals can also be vapor deposited onto appropriately sized substrates, including finely divided alumina, polystyrene beads, and even filter paper.[10]

These techniques all have merits. Electrodes offer control over double layer charge. To some extent electrode potential can be used to control adsorption or desorption of molecules. However, the need for two or three electrodes increases sample and container size. Further, electrode surfaces are notoriously heterogeneous and difficult to prepare reproducibly. Colloids can be prepared in advance and stored for some length of time. They can be mixed directly with samples or sprayed onto substrates deposited on inert surfaces, such as thin-layer chromatography plates. Adsorption on colloids is slow and the active forms are more aggregated than the freshly prepared colloid. Thus, SERS intensity builds slowly, but the process can be hastened by heating the colloid-sample mixture.[61] Commercial preparations of polystyrene beads and alumina can have small size dispersions. Vapor deposition onto such substrates can provide highly reproducible surfaces, which are defined by the substrates themselves. The surface preparation is more tedious than either solution chemistry or electrochemical roughening.

Although it is relatively easy to obtain SERS, quantitative measurements are difficult, in part because of the difficulty in preparing substrates reproducibly. In addition, because SERS depends on adsorption of molecules, the signal magnitude in multicomponent systems depends on competitive absorption. The most strongly adsorbed material will give the strongest signal, sometimes to the exclusion of other materials, which may even be present in higher concentrations.[62] There are no general solutions to this problem yet. Overcoating the SERS substrate with a thin, permeable permselective or ion-exchange coating can control competitive adsorption in some cases. This approach, which has been richly developed by electrochemists, has been shown to eliminate protein adsorption and allow the observation of neurotransmitter SERS from protein-rich matrices.[63]

Despite these problems, SERS provides the Raman spectroscopist with an easy route to highly sensitive spectroscopy. SERS should be considered, albeit cautiously, in instances when the normal Raman effect gives signals too weak to be usable.

6.4. Applications

Raman spectroscopy can be used in almost every area of chemistry and has many uses in materials science, atmospheric science, biology, and other fields as well. We can describe only a few selected applications here, emphasizing those that depend upon the recent advances we have described in Sections 6.2 and 6.3.

6.4.1. Applications of Fourier-Transform Raman Spectroscopy

Near-infrared excitation and FT measurement have generated new enthusiasm for Raman spectroscopy as a generally applicable spectroscopic probe. New examples appear regularly in the analytical literature, notably in *Applied Spectroscopy* and the *Journal of Raman Spectroscopy*. An entire issue of *Spectrochimica Acta* [46A (2), 1990] is devoted to FT Raman. We summarize only a few studies from diverse areas of chemistry, which demonstrate the ease of obtaining spectra of formerly difficult or intractable samples.

Levin and co-workers have demonstrated fluorescence-free Raman spectra of dispersions of dipalmitoylphosphatidylcholine (DPPC) bilayers in the presence of imbedded amphotericin B.[64] Amphotericin is a highly fluorescent polyene antifungal antibiotic. The dispersions contain relatively small amounts of DPPC/amphotericin, and signal averaging is required to obtain good signal-to-noise ratios. Care must be taken to cool the sample because the 2-W Nd–YAG power can heat the DPPC above the gel–liquid crystal transition temperature. The antibiotic is a putative membrane channel former and its presence perturbs the C—H stretching region of DPPC, in the 2,850–2,950 cm^{-1} region.

Bergin and Shurvell have shown that spectra of difficult industrial samples, including lubricating oils, polyurethane, and polyimide, are easily measured.[27] In the lubricant field, FT–Raman shows promise for identification of additives and their decomposition products. The intensities of C—S, P—S, and P=S stretches are strong in the Raman. These moieties are commonly found in additives, and characterization should prove feasible. As Bergin and Shurvell point out, fluorescence is not completely absent in industrial samples with 1.06-μm excitation. Although very few organic compounds emit above 1 μm, industrial products can contain traces of the transition metal compounds used as catalysts. The metal ions do fluoresce in the near infrared, and background subtraction, or even the admission of an occasional failure, may be necessary.

Observation of molecules adsorbed on zeolite catalysts is feasible.[65,66] Although FT–Raman is not particularly sensitive, the large surface area of zeolites allows observation of adsorbed molecules. Chemisorbed hydrogen-bonded pyridine and pyridinium ions can be identified in the ring breathing region (1,000–1,010 cm^1). Zeolites are favorable materials for such studies. They are nominally nonabsorbing at 1 μm, so that irradiation with 350–700 mW causes only minor sample heating.

Even foodstuffs are easily handled.[67] Survey spectra of saccharides and polysaccharides, including sucrose, Ficoll (poly-sucrose), maltose, lactose, and corn starch, are excellent, as are the spectra of carbohydrate-rich semolina, rice, and tapioca. Vegetable oils, margarine, and butter also give excellent spectra. FT–Raman provides a simple assay for the presence of unsaturated fat in these materials. Near-infrared spectroscopy is commonly used in foodstuff assays. Such methods could easily be displaced by FT–Raman. Except for the laser, the instrumentation is similar, and the Raman spectrum provides more certain identification and quantification of nutrients or contaminants.

6.4.2. Applications of the Raman Microscope and Microprobe

6.4.2.1. Semiconductors Semiconductor structural features, at least for current generation devices, are large enough for easy observation by visible light microscopy, including Raman microscopy. The Raman microprobe functions as nondestructive probe for several important properties of devices based on both silicon and gallium arsenide.[68] Phonon frequencies are a function of strain for both silicon and gallium arsenide structures. The silicon mode (~520 cm^{-1}) decreases in frequency with increasing stress, so that its frequency reports crystal strain at boundaries. The frequency range is only 1–2 cm^{-1}, so fairly high-resolution measurements are necessary. It is possible to construct rough cross-sectional maps, for example, across a region of 10–20 μm, which reveal the extent of strain across a feature. In polar semiconductors, such as gallium arsenide, gallium–aluminum arsenide, and zinc selenide, the free carrier plasma is a space charge with an electric field that can interact with LO phonon modes. If the plasmon and phonon frequencies are similar and the plasmon damping is small, two well-defined bands are observed. Peak intensity monitors carrier concentration. Raman scattering is also widely used to monitor damage by ion implantation in silicon as the decrease in intensity of a silicon crystalline lattice vibration. Quantitative measurements are possible by normalization to scattering intensity from an undamaged region.

       6.4.2.2. Geology   The Raman microprobe is an important nondestructive probe of the composition of fluid inclusions in minerals.[69,70] The composition of an inclusion offers some clues to the mechanism of the formation of a rock and can indicate the presence of economically valuable oil, gas, or mineral deposits nearby. Except for SO_4^{-2} and HS^-, most polyatomic anions are present at concentrations too low for Raman measurement. The pH at which CO_3^{-2} is the dominant ion (pH > 10.5) in $H_2O–CO_2$ fluids is higher than is normally reached. The major species is HCO_3^-, which is a weak scatterer, and for which the bands overlap those of the common hosts, silica and calcite. It is sometimes possible to identify the presence of monoatomic ions, including Na^+, K^+, Ca^{+2}, Mg^{+2}, and Fe^{+2} from the bands of their water complexes, normally in the range 350–600 cm^{-1}.

    The major use of the microprobe is to measure the permanent gases.[70] These may be present at pressures of several atmospheres. CO_2, CO, CH_4, C_2H_6, H_2, H_2S, SO_2, N_2, and O_2 are easily identified by their strong Raman spectra. It is possible to measure the relative composition (mole fraction) of the gases, but not the absolute amounts. The geometry of an inclusion is unknown, and as a practical matter, unknowable. Moreover, the host lattice usually has a high refractive index (>1.5) and may be birefringent. The crystal is a system of prisms of unknown size, angles, and orientation. Consequently, the fraction of the incident laser light that reaches the sample is unknown, and the collection efficiency is also unknown. Calibration with external samples is impossible. Because the bands used range from CO_2 1,388 cm^{-1} to H_2 4,156 cm^{-1}, it is necessary to correct for changes in instrument efficiency. Much work is done with argon ion 514.5 nm light, and fluorescence can sometimes be troublesome.

    Measurement of composition can be made with high precision, often approaching ±1%. However, the accuracy is limited by the accuracy of Raman cross sections and by the accuracy of the correction factors used to extrapolate to high-pressure conditions. Pasteris and co-workers have shown that the calculated mole fractions can vary by a factor of 2, depending on the extrapolations and corrections used, despite the excellent precision of the method.[69]

6.4.2.3. Biology, Medicine, and Clinical Chemistry The Raman microprobe has several applications in disease diagnosis,[71] although it remains a research tool rather than a routine clinical instrument. The Raman spectrum can identify inclusions of foreign materials in tissues. Examples include mineral particles in the lung tissues of silicosis victims, and fragments of prosthetic silicone implants that have migrated from their implantation sites. The compositions of gallstone or kidney stones are easily deduced from their micro-Raman spectra. The stones are commonly formed from calcium phosphate, cholesterol, bilirubin, uric acid, fatty acid

Figure 6.14 Raman microprobe spectra of white zinc stearate particles contained in bilirubin and cholesterol–bilirubin gallstones. The Raman spectrum of zinc stearate and the probable precursor dipalmitoyl–phosphatidyl choline (DPPC) are shown. (Reprinted from Ref. 72 with permission.)

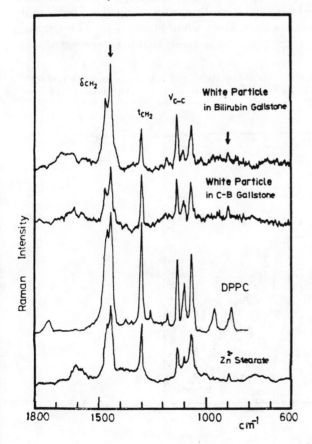

salts, or mixtures of these compounds. All have characteristic Raman spectra. Figure 6.14 shows the Raman microprobe spectrum of zinc stearate particles found in bilirubin and bilirubin–cholesterol gallstones. The stearate is probably a decomposition product of a phospholipid, such as dipalmityol–phosphatidylcholine. Infrared microscopy can provide similar information, albeit with greatly inferior spatial resolution. Both techniques are used in gallstone and kidney stone research.[72]

The Nelson group demonstrated that resonance Raman microprobe spectra could be obtained from the carotenoid pigments of individual bacteria.[73] They investigated freely moving bacteria, which remained in the field of view of the instrument long enough for spectral acquisition, even with a scanning spectrometer. Although this work demonstrated the feasibility of obtaining micro-Raman spectra from intact, living microorganisms, it was limited to carefully chosen systems containing highly colored pigments.

Using a microscope with extended-ultraviolet fluorite optics, Sureau and co-workers[74] have recently been able to obtain spectra from the DNA of a single living tumor cell. With a single grating spectrograph and multichannel detection, spectra can be obtained in 100 s, using 10 μW 257 nm from a frequency-doubled argon ion laser, as shown in Figure 6.15. The relative ease with which spectra can be obtained suggests that it is possible to use

Figure 6.15 Resonance Raman microprobe spectrum of DNA, excited with 10 mW of 257-nm light. (a) Nuclear DNA of T47D human mammary tumor cell, 100-s acquisition. (b) Solution sample (20 ml) DNA, 1 mg/ml, 10-sec acquisition. (Reprinted from Ref. 74 with permission.)

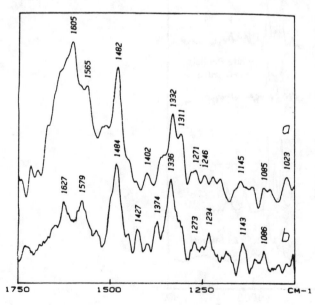

UV Raman spectroscopy as a probe of some dynamic events, such as metabolism, or of the effects of introduced materials, at the single-cell level.

6.4.2.4. Raman Microimaging

The pioneering imaging work of Delhaye and Dhamelincourt was confined to strongly scattering samples, typically inorganic crystals.[49] Technical advances have made application to interesting and important weak scatterers feasible. Using line imaging and mechanical scanning, Dauskardt and co-workers have studied the monoclinic–tetragonal phase transformations around a crack in partially stabilized zirconia.[75] Using the integrated intensity under the 181/192 cm^{-1} band pair as a measure of monoclinic zirconia and the intensity under the 264 cm^{-1} band as a measure of tetragonal zirconia, they were able to map fractional content of the monoclinic phase with a spatial resolution of about 27 μm (Figure 6.16). This resolution is obtainable with a good-quality 50-mm *f*/1 photographic lens, the light-gathering optic used in these experiments. The sample is a difficult one, because it is translucent and scatters light, thus reducing contrast and resolution. Nonetheless, the apparent half-width of the crack agrees well with that obtained by Nomarski interference microscopy.

Treado and coworkers[56] have employed Hadamard Raman microscopy to image edge plane microstructures in laser-modified highly ordered pyr-

Figure 6.16 Three-dimensional Raman plot of the morphology of the transformed zone ahead and in the wake of the crack tip in partially stabilized zirconia. (Reprinted from Ref. 75 with permission.)

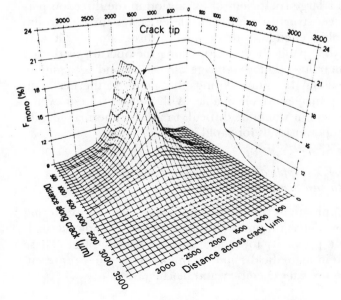

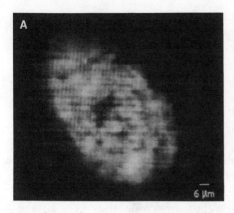
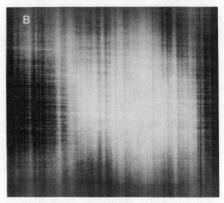

Figure 6.17 Hadamard transform Raman microscope image of edge plane defects in highly ordered pyrolytic graphite. 20×, 0.7 NA objective, 0.6 mm/pixel. (A) 1,360-cm^{-1} Raman image of edge plane microstructure, 10-min acquisition following 15-min thermal equilibration. (B) Out of focus 1,360-cm^{-1} Raman image, obtained with no prior thermal equilibration. (Reprinted from Ref. 56 with permission.)

olytic graphite (HOPG) electrodes. They have shown that the defects are uniformly distributed. The defects in a portion of a typical modified electrode are shown as bright regions in the Raman image of Figure 6.17. The image also shows dark regions, which are exposed unmodified basal plane graphite. The modified carbon is extremely fragile, and even careful handling damages it.

Figure 6.17 also illustrates one advantage of global illumination for imaging. Graphite has a large coefficient of expansion in the direction perpendicular to the sheet structures. Sufficient laser power for Raman imaging heats the graphite enough to defocus it. In this example, 250 mW causes the graphite to expand about 40 μm along the optical axis during image acquisition, blurring the lower image almost beyond recognition. Preequilibration for about 15 minutes brings the sample to thermal equilibrium, so that sharp images are obtainable. With a raster scan or even a line scan imaging system, it would be difficult to achieve thermal equilibrium at high enough power levels for rapid imaging.

6.4.3. Applications of Surface-Enhanced Raman Scattering

Applications of surface-enhanced Raman scattering (SERS) exploit its low detection limits and its sensitivity to molecules adsorbed on or in close proximity to a metal surface, usually silver. These properties are useful in many fields, and we present only a few representative examples from polymer chemistry and biochemistry.

6.4.3.1. Polymers SERS has been used to probe formation and structure of polymers. Both films and gel networks are amenable to SERS study. For example, cross-linked polyacrylamide polymers are widely used as electrophoresis gels for separation of proteins and nucleic acids. Although these gels are commonly prepared in advance and used afterward, their structure continues to change over a period of days or weeks.[76] With a colloidal silver probe weak SERS is initially visible, but this gradually disappears as the gel rearranges. This observation suggests that the silver colloids formed in the biological staining procedures are ultimately stabilized by steric factors. Although a pyridine probe shows that the SERS activity and thus the degree of aggregation of the colloids does not change over time, the colloids are not adsorbed on aged polyacrylamide.

Because SERS response is strongest from the layer in immediate contact with a metal, it has long been known SERS is useful for the study of thin polymeric films on metals. These studies are complicated both by laser-induced air oxidation of the polymer and by the effects of partial coverage.[77] If monolayers are formed, they may be susceptible to oxidation. Oxidation can be inhibited by overcoating with a second, thicker layer of polymer. If the polymer concentration in the spin-coating solution is less than about 0.1%, then the surface is incompletely coated. Overcoating with a second polymer completes the coating process, and a mixed spectrum is observed. However, if concentrated polymer solutions are used for spin-coating, then complete coverage is obtained, and only the spectrum of the layer in contact with the metal is observed. Because SERS easily detects less than 0.01 monolayers, it is a sensitive and simple diagnostic for flaws in monolayer films, as well as for damage to films. Perhaps the most promising application is study of adhesive systems. Boerio and co-workers have used SERS to study the bonding of acrylic adhesives to silver.[78] They find that saccharin, a component of the curing system, segregates at the surface and forms a silver salt. They suggest that this salt catalyzes the free-radical redox polymerization reaction, which accelerates curing of the polymer on metal.

6.4.3.2. Biological Molecules Much of the SERS of biological systems has been exploratory studies, which define the kinds of systems for which the technique is useful.[8,9] As that period closes, important substantive applications are appearing. In an elegant demonstration, Rohr and co-workers have used SERS detection to decrease the number of steps required in immunoassays.[79] The approach is to incorporate SERS-active functionalities into proteins or other components of the assay system. These are then allowed to bind to antibodies immobilized onto silver films. SERS, or even resonance-enhanced SERS, is then observed from the bound material. The results correlate well with detection techniques that use counting of tritium-labeled materials. However, the SERS readout requires no washing steps to remove excess reagents and, in addition, eliminates the use of radioactive materials. Although a fluorescence readout would also work, careful washing is required to eliminate unbound fluorophore. Because SERS signals are localized to surfaces, the need for washing is obviated.

Proteins adsorbed on silver and exposed to laser light may be easily denatured.[80] By working at 77 K, Rospendowski and co-workers have been able to avoid this problem with the light harvesting plant protein phytochrome.[81] At cryogenic temperatures the resonance-enhanced SERS of both the Z and E photoisomers of the biliverdin chromophore can be observed. Although the overall chromophore conformations are not very different, the Raman intensities of C—H bending modes are substantially different and change dramatically from 407 nm to 413 nm. These observations imply the presence of two nearly degenerate electronic excited states. This degeneracy could be caused by the protein matrix, which holds the chromophores in rigid positions that are somewhat different in the Z and E forms.

Adsorption of bacteriorhodopsin from the purple membranes of *H. halobium* onto colloidal silver or roughened electrodes does disrupt the structure and stop the polyene photoisomerization.[82] However, if the membrane is adsorbed onto an electrode that has not been roughened, it continues to function. Enhancement of the unperturbed resonance Raman spectrum is sufficient that the spectra are about $50\times$ stronger than from a suspension of the membranes, allowing a corresponding decrease in the concentration used. The signals from the kinetic intermediate maximize at the potential of zero charge for silver, probably because adsorption with the chromophore close to the surface maximizes at that point.

Although the electrodes have not been deliberately roughened, they are not smooth on the atomic scale. Surface enhancement occurs, but it is small compared to the effects on roughened electrodes. The trade-off between structural stability and enhancement is interesting and might also be applicable in other systems.

6.5. Conclusions

In this review we have been able to explore only a few of the recent developments in Raman spectroscopy. The interested reader is invited to browse through the journals and review series cited here for pointers to other active areas. Nonlinear techniques are described in the following chapter, and ultrafast time-resolved measurements are discussed in Chapter 10. The field is dynamic, and interest is spreading from the traditional communities of physicists, physical chemists, and physical biochemists to many other groups of users. Instrumental advances have largely eliminated the experimental problems of the technique. Spectral acquisition time is now routinely a matter of seconds or at most a few minutes and fluorescence problems have been largely conquered. Raman spectroscopy now stands among the techniques that any modern laboratory must include in its essential collection.

REFERENCES

1. D. A. Long, *Raman Spectroscopy* (McGraw-Hill, New York, 1977).

2. H. W. Schrötter and H. W. Klöckner, in *Raman Spectroscopy of Gases and Liquids*, A. Weber, ed., (Springer, Berlin, 1979).

3. A. B. Meyers and R. A. Mathies, in *Biological Applications of Raman Spectroscopy*, Vol. 2, T. G. Spiro, ed., (Wiley, New York, 1987), pp. 1–88.

4. S. A. Asher, *Ann. Rev. Phys. Chem.* **39,** 537 (1988).

5. R. L. Garrell, *Anal. Chem.* **61,** 401A (1989).

6. J. A. Creighton, in *Advances in Spectroscopy*, Vol. 16, R. J. H. Clark and R. E. Hester, eds. (Wiley, Chichester 1988), p. 37.

7. M. Moskovits, *Rev. Mod. Phys.* **57,** 783 (1985).

8. T. M. Cotton, in R. J. H. Clark and R. E. Hester, eds., *Advances in Spectroscopy*, Vol. 16 (Wiley, Chichester, 1988), p. 37.

9. R. F. Paisley and M. D. Morris, *Prog. Anal. Spectrosc.* **11,** 111 (1988).

10. T. VoDinh, *Chemical Analysis of Polycyclic Aromatic Compounds* (Wiley, New York, 1989).

11. M. Kerker, *Acc. Chem. Res.* **17,** 271 (1984).

12. P. Hildebrandt and M. Stockburger, *J. Phys. Chem.* **88,** 5935 (1984).

13. P. L. Flaugh, S. E. O'Donnell, and S. A. Asher, *Appl. Spectrosc.* **38,** 847 (1984).

14. M. M. Carrabba, K. M. Spencer, C. Rich, and R. D. Rauh, *Appl. Spectrosc.* **44,** 1558 (1990).

15. C. A. Murray and S. B. Dierker, *J. Opt. Soc. Am.* **3A,** 2157 (1986).

16. R. B. Bilhorn, J. V. Sweedler, P. M. Epperson, and M. B. Denton, *Appl. Spectrosc.* **41,** 1114 (1987).

17. R. B. Bilhorn, P. M. Epperson, J. V. Sweedler, and M. B. Denton, *Appl. Spectrosc.* **41,** 1125 (1987).

18. A. Weber, D. E. Jennings, and J. W. Brault, *Appl. Opt.* **25,** 284 (1986).

19. T. Hirschfeld and B. Chase, *Appl. Spectrosc.* **40,** 133 (1986).

20. D. J. Cutler, *Spectrochim. Acta* **46A,** 131 (1990).

21. S. F. Parker, K. P. J. Williams, P. J. Hendra, and A. J. Turner, *Appl. Spectrosc.* **42,** 796 (1988).

22. N. J. Everall and J. Howard, *Appl. Spectrosc.* **43,** 778 (1989).

23. E. N. Lewis, V. F. Kalaskinsky, and I. W. Levin, *Anal. Chem.* **60,** 2658 (1988).

24. D. D. Archibald, L. T. Lin, and D. E. Honigs, *Appl. Spectrosc.* **42,** 1558 (1988).

25. K. P. J. williams, *J. Raman Spectrosc.* **21,** 147 (1990).

26. R. G. Messerschmidt and D. B. Chase, *Appl. Spectrosc.* **43,** 11 (1989).

27. F. J. Bergin and H. F. Shurvell, *Appl. Spectrosc.* **43,** 516 (1989).

28. F. J. Bergin, *Spectrochim. Acta* **46A,** 153 (199).

29. A. P. Bohlke, J. D. Tate, J. V. Puakstelis, R. M. Hammaker, and W. G. Fately, *J. Mol. Struct. Theochem.* **20,** 471 (1989).

30. D. N. Waters, *Appl. Spectrosc.* **41,** 708 (1987).

31. M. Fujiwara, H. Hamaguchi, and M. Tasumi, *Appl. Spectrosc.* **40,** 137 (1986).

32. D. R. Porterfield and A. Campion, *J. Am. Chem. Soc.* **110,** 408 (1988).

33. J. M. Williamson, R. J. Bowling, and R. L. McCreery, *Appl. Spectrosc.* **43,** 372 (1989).

34. S. M. Angel and M. L. Myrick, *Anal. Chem.* **61,** 1648 (1989).

35. Y. Wang and R. L. McCreery, *Anal. Chem.* **61,** 2647 (1989).

36. Y. Nishimura, A. Y. Hirakawa, and M. Tsuboi, in R. J. H. Clark and R. E. Hester, eds., *Advances in Infrared and Raman Spectroscopy,* Vol. 5 (Wiley, Chichester, 1978), p. 217.

37. K. Bajdor, Y. Nishimura, and W. L. Peticolas, *J. Am. Chem. Soc.* **109,** 3514 (1987).

38. J. Teraoka, P. A. Harmon, and S. A. Asher, *J. Am. Chem. Soc.* **112,** 2892 (1990).

39. M. Baek, W. H. Nelson, and P. E. Hargraves, *Appl. Spectrosc.* **43,** 159 (1989).

40. S. D. Schwab and R. L. McCreery, *Anal. Chem.* **56,** 2199 (1984).

41. M. A. Leugers and R. D. McLachlan, *Proc. SPIE* **990,** 88 (1988).

42. M. L. Myrick, S. M. Angel, and R. Desiderio, *Appl. Opt.* **29,** 1333 (1990).

43. J. M. Bello and T. Vo-Dinh, *Appl. Spectrosc.* **44,** 63 (1990).

44. K. Newby, W. M. Reichert, J. D. Andrade, and R. E. Benner, *Appl. Opt.* **23,** 1812 (1984).

45. M. L. Myrick and S. M. Angel, *Appl. Spectrosc.* **44,** 565 (1990).

46. S. M. Angel and M. L. Myrick, *Appl. Opt.* **29,** 1350 (1990).

47. S. D. Schwab and R. L. McCreery, *Appl. Spectrosc.* **41,** 126 (1987).

48. B. Yang, and M. D. Morris, *Appl. Spectrosc.,* **45,** 512 (1991).

49. M. Delhaye and P. Dhamelincourt, *J. Raman Spectrosc.* **3,** 33 (1975).

50. G. J. Rosasco, in R. J. H. Clark and R. E. Hester, eds., *Advances in Infrared and Raman Spectroscopy,* Vol. 7 (Heyden, London, 1980), pp. 223–283.

51. S. Inoue, *Video Microscopy* (Plenum, New York, 1986).

52. F. Adar, *Microchem. J.* **38,** 50 (1988).

53. G. Turrell, *J. Raman Spectrosc.* **15,** 103 (1984).

54. C. Bremard, P. Dhamelincourt, J. Laureyns, and G. Turrell, *Appl. Spectrosc.* **39,** 1036 (1985).

55. P. J. Treado and M. D. Morris, *Appl. Spectrosc.* **44,** 1 (1990).

56. P. J. Treado, A. Govil, M. D. Morris, K. D. Sternitzke, and R. L. McCreery, *Appl. Spectrosc.* **44,** 1270 (1990).

57. D. K. Viers, G. M. Rosenblatt, R. H. Dauskardt, and R. O. Ritchie, in *Microbeam Analysis,* D. E. Newbury, ed. (San Francisco Press, San Francisco, 1988), p. 179.

58. M. Bowden, D. J. Gardiner, G. Rice, and D. L. Gerrard, *J. Raman Spectrosc.* **21,** 37 (1990).

59. D. J. Gardiner, C. J. Littleton, and M. Bowden, *Appl. Spectrosc.* **42,** 15 (1988).

60. R.-S. Sheng, L. Zhu, and M. D. Morris, *Anal. Chem.* **58,** 1116 (1986).

61. F. Ni, R.-S. Sheng, and T. M. Cotton, *Anal. Chem.* **62,** 1958 (1990).

62. J. J. Laserna, A. D. Campiglia, and J. D. Winefordner, *Anal. Chem.* **61,** 1697 (1989).

63. M. L. McGlashen, K. L. Davis, and M. D. Morris, *Anal. Chem.* **62,** 846 (1990).

64. E. N. Lewis, V. F. Kalisinsky, and I. W. Levin, *Appl. Spectrosc.* **42,** 1188 (1988).

65. P. J. Hendra, C. Passingham, G. M. Warnes, R. Burch, and D. J. Rawlence, *Chem. Phys. Lett.* **164,** 178 (1989).

66. R. Burch, C. Passingham, G. M. Warnes, and D. J. Rawlence, *Spectrochim. Acta* **46A,** 243 (1990).

67. J. Goral and V. Zichy, *Spectrochim. Acta* **46A,** 253 (1990).

68. S. Nakashima and M. Hangyo, *IEEE J. Quantum Electron.* **QE-25,** 965 (1989).

69. J. D. Pasteris, B. Wopenka, and J. C. Seitz, *Geochim. Cosmochim. Acta* **52,** 979 (1988).

70. J. Dubessy, B. Poty, and C. Ramboz, *Eur. J. Mineral.* **1,** 517 (1989).

71. Y. Ozaki, *Appl. Spectrosc. Rev.* **24,** 259 (1988).

72. H. Ishida, R. Kamoto, S. Uchida, A. Ishitani, K. Iriyama, E. Tsukie, K. Shibata, F. Ishirara, and H. Kameda, *Appl. Spectrosc.* **41,** 407 (1987).

73. R. A. Dalterio, W. H. Nelson, D. Britt, J. Sperry, and F. J. Purcell, *Appl. Spectrosc.* **40,** 271 (1986).

74. F. Sureau, L. Chinsky, C. Amirand, J. P. Ballini, M. Duquesne, A. Laigle, P. Y. Turpin, and P. Vigny, *Appl. Spectrosc.* **44,** 1047 (1990).

75. R. H. Dauskardt, D. K. Viers, and R. O. Ritchie, *J. Am. Ceram. Soc.* **72,** 1124 (1989).

76. A. M. Ahern and R. L. Garrell, *Langmuir* **4** 1162 (1988).

77. R. S. Venkatachalam, F. J. Boerio, P. G. Roth, and W. H. Tsai, *J. Polymer Sci. B* **26,** 2447 (1988).

78. F. J. Boerio, P. P. Hong, P. J. Clark, and Y. Okamoto, *Langmuir* **6,** 721 (1990).

79. T. E. Rohr, T. Cotton, N. Fan, and P. J. Tarcha, *Anal. Biochem.* **182,** 388 (1989).

80. N.-S. Lee, Y. Z. Hsieh, M. D. Morris, and L. M. Schopfer, *J. Am. Chem. Soc.* **109,** 1358 (1987).

81. B. N. Rospendowski, D. L. Farrens, T. M. Cotton, and P.-S. Song, *FEBS Lett.* **258,** 1 (1989).

82. I. R. Nabiev, G. D. Chumanov, and R. G. Efremov, *J. Raman Spectrosc.* **21,** 49 (1990).

7 *Nonlinear Raman Spectroscopy*

H. Berger, B. Lavorel, and G. Millot

7.1. Introduction

Raman spectroscopy has been revolutionized by the development of tunable laser sources that have facilitated the application of nonlinear techniques. Many forms of coherent Raman spectroscopy have been developed; the aim of this chapter is not to give a survey of them all but to point out the main characteristics of nonlinear Raman spectroscopy through two processes that are extensively used: *stimulated Raman spectroscopy (SRS) and coherent anti-Stokes Raman spectroscopy* (CARS).

The important quantity in any nonlinear optical phenomenon is the nonlinear molecular susceptibility. A particular emphasis is given to consideration of the third-order susceptibility, $\chi^{(3)}$, which is responsible for four-wave mixing processes, including CARS and SRS. Incoherent processes are briefly illustrated by discussing the hyper-Raman effect. It is shown how the intensities of nonlinear Raman signals can be obtained from Maxwell's equations where the nonlinear polarization acts as a source term. These calculations are based on classical methods that a nonspecialist can easily follow; the detailed quantum mechanical derivation of the nonlinear susceptibility has been avoided.

For those just starting in this field we discuss in detail the laser instrumentation. As far as applications are concerned, our examples deal essentially with gaseous systems because here it is possible to take the best advantage of narrow-linewidth lasers. This largely explains the great progress that has occurred during the last decade. At present, coherent Raman spectra can be obtained at microbar pressure with resolution limited only by Doppler broadening. Using an accurate laser wavemeter, Raman frequencies can be measured with an accuracy greater than 0.0005 cm^{-1}, a precision that approaches that obtained in infrared absorption spectroscopy. This fundamentally changes the role played by Raman spectroscopy in molecular structure studies. In the range of laser diagnostics, nonlinear Raman spectroscopy now constitutes a particularly powerful method is combustion media following the pioneering experiment of Taran and co-workers in 1973.[1] In is also shown how lineshape studies can provide very useful data in a wide range of applications in physics and physical chemistry.

7.2. Description of Nonlinear Raman Processes

The coherent Raman techniques discussed in this chapter have already been presented in several reviews.[2-4] In this section, we point out the essentials of the theory to explain the main features of coherent Raman processes. First, we must consider the electric polarization created by the incident radiation. The nonlinear response of matter is written as an expansion of the polarization density as a function of input fields:

$$P_{(r,t)} = P^{(1)}_{(r,t)} + P^{(2)}_{(r,t)} + \cdots + P^{(n)}_{(r,t)} \tag{7.1}$$

In fact, problems involving the interaction of an electromagnetic field with matter are generally considered in the frequency domain instead of the time domain. The relation between the field and the induced polarization is then simplified.

Let us consider the oscillating field of a single monochromatic laser in a point of the medium by the following real quantity:

$$E(r,t) = \frac{1}{2}E(\omega)\, e^{i(\mathbf{k}\cdot\mathbf{r}\, -\omega t)} + \text{c.c. (complex conjugate)} \tag{7.2}$$

In a similar way the induced polarization is written:

$$P(r,t) = \frac{1}{2}P(\omega)\, e^{i(\mathbf{k}\cdot\mathbf{r}\, -\omega t)} + \text{c.c.} \tag{7.3}$$

The relation between the complex amplitudes of the polarization and the fields then takes the form for the nth order term in equation (7.1), using SI units here and throughout the following:

$$P^{(n)}(\omega) = \left(\frac{1}{2}\right)^{n-1}\varepsilon_0\, \chi^{(n)}\left(-\omega;\, \omega_1,\, \omega_2,\, \ldots,\, \omega_n\right)$$
$$E_1(\omega_1)\, E_2(\omega_2) \cdots E_n(\omega_n) \tag{7.4}$$

with

$$\omega = \sum_n \omega_n$$

The susceptibility term $\chi^{(n)}$ is a rank $(n + 1)$ tensor. The linear and nonlinear susceptibilities characterize the optical properties of the medium. We mention in passing that the convention of Maker and Terhune,[5] which entails inclusion of the constant $\varepsilon_0 \left(\frac{1}{2}\right)^{n-1}$ in the definition of $\chi^{(n)}$, is not used in this chapter. In Eesley's notation, the constant ε_0 is incorporated in $\chi^{(n)}$.[2] We consider that it is better to have, for the higher terms of the polarization, an expression similar to the familiar expression $P^{(1)}(\omega) = \varepsilon_0 \chi^{(1)} E(\omega)$.

Terms of successively higher order in equation (7.4) are of diminishing magnitude. So before the arrival of high-power lasers in the 1960s, the first term was the main origin of observed phenomena. The first nonlinear term $\chi^{(2)}$ is responsible for harmonic generation (first observed in 1961[6]) and optical rectification[7,8]; $\chi^{(2)}$ and all high-order even terms vanish in isotropic media such as gases and liquids and in centrosymmetric crystals.

By contrast, all matter provides nonvanishing contributions to $\chi^{(3)}$ and higher-order odd terms. The most important coherent Raman processes are described by the third-order nonlinear susceptibility $\chi^{(3)}$. Moreover, $\chi^{(3)}$ can be regarded as responsible for many other forms of spectroscopy, including two-photon absorption and polarization spectroscopy. The fifth-order nonlinear susceptibility is associated with hyper-Raman spectroscopy.

The ith component of the polarization can be written as

$$P_i^{(n)}(\omega) = \Sigma_{\alpha_1 \cdots \alpha_n} \left(\frac{1}{2}\right)^{n-1} \varepsilon_0 \chi_{i\alpha_1\alpha_2 \cdots \alpha_n}^{(n)} (-\omega; \omega_1, \omega_2, \cdots, \omega_n)$$
$$E_{\alpha_1}(\omega_1) \times E_{\alpha_2}(\omega_2) \cdots \times E_{\alpha_n}(\omega_n) \quad (7.5)$$

The frequency values of photons that are annihilated are traditionally written in the argument of $\chi^{(n)}$ with positive signs and the created photons with negative signs. (See below the expressions for CARS and SRS.) The summation only includes distinguishable terms of polarization frequency pairs (α_i, ω_i). Notice that the ordering of frequency terms and polarization indices is arbitrary. $\chi^{(n)}$ is invariant under the $n!$ permutations of pairs (α_i, ω_i); if q pairs are identical, $q!$ are indistinguishable. Thus, defining $D = n!/q!$ as a factor representing the degeneracy, equation (7.5) can be written as

$$P_i^{(n)}(\omega) = D \left(\frac{1}{2}\right)^{n-1} \varepsilon_0 \chi_{i\alpha_1 \cdots \alpha_n}^{(n)} (-\omega; \omega_1, \cdots, \omega_n)$$
$$E_{\alpha_1}(\omega_1) \times \cdots \times E_{\alpha_n}(\omega_n) \quad (7.6)$$

7.2.1. Third-Order Nonlinear Processes

The $\chi^{(3)}$ term is a fourth-rank tensor having 81 elements, but fortunately, the symmetry properties of system significantly reduce the number of independent components. The nonzero elements for different crystals are given in Refs. 3 and 9. For isotropic media (liquids and gases) where the majority of experiments have been performed, only three terms are independent:

$$\chi_{1111} = \chi_{2222} = \chi_{3333} = \chi_{1122} + \chi_{1212} + \chi_{1221} \tag{7.7}$$

and

$$\chi_{1122} = \chi_{1133} = \chi_{2233} = \chi_{2211} = \chi_{3311} = \chi_{3322}$$
$$\chi_{1212} = \chi_{1313} = \chi_{2323} = \chi_{2121} = \chi_{3131} = \chi_{3232}$$
$$\chi_{1221} = \chi_{1331} = \chi_{2332} = \chi_{2112} = \chi_{3113} = \chi_{3223} \tag{7.8}$$

To take the polarization of input fields into account, the unit vectors e_i can be introduced in equation (7.6). The induced polarization corresponding to four-wave mixing, Figure 7.1a, is thus written

$$\begin{aligned}
\mathbf{P}^{(3)}(\omega,\mathbf{r}) = \frac{D}{4}\,\varepsilon_0 \Big\{ &\chi^{(3)}_{1122}(-\omega;\,\omega_1,\,-\omega_2,\,\omega_3)\,\mathbf{e}_1\,(\mathbf{e}_2\cdot\mathbf{e}_3) \\
&+ \chi^{(3)}_{1212}(-\omega;\,\omega_1,\,-\omega_2,\,\omega_3)\mathbf{e}_2(\mathbf{e}_1\cdot\mathbf{e}_3) \\
&+ \chi^{(3)}_{1221}(-\omega;\,\omega_1,\,-\omega_2,\,\omega_3)\mathbf{e}_3(\mathbf{e}_1\cdot\mathbf{e}_2)\Big\} \\
&E(\omega_1)\,E^*(\omega_2)\,E(\omega_3)\,e^{i(\mathbf{k}_1 - \mathbf{k}_2 + \mathbf{k}_3)\cdot\mathbf{r}}
\end{aligned} \tag{7.9}$$

From this equation are deduced all forms of four-wave processes, including SRS and CARS. In practice, the third-order nonlinear processes are generated when two lasers at frequencies ω_1 and ω_2 are focused in a sample (see Figure 7.1b.) When the frequency difference $\omega_1 - \omega_2$ corresponds to the frequency of a molecular transition, a nonlinear polarization appears in the medium and is responsible for the modification of interacting waves (SRS) or the creation of new waves (CARS). The different methods for monitoring the interaction are described subsequently.

7.2.1.1. Stimulated Raman Spectroscopy (SRS)

The stimulated Raman effect is often considered as the process where a strong laser is sent into a medium and the Stokes photons produced by spontaneous Raman scattering create an autostimulated effect, leading to a strong Stokes wave.[10,11] Raman lasers are based on this principle. However, the stimulated Raman spectroscopy described here concerns the mixing of two lasers at frequencies ω_1 and ω_2 [12,13]$(\omega_1 > \omega_2)$. The second laser stimulates the scattering at the frequency ω_2. (See Figure 7.2.) The generated Raman wave at frequency ω_2 has the same properties as the ω_2 laser wave (frequency, direction, polarization). During the process a gain is produced at the frequency ω_2, whereas photons of frequency ω_1 are annihilated. As a consequence there are two ways for the experimenter to measure the interaction, either by measuring the gain at the frequency ω_2, which is called

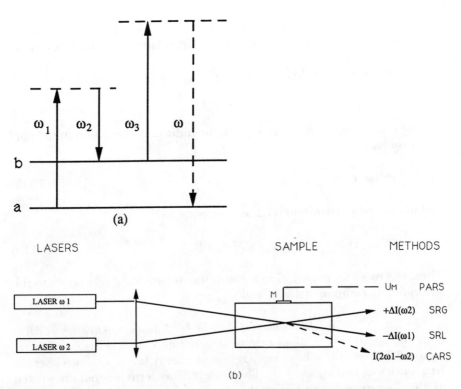

Figure 7.1 (a) Schematic diagram representing the four-wave mixing process: a polarization appears at the frequency $\omega = \omega_1 - \omega_2 + \omega_3$. (b) Third-order nonlinear Raman processes generated with two lasers (M: sensitive microphone)

Raman gain spectroscopy, or by measuring losses at the frequency ω_1, this technique being called *Raman loss spectroscopy* or *Inverse Raman spectroscopy.* The probe laser must be very stable in intensity; generally, a CW ion laser is chosen as a probe, whereas a pulsed laser serves as the pump beam.

In stimulated Raman gain spectroscopy the nonlinear polarization at the frequency ω_2 is of interest, and the responsible susceptibility term is

Figure 7.2 Stimulated Raman process with two lasers at frequency ω_1 and ω_2.

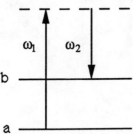

$\chi^{(3)}(-\omega_2; \omega_1, -\omega_1, \omega_2)$. Let us denote by $E_1 = E(\omega_1)$ and $E_2 = E(\omega_2)$ the amplitudes of the two interacting waves. The relation (7.9) then becomes

$$\mathbf{P}^{(3)}(\omega_2,\mathbf{r}) = \frac{6}{4}\varepsilon_0\Big\{\chi^{(3)}_{1122}\,\mathbf{e}_1(\mathbf{e}_1\cdot\mathbf{e}_2) + \chi^{(3)}_{1212}\,\mathbf{e}_1(\mathbf{e}_1\cdot\mathbf{e}_2) + \chi^{(3)}_{1221}\,\mathbf{e}_2\Big\}$$

$$|E_1|^2 E_2\, e^{i(\mathbf{k}_2\cdot\mathbf{r})} \quad (7.10)$$

Taking into account the relation (7.7), the induced polarization in the parallel case ($\mathbf{e}_1\|\mathbf{e}_2$) is

$$\mathbf{P}^{(3)}(\omega_2,\mathbf{r}) = \frac{6}{4}\,\varepsilon_0\chi^{(3)}_{1111}\,\mathbf{e}_2\,|E_1|^2\,E_2\,e^{i(\mathbf{k}_2\cdot\mathbf{r})} \quad (7.11a)$$

and in crossed polarization ($\mathbf{e}_1 \perp \mathbf{e}_2$), we obtain

$$\mathbf{P}^{(3)}(\omega_2,\mathbf{r}) = \frac{6}{4}\,\varepsilon_0\chi^{(3)}_{1221}\,\mathbf{e}_2\,|E_1|^2\,E_2\,e^{i(\mathbf{k}_2\cdot\mathbf{r})} \quad (7.11b)$$

Thus, the two SRS polarization measurements provide direct access to the susceptibility components χ_{1111} and χ_{1221}.

7.2.1.2. Coherent Anti-Stokes Raman Spectroscopy (CARS)

Only the commonly applied case of degenerate CARS is discussed here. Figure 7.3 shows the schematic diagram of energy levels with two lasers at frequencies ω_1 and ω_2, $\omega_1 > \omega_2$. The susceptibility term is generally written in the literature by ordering the frequencies as follows: $\chi^{(3)}(-\omega; \omega_1, \omega_1, -\omega_2)$. As we have a frequency degeneracy, the factor D equals 3; moreover, $\chi^{(3)}_{1122}(-\omega; \omega_1, \omega_1, -\omega_2) = \chi^{(3)}_{1212}(-\omega; \omega_1, \omega_1, -\omega_2)$, because of permutation invariance. Then the induced polarization at the anti-Stokes frequency $\omega = 2\omega_1 - \omega_2$ is written as follows:

$$\mathbf{P}^{(3)}(\omega,\mathbf{r})^{\,\text{CARS}} = \frac{3}{4}\,\varepsilon_0\Big\{2\chi^{(3)}_{1122}\,\mathbf{e}_1\,(\mathbf{e}_1\cdot\mathbf{e}_2) + \chi^{(3)}_{1221}\,\mathbf{e}_2\Big\}$$

$$E_1^2\,E_2{}^*\,e^{i(2\mathbf{k}_1-\mathbf{k}_2)\cdot\mathbf{r}} \quad (7.12)$$

Figure 7.3 CARS excitation with two lasers (ω_1 and ω_2).

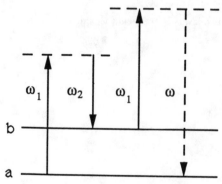

For the parallel case, $(\mathbf{e}_1 \parallel \mathbf{e}_2)$, we have

$$\mathbf{P}^{(3)}(\omega,\mathbf{r})^{\text{ CARS}} = \frac{3}{4} \varepsilon_0 \left\{ \chi^{(3)}_{1111} \mathbf{e}_2 \right\} E_1^2 E_2^* \, e^{i(2\mathbf{k}_1 - \mathbf{k}_2) \cdot \mathbf{r}} \qquad (7.13a)$$

and for the perpendicular case $(\mathbf{e}_1 \perp \mathbf{e}_2)$:

$$\mathbf{P}^{(3)}(\omega,\mathbf{r})^{\text{ CARS}} = \frac{3}{4} \varepsilon_0 \left\{ \chi^{(3)}_{1212} \mathbf{e}_2 \right\} E_1^2 E_2^* \, e^{i(2\mathbf{k}_1 - \mathbf{k}_2) \cdot \mathbf{r}} \qquad (7.13b)$$

In a similar way, a polarization appears at the Stokes frequency $2\omega_2 - \omega_1$ (coherent Stokes spectroscopy). The corresponding relations are obtained by exchanging ω_1 and ω_2. In conclusion, the expressions for the induced polarization are very similar for SRS and CARS except for the phase term, which is at the origin of the phase-matching condition in CARS (see Section 7.2.4.2).

7.2.2. Fifth-Order Processes: Hyper-Raman Effects

Taking into account the laser intensities usually available, the contribution of the fifth-order polarization term is weaker than the contribution of the cubic term. In these conditions the hyper-Raman effect that is due to the fifth-order polarization is more difficult to observe, and this kind of spectroscopy has only been used in a few investigations.

7.2.2.1. Spontaneous Effect The spontaneous process is represented by the energy diagram of Figure 7.4a. Two pump photons ω_1 are annihilated and a photon is emitted at the frequency $\omega = 2\omega_1 - \omega_{ba}$. The spontaneous hyper-Raman effect was discovered in 1965.[14] The interest of such a process results from the selection rules, which are different from SRS and CARS. In the hyper-Raman process, we have a three-photon

Figure 7.4 (a) Spontaneous hyper-Raman process. (b) Stimulated hyper-Raman process. (c) Spatial orientation of the wave vectors to eliminate Doppler broadening.

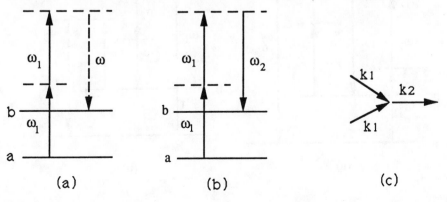

resonance and the transition $a \rightarrow b$ must satisfy the parity rule $u \leftrightarrow g$. Some vibrational bands that are forbidden in a SRS or CARS process can be allowed in the hyper-Raman process. One can say that the hyper-Raman technique is useful when the infrared spectroscopy or SRS and CARS do not work. For example, the fundamental $\nu_6(t_{2u})$ band of SF_6, forbidden in dipolar absorption and Raman scattering, could be investigated by means of this high-order Raman process. The C_6H_6 molecule has nine modes of vibration that are inactive using normal spectroscopic techniques; six of them are allowed in the hyper-Raman effect ($2b_{1u}$, $2b_{2u}$, and $2t_{2u}$).

7.2.2.2. Coherent Hyper-Raman Effects The first observation of autostimulated hyper-Raman scattering appears to be that of Yatsiv et al. in 1968.[15] The stimulated hyper-Raman process with two lasers of frequencies ω_1 and ω_2, Figure 7.4b, offers the possibility of a Doppler-free resonance if the spatial orientations of the wave vectors correspond to Figure 7.4(c).[16] Such a situation cannot be realized in a two-photon Raman resonance. Moreover, to our knowledge, stimulated hyper-Raman gain/loss techniques have not been used to provide spectroscopic information.

Concerning six-wave-mixing processes, one often has in mind a process equivalent to CARS, as represented in Figure 7.5(a); a resonance appears at the frequency $2\omega_1 - \omega_2$ and the signal at $4\omega_1 - \omega_2$. The selection rules are identical to those of the spontaneous hyper-Raman effect. To our knowledge, no spectrum has been recorded via this process, though the six-wave mixing illustrated in Figure 7.5b is responsible for observation of the coherent spectra of H_2 and N_2 in a recent study.[17] The main characteristic is the possibility to have a four-photon resonance at $2\omega_1 - 2\omega_2$, which means that $\omega_1 - \omega_2 = \omega_{ab}/2$. This process is called coherent subharmonic

Figure 7.5 Six-wave mixing processes: (a) with a three-photon resonance; (b) with a four-photon resonance at $\omega_1 - \omega_2 = \omega_{ba}/2$.

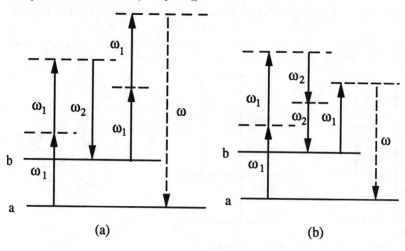

(a)   (b)

anti-Stokes Raman scattering (CSARS), because resonance appears at the subharmonic of the usual molecular resonance ($\omega_1 - \omega_2$).

It is worth noting that $\chi^{(5)}$, which appears in the hyper-Raman process, is proportional to $(\delta\beta/\delta q)2$, where β is the first hyperpolarizability and q represents a generalized normal vibrational coordinate, whereas $\chi^{(5)}$, which appears in CARS, is proportional to $(\delta\beta/\delta q)(\delta\gamma/\delta q)$, γ being the second hyperpolarizability. The latter process thus represents the first time that a Raman experiment allows the assess to measurement of the second hyperpolarizability.

7.2.3. Wave Equations in Nonlinear Optics

Following the phenomenological description of Raman processes in terms of nonlinear susceptibility, a quantitative description consists in calculating the intensities of the coherent Raman signals. The basis of the calculation starts from Maxwell's equations, from which the well-known wave equation for a nonmagnetic homogeneous medium is derived:

$$\nabla \times (\nabla \times \mathcal{E}) + \frac{1}{c^2}\frac{\partial^2 \mathcal{E}}{\partial t^2} = -\frac{1}{c^2\varepsilon_0}\frac{\partial^2 P}{\partial t^2} \tag{7.14}$$

where $P = P^L + P^{NL}$. We assume that both $\mathcal{E}$ and P can be written as a sum of plane waves:

$$\mathcal{E}(r,t) = \frac{1}{2}\sum_n \mathcal{E}(r, \omega_n)\, e^{-i\omega_n t} \tag{7.15}$$

$$P(r,t) = \frac{1}{2}\sum_n P(r, \omega_n)\, e^{-i\omega_n t}$$

It is usual to write equation (7.14) in the frequency domain for each frequency component ω_n and to extract the linear term $P^L(r,\omega) = \varepsilon_0(n^2 - 1)\,\mathcal{E}(r, \omega)$. We thus obtain

$$\nabla \times (\nabla \times \mathcal{E})(\omega_n, r) - \frac{n^2\omega^2}{c^2}\,\mathcal{E}(\omega_n, r) = \frac{1}{c^2\varepsilon_0}\,\omega_n^2\,P^{NL}(\omega_n, r) \tag{7.16}$$

where only the nonlinear polarization appears. Equation (7.16) indicates that the n waves are nonlinearly coupled via the nonlinear polarization, leading to an energy transfer among them.

For simplification, we assume propagation of the fields along the z-axis with slowly varying amplitude:

$$\mathcal{E}(\omega_n, z) = E(\omega_n, z)\, e^{ik_n z}\, e \tag{7.17}$$

The substitution of (7.17) into (7.16) yields the evolution equation:

$$- \left(\frac{\partial^2 E(\omega_n, z)}{\partial z^2} + 2ik_n \frac{\partial E(\omega_n, z)}{\partial z} - k_n^2 E(\omega_n, z) \right)$$
$$- \frac{n^2 \omega^2}{c^2} E(\omega_n, z) = \frac{1}{c^2 \varepsilon_0} \omega^2 P^{\mathrm{NL}}(\omega_n, z) \quad (7.18)$$

In this expression the second derivative can be neglected if we assume that the amplitude $E(\omega_n, z)$ varies slowly as a function of z. For n frequencies ω_n interacting in the medium, we have n equations coupled via the nonlinear term P^{NL}:

$$- 2ik_n \frac{\partial E(\omega_n, z)}{\partial z} = \frac{1}{c^2 \varepsilon_0} \omega^2 P^{\mathrm{NL}}(\omega_n, z) \quad (7.19)$$

Fortunately, in most of the cases, the intensities of incident waves are only weakly modified, and each equation associated with component ω_n can be solved analytically. In the following, we will denote $E(\omega_n) = E_n$ and the refractive index $n(\omega_n) = n_n$.

7.2.3.1. SRS

If we assume that the fields have the same polarization, the two coupled equations in the SRS process are

$$\frac{\partial}{\partial z} E_2(z) = \frac{3}{4} \frac{ik_2}{n_2^2} \chi_{1111}^{(3)} (-\omega_2; \omega_1, -\omega_1, \omega_2) |E_1(z)|^2 E_2(z) \quad (7.20a)$$

$$\frac{\partial}{\partial z} E_1(z) = \frac{3}{4} \frac{ik_1}{n_1^2} \chi_{1111}^{(3)} (-\omega_1; \omega_2, -\omega_2, \omega_1) |E_2(z)|^2 E_1(z) \quad (7.20b)$$

These equations can be solved by assuming E_1 is a constant in (7.20a) and E_2 is a constant in (7.20b). We then obtain

$$E_2(z) = E_2(z = 0) \exp \left\{ \frac{3ik_2 z}{4n_2^2} \chi_{1111}^{(3)} (-\omega_2; \omega_1, \right.$$
$$\left. -\omega_1, \omega_2) |E_1|^2 \right\} \quad (7.21a)$$

$$E_1(z) = E_1(z = 0) \exp \left\{ \frac{3ik_1 z}{4n_1^2} \chi_{1111}^{(3)} (-\omega_1; \omega_2, \right.$$
$$\left. -\omega_2, \omega_1) |E_2|^2 \right\} \quad (7.21b)$$

In a resonance region, $\chi^{(3)}$ can be written as the sum of two contributions, a resonant contribution χ_{R}, which is frequency-dependent, and a nonresonant contribution χ_{NR} (generally a real quantity) due to all other transitions that are off-resonance:

$$\chi^{(3)} = \chi_{\mathrm{NR}}^{(3)} + \chi_{\mathrm{R}}^{(3)} \quad \text{with } \chi_{\mathrm{R}}^{(3)} = \chi'^{(3)} + i\chi''^{(3)} \quad (7.22)$$

Only the imaginary part of the nonlinear susceptibility $\chi''^{(3)}$ is responsible for amplitude changes of the probe laser. We finally have

$$E_2(z) = E_2(0) \exp\left\{\frac{-3k_2z}{4n_2^2} \chi''^{(3)}_{1111}(-\omega_2; \omega_1,\right.$$

$$\left. -\omega_1, \omega_2) |E_1|^2\right\} \quad (7.23a)$$

$$E_1(z) = E_1(0) \exp\left\{\frac{-3k_1z}{4n_1^2} \chi''^{(3)}_{1111}(-\omega_1; \omega_2,\right.$$

$$\left. -\omega_2, \omega_1) |E_2|^2\right\} \quad (7.23b)$$

Explicit expressions for the nonlinear susceptibilities can be derived from a quantum mechanical treatment,[9,18] which is not discussed here. As an example, we give the following expression for $\chi^{(3)}(-\omega_2; \omega_1, -\omega_1, -\omega_2)$ cast in SI units[19]:

$$\chi^{(3)}(-\omega_2; \omega_1, -\omega_1, -\omega_2)$$
$$= \chi^{(3)}_{NR} + \sum_{ab} \frac{16\pi^3 \varepsilon_0 c^4}{3h\omega_2^4} \left(\frac{d\sigma}{d\Omega}\right)_{ba} \frac{N(\rho_{aa}^0 - \rho_{bb}^0)}{\omega_{ba} + \omega_2 - \omega_1 + i\Gamma_{ba}} \quad (7.24a)$$

$$\chi^{(3)}(-\omega_1; \omega_2, -\omega_2, -\omega_1)$$
$$= \chi^{(3)}_{NR} + \sum_{ab} \frac{16\pi^3 \varepsilon_0 c^4}{3h\omega_2^4} \left(\frac{d\sigma}{d\Omega}\right)_{ba} \frac{N(\rho_{aa}^0 - \rho_{bb}^0)}{\omega_{ba} + \omega_2 - \omega_1 - i\Gamma_{ba}} \quad (7.24b)$$

with $\quad \left(\frac{d\sigma}{d\Omega}\right)_{ba} = \left|\left\{\frac{1}{4\pi\varepsilon_0}\frac{\omega^2}{c^2}\right\} |<a|\alpha|b>|\right|^2 \quad (7.25)$

These equations make a connection between the macroscopic susceptibility, the usual differential cross section $(d\sigma/d\Omega)_{ba}$, and the polarizability matrix element $<a|\alpha|b>$ of the transition $a \rightarrow b$. Here N is the density of molecules and $N(\rho_{aa}^0 - \rho_{bb}^0)$ represents the difference of population between the initial and final states; Γ_{ba} is the half-width at half-maximum of the Raman transition.

It transpires that in equation (7.24), $\chi''^{(3)}_{1111}(-\omega_2; \omega_1, -\omega_1, \omega_2) = \chi''^{(3)SRG}$ has a negative value, so that the ω_2 wave intensity is increased. On the contrary, $\chi''^{(3)}_{1111}(-\omega_1; \omega_2, -\omega_2, \omega_1) = \chi''^{(3)SRL}$ is positive, leading to losses for the ω_1 field. As the gains or losses are normally small for most stimulated experiments, the exponential term can be developed and only the first two terms of expression (7.23) are kept:

$$E_2(z) = E_2(0)\left(1 - \frac{3k_2z}{4n_2^2} \chi''^{(3)SRG} |E_1|^2\right) \quad (7.26)$$

The time average intensity of an electromagnetic wave at frequency ω is $I(\omega) = \frac{1}{2}n(\omega)c\varepsilon_0|E_\omega|^2$ and the intensity of the ω_2 wave after the interaction in the medium can thus be written as

$$I_2(z) = I_2(0) \left(1 - \frac{3\omega_2 z}{\varepsilon_0 c^2 n_1 n_2} \chi''^{(3)\text{SRG}} I_1 \right) \tag{7.27}$$

For an interaction length L, it is usual to define the gain G as

$$G^{\text{SRG}} = \frac{I_2(L) - I_2(0)}{I_2(0)} = \frac{-3\omega_2 L}{\varepsilon_0 c^2 n_1 n_2} \chi''^{(3)\text{SRG}} I_1 \tag{7.28}$$

In stimulated Raman loss (SRL) spectroscopy, the losses per unit of length are defined in a similar way. The negative gain G is then

$$G^{\text{SRL}} = \frac{-3\omega_1 L}{\varepsilon_0 c^2 n_1 n_2} \chi''^{(3)\text{SRL}} I_2 \tag{7.29}$$

7.2.3.2. CARS

Of the three coupled equations for the CARS process, only the wave equation describing the evolution of the created frequency $\omega_3 = 2\omega_1 - \omega_2$ is of interest if one assumes, as for SRS, that the intensities of the laser beams at frequencies ω_1 and ω_2 are modified only weakly during the interaction. The starting equation for the new wave ω_3 is derived from equation (7.19) with the inclusion of appropriate phase terms:

$$2ik_3 \frac{\partial E_3(z)}{\partial z} e^{ik_3 z} = -\frac{1}{c^2 \varepsilon_0} \omega_3^2 \varepsilon_0 \frac{6}{4} \chi^{(3)\text{CARS}} E_1^2 E_2^* e^{(2k_1 - k_2)z} \tag{7.30}$$

with $n_3 = ck_3/\omega_3$. Integrating equation (7.30) with $E_3(z = 0) = 0$, we obtain

$$E_3 = \frac{3}{4} \frac{1}{\Delta k} \frac{\omega_3}{n_3 c} (e^{i(\Delta k z)} - 1) \chi^{(3)\text{CARS}} E_1^2 E_2^* \tag{7.31}$$

with $\Delta k = 2k_1 - k_2 - k_3$. Finally, the CARS intensity $I_3 = \frac{1}{2} n_3 c \varepsilon_0 |E_3|^2$ is written as

$$I_3 = \frac{9\omega_3^2}{16 n_1^2 n_2 n_3 \varepsilon_0^2 c^2} |\chi^{(3)\text{CARS}}|^2 I_1^2 I_2 z^2 \text{sinc}^2 \left(\frac{1}{2} z\Delta k \right) \tag{7.32}$$

From this equation, a number of remarks can be made. First, the CARS signal is also proportional to the modulus square of the nonlinear susceptibility χ^3; the consequences of this will be discussed in Section 7.2.4.3. The CARS intensity is also proportional to the square of the well-known sinc function (sinc $x = x^{-1} \sin x$). We know that the principal maximum is obtained if the argument is zero—that is, $\Delta k = 0$. This is the phase-matching condition. In SRS we do not have such a condition on the wave vectors. In the case of collinear fields, the CARS interaction is coherent over a length

z_c, called *coherence length*, which depends on the dispersion of the medium:

$$\Delta k = 2k_1 - k_2 - k_3 = \frac{1}{c}[2n_1\omega_1 - n_2\omega_2 - n_3(2\omega_1 - \omega_2)]$$

$$= \frac{1}{c}[2\omega_1(n_1 - n_3) - \omega_2(n_2 - n_3)] \qquad (7.33)$$

The generated signal ω_3 reaches the first maximum value at $z_c = \pi/\Delta k$. For gases the coherence length is in the centimeter range at atmospheric pressure so that a collinear mixing of the waves may be easily realized. For a condensed medium z_c would be too small for a collinear geometry and a crossed-beam geometry corresponding to the vector condition $\Delta k = 2\mathbf{k}_1 - \mathbf{k}_2 - \mathbf{k}_3 = 0$ is generally used.

7.2.4. Characteristics of Coherent Raman Spectroscopy

7.2.4.1. Common Properties Among the features of coherent Raman techniques that are distinct from those of spontaneous Raman spectroscopy we notice the following:

1. The output signal is coherent. This point is easily understood from the theory of stimulated emission. The coherent signal results from an interaction of matter with a mode of the electromagnetic field. Consequently, the Raman signal has exactly the properties of this mode (direction, polarization, frequency).

2. The instrumental spectral resolution is limited only by the linewidth of the input laser source. So, by using very-narrow-linewidth lasers, a resolution better than 10^{-4} cm^{-1} could be reached. In fact, the resolution is often limited by Doppler broadening, though this can be minimized in the forward scattering configuration. (See Section 7.5.1.1.)

3. Spectra are often derived by scanning the frequency of one laser. The response of the sample is recorded when successive molecular resonances appear. Spectroscopy is achieved without a spectrometer!

4. A high spatial resolution can be obtained because the Raman signal is generated in the beam overlap region.

5. Another feature is the high fluorescence rejection capability.

7.2.4.2. Phase-Matching Condition in CARS The standard scheme shown in Figure 7.6 corresponds to the case where the two photons ω_1 have the same direction. The adjustment of the angle θ between the beams is obtained by changing the distance d between the parallel incident beams ω_1 and ω_2. Obviously, if a scanning of 100 cm^{-1} (or more) is required, it is necessary to adjust the θ angle continuously to have the maximum signal. The CARS signal is generated in a direction off from the laser beams, making separation of the strong pump beams ω_1 and ω_2 easy. In the

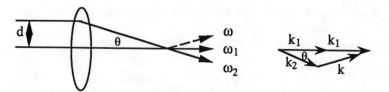

Figure 7.6 Triangular phase matching in CARS.

triangular phase-matching arrangement the angle θ is unique. To be able to change the crossing angle, an arrangement called BOXCARS[20,21] can be used. In the configuration represented in Figure 7.7a the wave vector arrangement takes the form of a "box." Such a configuration is best adopted if an improvement in spatial resolution is required; however, in this geometry the generated CARS signal has approximately the same direction as one ω_1 beam. This can be a drawback if a weak CARS signal is to be extracted. A clever solution is obtained with the folded BOXCARS geometry (Figure 7.7b), where half of the phase-matching diagram is rotated by $\frac{\pi}{2}$ radians.

7.2.4.3. Influence of Susceptibility Terms on the CARS Signal

The CARS signal is governed by the square of the nonlinear susceptibility. Consequently, the CARS intensity has four contributions as

$$|\chi^{(3)}|^2 = (\chi_{NR} + \chi' + i\chi'')^2 = \chi_{NR}^2 + 2\chi'\chi_{NR} + \chi'^2 + \chi''^2$$
$$= \chi_{NR}^2 + 2\chi'\chi_{NR} + |\chi_R|^2 \quad (7.34)$$

The nonresonant χ_{NR} is assumed to be real if no electronic resonance exists and so gives a constant contribution. The second term has a dispersive character and strongly distorts the CARS profile of an isolated line in dense media. If we consider two (or more) closely spaced lines (at frequencies ω and ω'), interference effects take place between the real parts of the resonant susceptibility $\chi'_R(\omega)$ and $\chi'_R(\omega')$; cancellation of the real terms χ' leads

Figure 7.7 (a) BOXCARS phase-matching. (b) Folded BOXCARS.

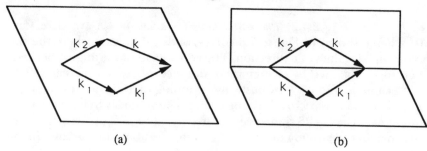

(a)   (b)

to an asymmetry in the profile and a slight shift of the maxima, increasing the apparent separation of the lines.[22]

7.3. Experimental Techniques

7.3.1. Laser Systems for Coherent Raman Spectroscopy

Our aim in this section is not to describe all devices but only to give a description of typical arrangements. It seems obvious that pulsed lasers play an important role for investigations of nonlinear processes. Indeed, to increase the signal-to-noise ratio, it is easier most of the time to increase the signal than to reduce the noise!

7.3.1.1. Single-Mode Injected Nd–YAG Laser In the past five years, the injection-seeding technique of pulsed laser instrumentation has seen significant development, in particular with commercial injected Nd–YAG lasers. The beam of a CW single-mode "mini" YAG pumped by a high-power diode-laser is injected into the cavity of a Q-switched Nd–YAG laser. If the injected radiation is near the frequency of a mode of the pulsed laser, then the pulsed laser can operate at high power in a single axial mode.

To ensure single-mode output, the injected power need only exceed the power loss due to spontaneous emission. A Faraday rotator generally isolates the CW laser from the pulsed laser beam, avoiding perturbation of the single-mode CW laser operation. The effect of injection seeding on the temporal pulse shape of the oscillator output is spectacular. If the length of the oscillator cavity is resonant with the injected frequency, then the shape of the pulse is smoothed. The smoother the pulse is, the shorter is the build-up time. Active stabilization of the slave cavity is achieved by minimizing the pulse buildup time, so that thermal drifts of the oscillator can be compensated. The frequency-doubled wave at 532 nm today constitutes a very convenient monochromatic high-power source for nonlinear Raman spectroscopy. In the near future, commercial injected Ti-sapphire lasers should be available in the 700–1,000-nm range.

7.3.1.2. Single-Mode Amplified Dye Laser A second possibility for a monochromatic high-power tunable source is to amplify a CW dye laser in the megawatt range. To do that, the beam of a single-mode CW dye laser is passed through three or four stages of an amplifier system pumped by a frequency-doubled Nd–YAG laser. From a CW beam of about 10–50 mW, pulses of more than 1 MW power can be obtained. The linewidth is determined by the Fourier transform of the temporal pulse. Such a system is commercially available. In the future, this technique could be employed to amplify the CW output of single-mode diode lasers, the latter replacing dye lasers in some frequency ranges.

7.3.1.3. Injected Flashlamp–Pumped Dye Laser   For the pulsed laser described in the previous section, the spectral linewidth (FWHM) is in the order of 0.0022 cm^{-1} for a 12-ns pulse duration. The linewidth could be reduced by working with a pulse length of 300–400 ns; that is possible with flashlamp–pumped dye lasers. A CW single-mode dye laser is injected into a pulsed laser cavity whose optical length is adjusted to the injected frequency. A laser linewidth of 6 MHz (0.0002 cm^{-1}) has in fact been reported.[23] The peak power is in the 10-kW range.

7.3.1.4. Ion Lasers   Ion lasers play an important role in Raman spectroscopy. In CARS, most of the devices use pulsed lasers, though CARS spectrometers with ion laser pumps are used for investigations in condensed media. We can note an exception concerning gas studies. At Munich University the gas sample is put inside the cavity of a single-mode CW Ar$^+$ laser.[24]

In SRS, ion lasers are used as the probe; this is due to the remarkable output stability of the ion laser in comparison with a dye laser (typically better by three orders of magnitude). So the amplitude noise is minimized and the shot noise limit can be attained. The Ar$^+$ laser serves as probe source in Raman loss spectroscopy, whereas the Kr$^+$ is used in Raman gain spectroscopy. The usual jitter of ion lasers (about 20 MHz) can be reduced to 1 MHz by realizing an active stabilization of the laser cavity, as in Figure 7.8. The frequency of the laser is stabilized by locking it to the side of a fringe of an external Fabry–Perot interferometer. The intensity variation detected by a photodiode generates an error signal that drives the laser cavity end-mirror mounted on a piezoelectric transducer. Moreover, in the case of Ar$^+$ laser, if long-term frequency stability is required, a small part of the argon beam is sent into an I_2 absorption cell, and after detection, the

Figure 7.8 Active stabilization of the Ar$^+$ laser to an iodine absorption line.

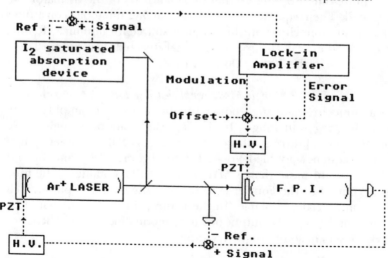

laser frequency is locked to the frequency of a hyperfine component of $P(13)$ or $R(15)$ in the 43–0 band. The frequencies of the hyperfine structure are given in Ref. 25.

7.3.2. Typical Apparatus in Coherent Raman Spectroscopy

7.3.2.1. The Quasi-CW Spectrometer for SRS The best compromise to have a high signal-to-noise ratio in SRS consists in associating a high-power pulsed laser as pump beam with a CW laser as probe. That explains the term *quasi-CW* given to such a device by A. Owyoung, a pioneer of stimulated Raman spectroscopy.[26] For the pulsed laser the solution of amplifying a CW dye laser (Section 7.3.1.2) is generally used. The probe laser is an Ar^+ or Kr^+ laser. If a single-mode Nd–YAG laser (or another frequency-fixed laser) were to be used as pump beam, it would be necessary to take a CW dye laser as probe beam. This possibility is generally not adopted because dye lasers are too noisy in comparison with ion lasers. A typical device developed in our laboratory is shown in Figure 7.9.

The CW probe laser is modulated at the repetition rate of the pulsed laser to form 50–100 μs pulses to avoid saturation or destruction of the photodiode; this allows the probe laser to work at a high power, which can reach 1 W. The modulation can be achieved with electro-optic or acousto-optic modulation, but a mechanical chopper is also very convenient and moreover allows 100% modulation. The two lasers are focused in the gas sample either in a collinear or in a crossed configuration. The latter is used to study N_2 or O_2 to avoid generation of a Raman signal outside the Raman cell. A loss factor of about 3 is observed with a crossed beam geometry.

After the interaction, the CW probe beam is separated from the pulsed beam by a dispersive system (or dichroic mirror) before being sent on to a fast photodiode. (EGG-FND 100 is often used.) This has a dynamic range of seven decades and a rise-time of less than 1 ns. A high-frequency amplifier isolates the Raman gain/loss (12 ns) in the probe laser pulse (50–100 μs). The ratio of the Raman signal to pump laser peak power allows the fluctuations of the pulsed pump beam to be taken into account. Finally, the signal is averaged in a boxcar and stored on disk. The problem of frequency calibration is discussed later.

7.3.2.2. Scanning CARS Apparatus A typical CARS device developed at the Oregon State University is shown in Figure 7.10.[27] A narrow frequency-doubled Nd–YAG laser at 532 nm is used as source at the frequency ω_1 in the Raman process and serves to pump the dye laser necessary in the experiment. About 20 mJ of the pump is used for the ω_1 source, and the rest is used for pumping the dye laser. This is a conventional dye laser with an intracavity Fabry–Perot etalon to provide a narrow bandwidth (≈ 0.05 cm^{-1}). The energies of pulses in the ω_2 beam are in the 2–10-mJ range. Complete CARS systems are now commercially available.

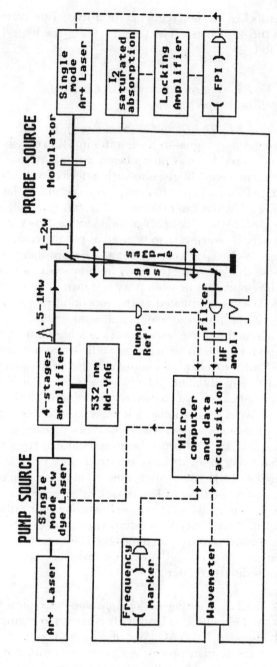

Figure 7.9 Schematic diagram of the stimulated Raman device at Dijon University.

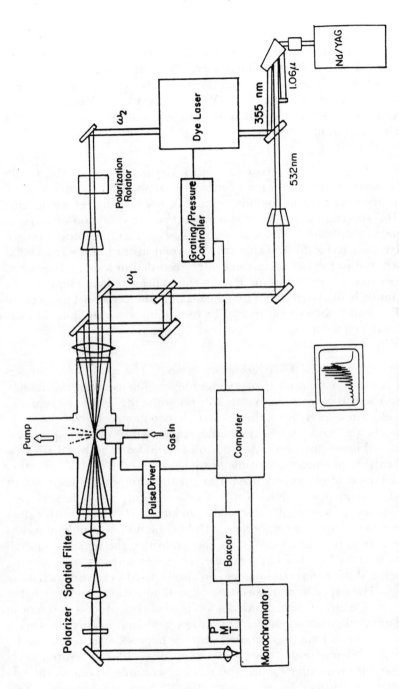

Figure 7.10 Typical CARS apparatus at Oregon State University (Reproduced from Ref. 27 with permission).

That could explain the success of the CARS technique over SRS, although the latter may often be more appropriate in particular for studies of condensed media. In CARS, the signal is generated at a frequency different from that of either input laser, so that the detector can be a photomultiplier sampled by a boxcar averager. In a CW CARS device, the Raman signal is weaker and is often detected with a cooled photomultiplier and integrated in a photon counting system.

7.3.2.3. Multiplex CARS Apparatus   Scanning CARS measurements are not often used for industrial applications such as temperature measurement in combustion media, because the recording time of the full spectrum is too long with respect to the evolution of physical phenomena (explosions, chemical reactions). For such purposes, a broadband dye laser (typically 150–200 cm^{-1}) is used instead of a narrowband dye laser, and so the full spectrum can be recorded in a single laser shot. A monochromator disperses the Raman signal prior to being sent into an optical multichannel detector. This consists of a photocathode, an electron intensifier, and a photodiode array. The resolution is mainly limited by the diode array ($\approx$0.4 cm^{-1}).

7.3.2.4. Hyper-Raman Devices   The spontaneous hyper-Raman effect is many orders of magnitude less intense than the Raman process. A way to overcome this difficulty is to tune the laser to be resonant with an electronic level. Then the resonant hyper-Raman spectrum has approximately the same intensity as the spontaneous Raman spectrum. Thanks to this enhancement, the first observation of a resolved rovibrational structure in ammonia became possible in 1987.[28] The UV excitation of the electronic level was produced from the third anti-Stokes Raman shift in H_2 of a frequency-doubled Rh6G dye laser (5 mJ per pulse). Hyper-Raman scattering is generally observed at an angle of 90° from the incident laser direction. As for any spontaneous phenomenon, a double monochromator is required. To observe such low intensities, the signal is normalized and averaged over a large number of pulses with photon-counting equipment. Multichannel techniques can also be used to record such weak processes. The hyper-Raman spectrum of ammonia had to be recorded at a pressure of 8 atm, which shows the limits of such a process as a tool in investigations of gaseous systems. Only a few gases and liquids but a great number of crystals have been studied with the hyper-Raman technique.

The experimental setups described earlier for CARS can be used to record coherent high-order Raman spectra. For example, in one of the few experiments performed to date via the CSARS process, pulses of a Nd–YAG laser (30 mJ) were simply mixed collinearly with pulses from a dye laser (1–2 mJ) in the Raman cell with a detection system as for CARS.[17] The spectra of the Q branches of H_2 and N_2 were recorded at 400 mbar and 900 mbar, respectively.

7.3.3. Polarization CARS

The nonresonant background susceptibility χ^{NR} limits CARS sensitivity for the observation of weak Raman resonances; that explains the great interest of developing a polarization technique to suppress this background. An excellent review of this technique and numerous applications are given in Ref. 29. The basic idea of polarization CARS lies in the fact that the Raman signal arising from Raman resonant (P^R) and nonresonant (P^{NR}) response of the medium generally have different polarizations. Therefore, the nonresonant contribution can be eliminated by means of a polarization analyzer. The explanation is based on the symmetry properties of the different components of $\chi^{(3)}$. As shown in Section 7.2.1.2, $\chi^{(3)}_{1122} = \chi^{(3)}_{1212}$, and $\chi^{(3)}_{1111} = 2\chi^{(3)}_{1122}; + \chi^{(3)}_{1221}$. Moreover, at resonance, an additional relation (Kleinman's relation[30]) can be considered:

$$\chi^{NR}_{1122} = \chi^{NR}_{1212} = \chi^{NR}_{1221} = \left(\frac{1}{3}\right)\chi^{NR}_{1111} \tag{7.35}$$

So one can define two depolarization ratios:

$$\rho_R = \frac{\chi^R_{1221}}{\chi^R_{1111}} \quad \text{and} \quad \rho_{NR} = \frac{\chi^{NR}_{1221}}{\chi^{NR}_{1111}} \tag{7.36}$$

Let us assume that the E_1 field is parallel to the x-axis and that the E_2 field makes an angle Φ with it. The Cartesian components of $P^{(3)}$ follow from equation (7.10):

$$P_x^{(3)} = \frac{3}{4}\varepsilon_0 (2\chi^{(3)}_{1122} + \chi^{(3)}_{1221}) \cos \Phi \, E_1^2 E_2^* \, e^{i\Delta kz}$$

$$= \frac{3}{4}\varepsilon_0 (\chi^{(3)R}_{1111} + \chi^{(3)NR}_{1111}) \cos \Phi \, E_1^2 E_2^* \, e^{i\Delta kz} \tag{7.37a}$$

$$P_y^{(3)} = \frac{3}{4}\varepsilon_0 \chi^{(3)}_{1221} \sin \Phi \, E_1^2 E_2^* \, e^{i\Delta kz}$$

$$= \frac{3}{4}\varepsilon_0 (\chi^{(3)R}_{1221} + \chi^{(3)NR}_{1221}) \sin \Phi \, E_1^2 E_2^* \, e^{i\Delta kz} \tag{7.37b}$$

The resonant term P^R and nonresonant term P^{NR} form angles α and β with respect to the field E_1:

$$\tan \alpha = \frac{P_y^{NR}}{P_x^{NR}} = \rho_{NR} \tan \Phi \tag{7.38a}$$

$$\tan \beta = \frac{P_y^R}{P_x^R} = \rho_R \tan \Phi \tag{7.38b}$$

$$\tan (\beta - \alpha) = \rho_{NR} \tan \Phi \, \frac{\left(\dfrac{\rho_R}{\rho_{NR}}\right) - 1}{(1 + \rho_R \, \rho_{NR} \tan^2 \Phi)} \tag{7.38c}$$

It can be noted that ρ_R and ρ_{NR} are always equal if $\Phi = 0$ or $90°$—that is, in usual CARS configurations. If $\rho_R \neq \rho_{NR}$, then the ρ_{NR} contribution can be canceled by crossing the background signal with the analyzer. A spectacular demonstration of background suppression using this technique is shown in Figure 7.11. Obviously, a reduction in the resonant signal intensity goes along with the nonresonant signal suppression. Polarization CARS has been performed in scanning as well as in multiplex mode.

7.3.4. Resonant CARS

A resonant enhancement of the order of 10^2–10^4 can be obtained when one or both intermediate levels correspond to excited electronic states. For gaseous systems, only a few molecules have been studied under these conditions because their electronic states often have too high

Figure 7.11 Square root of the CARS signal from a methane–air flame in the region of the CO Q branch. (a) Background-free spectrum. (b) Conventional CARS spectrum with all polarizations parallel. (Reproduced from Ref. 30 by permission.)

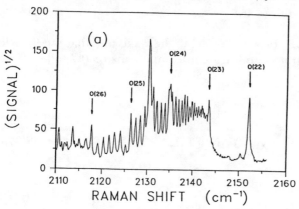

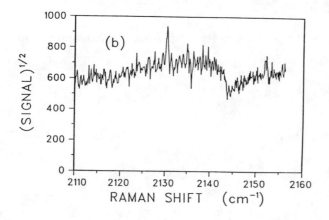

an energy to be reached with commercially available lasers. Discrete state resonances have been reported in I_2, C_2, NO_2, OH, and CS_2. However, many resonance-enhanced spectra have been reported for liquids. Resonance CARS is carried out by holding ω_1 near a one-photon absorption ($\omega_1 = \omega_{na}$) while ω_2 is scanned. In Figure 7.12a, the CARS signal is enhanced and corresponds to the usual resonance condition $\omega_1 - \omega_2 = \omega_{ba}$. In Figure 7.12b, $\omega = \omega_{n'a} = 2\omega_1 - \omega_2$, and thus $\omega_1 - \omega_2 = \omega_{n'a} - \omega_1$. The latter depends on ω_1; this can be a way to identify the lines. Investigations in the visible region are easy with tunable dye lasers, but for most molecules tunable lasers in the UV range would be required. Some electronic resonances have been studied with excimer lasers or visible dye lasers Raman-shifted in hydrogen to provide tunable radiation near 200 nm.[31]

7.3.5. Ionization-Detected Stimulated Raman Spectroscopy

Ionization-detected stimulated Raman spectroscopy is a double-resonance technique developed by Esherick and Owyoung. It is illustrated in Figure 7.13. In this two-step process, molecules are first pumped to the $v = 1$ vibrational state by SRS and then selectively ionized via a multiphoton process (two-photon in the case of C_6H_6[32]). The ionization process is enhanced if the UV laser excitation is resonant with an electronic transition. The tunable UV radiation is provided by a commercially available frequency-doubled dye laser. This new technique, called *ionization-detected stimulated Raman spectroscopy* (IDSRS), allows a sensitivity enhancement of two or three orders of magnitude because ions can be de-

Figure 7.12 Resonant CARS conditions: (a) $\omega_1 - \omega_2 = \omega_{ba}$; (b) $\omega_1 - \omega_2 = \omega_{n'a} - \omega_1$.

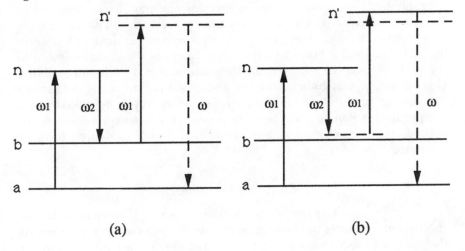

(a)   (b)

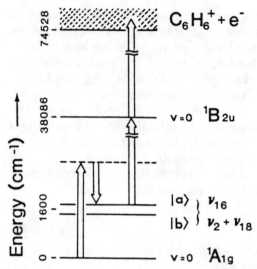

Figure 7.13 Diagram illustrating the excitation scheme in the double-resonance process (IDSRS)

tected with higher sensitivity than photons. Ions are collected by simple parallel-plate electrodes. Nitric oxide has been studied with this technique at the very low pressure of 0.001 torr.[33]

7.3.6. *Photoacoustic Raman Spectroscopy (PARS)*

In photoacoustic Raman spectroscopy (PARS) the energy deposited in the sample by the stimulated Raman process is detected by sensitive acoustic methods. During the coherent SRS process, the molecules are pumped on a rovibrational level; collision-induced relaxation to translational energy leads to a pressure increase that can be detected by a microphone. Photoacoustic Raman spectroscopy was first introduced by Barrett and Berry with CW laser excitation of methane.[34] The use of high-power pulsed laser sources gave great improvements in the PARS sensitivity.[35] This method is limited by the requirement of a static sample and by the pressure necessary to transfer the acoustic signal to the microphone. One advantage that comes from the nonoptical nature of the detection is the absence of the Rayleigh line, because no energy is deposited in the sample at zero Raman shift. So pure rotational Raman spectra can be recorded by the PARS technique without interference caused by Rayleigh scattering.[35]

7.3.7. *Frequency Measurements*

The use of tunable laser sources in nonlinear Raman spectroscopy requires the calibration of laser frequencies with accuracy. Although not all spectroscopy experiments require very accurate determination of laser frequencies, most spectroscopists have laboratory wavemeters with an accuracy limited only by the linewidth of the laser. The methods differ according to the type of laser (CW or pulsed).

7.3.7.1. Continuous Lasers

A traveling double Michelson interferometer is generally used. The unknown frequency of a continuous laser is compared to a frequency reference of a stabilized laser which is commonly a red He–Ne laser stabilized on a hyperfine component of I_2. In the device developed in our laboratory the mirrors of a conventional Michelson interferometer have been replaced by corner-cube reflectors so that the light in the moving arm is always reflected in the incident direction.[36,37] The system is built so that the path difference is identical for each of the two beams. The traveling corner-cube moves vertically in an evacuated cylinder over a distance of 1 m. The duration of the measurement is relatively short (around 3 s) with respect to devices referenced in the literature. The wavemeter has been tested by measuring the frequency of a single-mode Ar^+ laser stabilized on a hyperfine component of I_2. A standard deviation of 3 MHz was attained; this means a relative accuracy of about 6×10^{-9} in the visible region.

In practice, only the initial and the final wavenumbers of each scan are determined using the wavemeter, with the rest being interpolated from transmission fringes of a temperature-stabilized confocal Fabry–Perot interferometer. This etalon generates frequency markers as the dye laser is scanned; then Raman frequencies are obtained by measuring their positions with respect to these markers, using linear interpolation between fringes. Absolute Raman shifts can be determined with an accuracy of about 15 MHz, depending on the signal-to-noise ratio of the Raman spectrum.

7.3.7.2. Pulsed Lasers

For pulsed lasers, static interferometers such as Fizeau or Fabry–Perot, or a combination of both, are used. Figure 7.14 shows a typical setup.[38] The laser beam is simultaneously sent through a system of three temperature-stabilized Fabry–Perot interferometers (FPI) having different free spectral ranges and with ratios of about 20. The fringes are imaged onto diode arrays that are read by a computer. In the past, such devices required a computation time of about 20 s to analyze the fringe pattern. Today this is unacceptable for pulsed lasers, which may be working at a repetition rate of 30 Hz. So it is necessary to develop fast fringe-fitting algorithms to calibrate the frequency. A device based on a flat FPI recently developed in our laboratory shows the possibility of measuring the frequency at close to a 30-Hz repetition rate.[39] The calibration is achieved using different frequencies of a dye laser, simultaneously measured with the traveling Michelson interferometer described earlier. The accuracy is about 20 MHz in the pulsed regime.

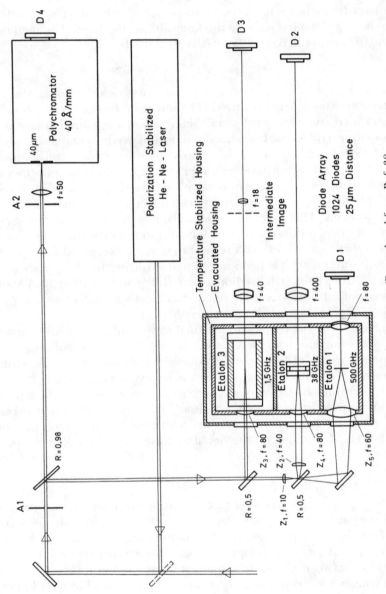

Figure 7.14 Optical arrangement for a pulsed wavemeter. (Reproduced from Ref. 38 with permission.)

7.4. Applications to the Study of Molecular Structures

As shown at the beginning of this chapter, the third-order nonlinear susceptibility $\chi^{(3)}$ has a resonance when the value of $\omega_1 - \omega_2$ (the difference of the laser frequencies) is equal to ω_{ba}, the frequency of an allowed Raman transition. The set of Raman frequencies $\{\omega_{ba}\}$ is characteristic of the molecule probed. If we consider the vibrational and rotational spectra of molecular gases, the set $\{\omega_{ba}\}$ corresponds to transitions between rotational and vibrational levels. Measurement of the frequencies of the resonance peaks provides information about these energy levels. The information that can be extracted from a coherent Raman spectrum (SRS, CARS) and that available from a spontaneous Raman spectrum are in principle of the same nature. However, only modest resolution can routinely be achieved in spontaneous spectroscopy (0.05–0.5 cm^{-1}), whereas nonlinear techniques allow high resolution, typically to a few thousandths of a cm^{-1}. This feature follows from the fact that in nonlinear Raman spectroscopy the factor limiting the experimental resolution is not the monochromator, but the spectral linewidth of the lasers. So, very close resonance peaks or Raman lines can be separated and hence the spectra of many complex molecules have been resolved for the first time. In the following, the contribution of nonlinear Raman spectroscopy to the studies of some linear and spherical top molecules are given as striking examples. A review of other contributions can be found in Ref. 13.

7.4.1. Spectroscopy of Gas-Phase Molecules

SRS and CARS techniques have demonstrated high spectral resolution capabilities in gaseous systems. The choice between these techniques must take into account the signal-to-noise ratio, and this depends on the lasers available for the experiment. In any case, the weak Raman cross section remains a limit for Raman investigations; so, among the possibilities for enhancing the sensitivity, the SRS technique, which is not phase-matched, offers the possibility of a very long interaction length thanks to a multiple-pass cell.[40–43] In Ref. 43 the multipass optics consists of two approximately concentric mirrors with a 50-cm curvature radius. The laser beams enter and exit the cell through small holes in the substrates (Figure 7.15). So the sensitivity is significantly increased by a factor of 25–30 after 40–50 passes. As a striking example, Figure 7.16 shows the weak $2\nu_2$ band of CO_2 at 1285 cm^{-1}.[43] This band was not completely resolved by CARS in a previous study. Thanks to the Raman cell it is possible to work in SRS in the 0.1–1 torr pressure range. Thus, Raman spectra can actually be recorded in the Doppler limit, providing new information on molecular structures.

7.4.1.1. Linear Molecules   A homonuclear diatomic molecule displays no dipole-allowed spectra in the infrared because of its symmetry, so Raman investigations are of a great interest. Let us take as an example the nitrogen molecule. Apart from a weak quadrupolar absorption spectrum,[44] a lot of information about rotational–vibrational levels in the electronic ground state of this molecule has come from Raman studies. The high resolution achieved by nonlinear Raman spectroscopy has the potential to give very accurate measurements of such molecular spectra.

The vibrational energy levels of a diatomic molecule such as N_2 can be described by the term values of an anharmonic oscillator[45]:

$$G(v) = \omega_e \left(v + \frac{1}{2} \right) - \omega_e x_e \left(v + \frac{1}{2} \right)^2 + \omega_e y_e \left(v + \frac{1}{2} \right)^3 + \cdots \quad (7.39)$$

where ω_e is the fundamental vibration wavenumber; $\omega_e x_e$, $\omega_e y_e$, ... are anharmonic constants; and v is the vibrational quantum number. The rotational structure of a given vibrational level v is expressed through the following formula:

$$F_v(J) = B_v J(J + 1) - D_v J^2 (J + 1)^2 + \cdots \quad (7.40)$$

Figure 7.15 Illustration of the ray trajectories in the multipass cell. (Reproduced from Ref. 43 with permission.)

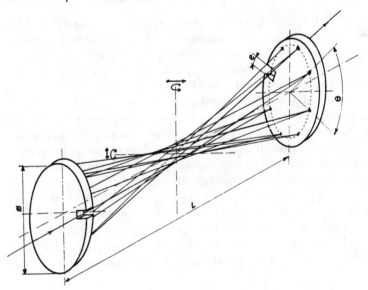

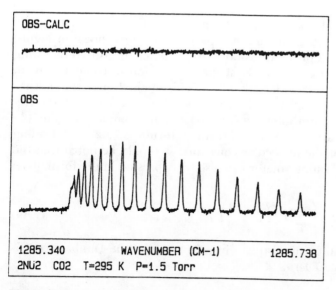

OBS-CALC

OBS

| 1285.340 | WAVENUMBER (CM-1) | 1285.738 |

2NU2 CO2 T=295 K P=1.5 Torr

Figure 7.16 Multipass Raman spectrum of the $2v_2$ band of CO_2 at 1.5 torr (lower trace). The residual of the Voigt profile fit is shown (upper trace).

where B_v is the rotational constant, D_v is the centrifugal distortion constant, and J is the rotational quantum number. B_v and D_v are generally written as expansions of $\left(v + \dfrac{1}{2} \right)$:

$$B_v = B_e - \alpha_e \left(v + \frac{1}{2} \right) + \gamma_e \left(v + \frac{1}{2} \right)^2 + \cdots \tag{7.41a}$$

$$D_v = D_e + \beta_e \left(v + \frac{1}{2} \right) + \cdots \tag{7.41b}$$

The Raman Q branch corresponds to transitions with the selection rules $\Delta v = +1$ and $\Delta J = 0$. The wavenumber of such transitions is therefore

$$v_{v \to v+1}(J) = \Delta G_{(v \to v+1)} + \Delta B_v J(J + 1)$$
$$- \Delta D_v J^2 (J + 1)^2 + \cdots \tag{7.42}$$

with

$$\Delta G_{(v \to v+1)} = \omega_e - 2\omega_e x_e (v + 1)$$
$$+ \omega_e y_e \left(3v^2 + 6v + \frac{13}{4} \right) + \cdots \tag{7.43}$$

and $\Delta B_v \approx -\alpha_e + 2\gamma_e (v + 1)$; $\Delta D_v \approx \beta_e$. The recording of the Raman Q branch of the fundamental band ($v = 0 \to v = 1$) and hot bands ($v \neq 0$) of N_2 have been very useful in determination of the molecular constants ω_e, $\omega_e x_e$, $\omega_e y_e$, ... and α_e, β_e, γ_e, ...[46–48]

The experimental spectra have been obtained by using either the CARS[46,47] or the SRS[48] technique. A typical spectrum is shown in Figure 7.17. The frequencies of the transitions are generally measured by fitting the experimental trace with a lineshape, taking into account the main causes of broadening. (See Section 7.5.1). This method gives the best results with very accurate measured frequencies. But in a few cases, the frequency of the maximum intensity can be used in a simple way. The observed frequencies are then analyzed through formulas 7.42–7.43, leading to determination of the molecular constants. It should be noted that the term values of a vibrating rotator can also be expressed in the form given by Dunham[49]:

$$T = G + F = \sum_{kl} Y_{kl} \left(v + \frac{1}{2} \right) J^l (J + 1)^l \qquad (7.44)$$

Therefore, the results can also be given in terms of the Dunham coefficients Y_{kl}. (See Table 7.1.)

7.4.1.2. Spherical Top Molecules Using coherent techniques, the Q branches of many spherical tops (e.g., $^{12}CH_4$, $^{13}CH_4$, $^{12}CD_4$, $^{13}CD_4$, SiH_4, GeH_4, CF_4, and SF_6) have been resolved. In this section, we focus on a typical example, $^{13}CD_4$.[50] An XY_4 molecule with tetrahedral symmetry is characterized by four normal vibrational modes corresponding to symmetry species in the T_d point group[51]: two stretching modes, $v_1(a_1)$ and $v_3(t_2)$, and two bending modes $v_2(e)$ and $v_4(t_2)$. In methane, the normal modes of each type of vibration (the two stretching modes and the two bending modes) are close in energy, and the stretching mode energy is twice that of the bending modes. So the harmonic and combination levels $2v_2$, $2v_4$, and $v_2 + v_4$ fall near the v_1 and v_3 levels. These vibrational levels v_1, v_3, $2v_2$, $2v_4$, $v_2 + v_4$ form a polyad of interacting bands usually called a *pentad*. A stimulated Raman spectrum of three spectral regions, including the bands $v_1 (a_1)$, $(v_2 + v_4) (t_1 + t_2)$, $2v_4(a_1)$, and $2v_4(a_1)$ has been recorded over 50 cm^{-1}.[50] A part of this spectrum is shown in Figure 7.18. The gas pressure was about 30 torr except for the weak $2v_2(a_1)$ band for which the

Table 7.1 Molecular Constants and Dunham Coefficients of N_2 (cm^{-1}).

Dunham Coefficient	Corresponding Molecular Constant	Value
Y_{11}	$(-\alpha_e)$	-0.017249
Y_{21}	(γ_e)	-3.24×10^{-5}
Y_{10}	(ω_e)	$-2,358.535$
Y_{20}	$(-\omega_e x_e)$	-14.3074
Y_{30}	—	-4.98×10^{-3}
Y_{40}	—	-1.22×10^{-4}

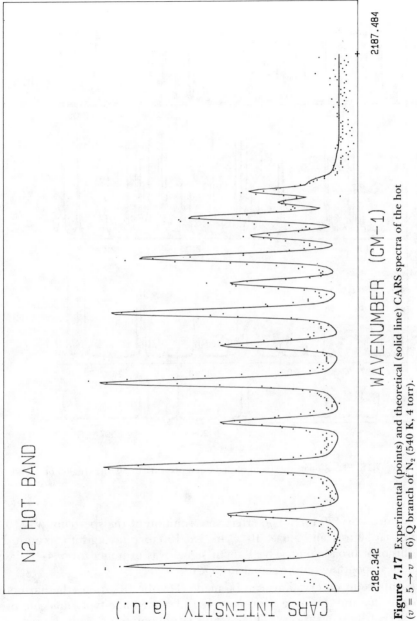

Figure 7.17 Experimental (points) and theoretical (solid line) CARS spectra of the hot ($v = 5 \rightarrow v = 6$) Q branch of N_2 (540 K, 4 torr).

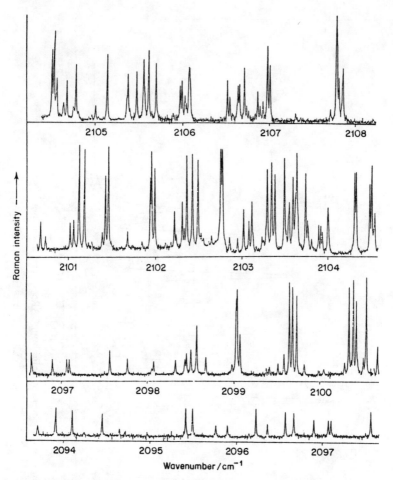

Figure 7.18 SRS Raman spectrum of $^{13}CD_4$ at a pressure of 30 torr (region of ν_1 and $\nu_2 + \nu_4$).

pressure was 150 torr. The effective resolution of the spectrum was 10^{-2} cm^{-1} at 30 torr and 3.5×10^{-2} cm^{-1} at 150 torr, taking into account the collisional broadening. The Raman signal was averaged over 4–32 laser shots, depending on the signal strength.

Important observations can be made: For the first time, the harmonic bands $2\nu_2$ and $2\nu_4$ have been observed in a methane-type molecule by a coherent Raman process. These bands strongly interact with the ν_1 band, and their intensity is enhanced. The second observation concerns the great number of measured spectral lines, which was more than 300. The accuracy of these absolute frequency measurements was less than 10^{-3} cm^{-1} in most cases. Such a study demonstrates the possibilities of stimulated Raman spectroscopy: Observation of relatively weak bands and numerous measurements of resolved lines. The signal enhancement provided by a multipass cell can also improve the study of weak bands.

As mentioned earlier, the five bands interact within the pentad, and the analysis of the data needs a model taking into account the different interactions. An effective Hamiltonian developed in terms of tensorial irreducible operators[52,53] up to the third order of approximation has been used.[50] By combining Raman and IR data, a simultaneous analysis of the pentad has been achieved. Thanks to their accuracy, the Raman data have for the first time provided a significant contribution to the determination of the Hamiltonian. The main interactions responsible for the observed perturbations were the strong second-order Coriolis interaction between $v_1(a_1)$ and the t_1 level of $(v_2 + v_4)$, and the Fermi interaction between $v_1(a_1)$, $2v_2(a_1)$, and $2v_4(a_1)$. The Raman data were less numerous compared to the infrared data, but they yielded important complementary information, in particular on the a_1 levels and their interactions with other levels.

7.4.2. Studies of Nonequilibrium Systems

The Raman spectral signature of molecules gives information about their energy levels, as shown in the previous section. One can also take advantage of the spectral signature to identify species in reactive processes or to measure state populations of species involved in such processes and changes. Indeed, information on populations is contained in the $\chi^{(3)}$ susceptibility. [$\chi^{(3)}$ is proportional to the difference of population between the states a and b. See equation (7.24).] CARS, for example, has been used to measure vibrational populations of the electronic ground state of nitrogen.[54] By recording spectra of the fundamental and hot bands up to $v = 14$ in a low-pressure DC glow discharge, the authors were able to extract vibrational populations. Non-Boltzmann behavior of the vibrational distribution becomes evident with a vibrational parameter of about 4,000–5,000 K and a rotational temperature of about 550 K.

On the contrary, in a supersonic free jet expansion, important cooling is achieved and produces a substantial lowering of rotational and vibrational temperatures. By comparison with calculated spectra it is possible to deduce the rotational temperature with an accuracy of a few degrees; such information on N_2 or O_2 is available only from Raman spectra. Moreover, one of the advantages of molecular jets is their simplification of gas-phase spectra, especially of heavy molecules that have low-energy vibrational modes thermally populated at room temperature; in a free expansion jet, the hot-bands are eliminated. Moreover, small van der Waals complexes can appear under jet conditions in concentrations sufficient for detection by coherent Raman techniques. Figure 7.19 shows an illustrative example of these two features obtained with a supersonic expansion jet by SRS.[27] The nonlinear Raman methods have the potential, under appropriate expansion conditions, to provide valuable information about the structure

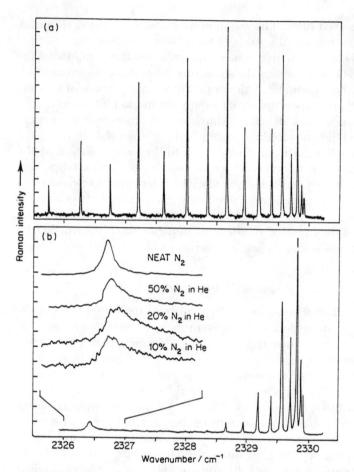

Figure 7.19 High-resolution SRS spectra of nitrogen. (a) Static sample at 50 torr, 290 K. (b) Same spectrum cooled to around 40 K in a free jet expansion. The weak feature at 2,326.4 cm^{-1} is due to N_2 clusters. The inset shows the broadening caused by added the driving gas. (Reproduced from Ref. 27 with permission.)

and dynamical properties of small molecular aggregates. A review of such studies on hydrogen-bonded clusters and van der Waals clusters can be found in Ref. 27.

7.4.3. Photochemical Applications

The use of pulsed lasers in CARS spectroscopy offers a high degree of temporal resolution (typically 10 ns) and the possibility of studying transient species generated by photolysis. In this way, the methyl radical CH_3 has been investigated by Holt and co-workers.[55,56] After illumination of a sample of diazomethane $CH_3N = NCH_3$ by a 355-nm laser pulse, a delayed CARS experiment was used to study the photoproducts N_2 and CH_3. The last undetermined vibrational frequency of the ν_1 band of the methyl radical was found at 3,004.8 cm^{-1}, and some rovibrational constants were derived from the experimental spectrum. Measurement of the vibrational distribution of the N_2 product deduced from CARS intensities has also brought new ideas about the photodissociation process of the diazomethane molecule. Other studies of photofragments by CARS spectroscopy can be found in Ref. 27.

7.5. Raman Lineshape Studies and Applications in Diagnostic Methods

Besides the initial object of nonlinear Raman spectroscopy as a tool in the determination of molecular energy levels and transition probabilities, there are several other scientific and technical applications. We focus our attention on a few examples. First, we describe the spectral lineshape studies that are essential for the main practical applications and that provide information about intermolecular forces. In a second part, we illustrate the important role of nonlinear Raman spectroscopy in diagnostic methods.

7.5.1. Spectral Lineshape Studies

To calculate SRS and CARS spectra in detail, it is essential to have the exact lineshapes for the observed transitions. The more exact this lineshape is the more accurate the physical parameters are (e.g., temperature, species concentrations, velocity) that may be extracted from the spectra. Many physical effects contribute to the spectral shape of a Raman spectrum. We only consider here the case of the pure isotropic Q branch ($\Delta v = 0$, $\Delta J = 0$). In this case, we must consider four types of molecular motion or collision:

1. The translational motion responsible for Doppler broadening;
2. The elastic collisions that change the velocity of the molecules and that are responsible for Dicke narrowing;
3. The inelastic collisions that induce pressure broadening and spectral narrowing (also called motional narrowing);
4. The elastic–vibrational phase-perturbing collisions that bring about pressure-broadening.

These narrowing and broadening mechanisms strongly depend on the gas pressure and on the molecular collisional system. Four pressure regions must be considered.

7.5.1.1. Very-Low-Pressure Region   The first region relates to very low pressure where a small number of collisions exist and
where the main contributions to lineshape are the random molecular velocities with respect to a fixed frame of reference: this corresponds to Doppler broadening. For quasi-collinear beams in forward scattering, the Doppler width is given by

$$\Gamma_D = \frac{\omega_j}{c} \sqrt{\frac{2kT \ln 2}{m}} \tag{7.45}$$

where Γ_D is the half-width at half-maximum (HWHM) expressed in wavenumbers, m is the molecular mass, T is the temperature, and $\omega_j = \omega_{ba}$ is the
transition frequency (in cm^{-1}). The Doppler halfwidth at $1/e$ of the peak
intensity is

$$\Gamma'_D = \frac{\Gamma_D}{\sqrt{\ln 2}} \tag{7.46}$$

The third-order complex susceptibility $\chi^{(3)}$ can be expressed in terms of the
complex error function $W_{(z)}$, whose general expression is

$$W(z) = \frac{1}{\pi} \int_{-\infty}^{+\infty} \frac{e^{-t^2} dt}{y - i(x - t)} \tag{7.47}$$

with $z = x + iy$ $(y > 0)$. Here x is the normalized frequency separation
from line center and $y = \Gamma/\Gamma'_D$, where Γ is the collisional halfwidth at half-
maximum; this is related to the linebroadening coefficient γ by $\gamma = \Gamma/p$,
where p is the pressure. When $y = 0$, the susceptibility $\chi^{(3)}$ is given by

$$\chi^{(3)} \propto \frac{\sqrt{\pi}}{\Gamma'_D} \sum_j i \, \alpha_j \, \Delta\rho_j \, W(-x_j) \tag{7.48}$$

where subscript j denotes a specific transition, $x_j = [\omega_j - (\omega_1 - \omega_2)]/\Gamma'_D$, α_j
is the polarizability matrix element of the specific transition j, and $\Delta\rho_j$ is the
difference in population of the energy levels connected by this transition.
The complex error function W is easily computed by a FORTRAN subrou-
tine given by Humlicek.[57] In this limit of zero homogeneous linewidth (y
$= 0$), the real part of W is purely gaussian: $D(x) = e^{-x^2}$. This is precisely
the lineshape obtained in forward spontaneous Raman scattering and stimulated Raman spectroscopy (SRS). In the case of CARS lines, the simple
expression $D(x)$ is incorrect because of the additional real part of $\chi^{(3)}$.

7.5.1.2. Low-Pressure Region   The second-pressure region corresponds to very low to medium pressures, where both random
thermal motion and molecular collisions are significant. Molecular collisions induce two physical effects. The most important is the pressure

broadening and shifting of the isolated lines. The second, which is significant under certain conditions, is the effect of velocity-changing collisions called Dicke narrowing.[58] If Dicke narrowing is neglected, $\chi^{(3)}$ is then obtained using equation (7.48), where x_j is substituted by z_j^*, with $z_j = x'_j + iy_j$; $x'_j = x_j - s_j$; $s_j = \Delta_j/\Gamma'_D$ and $y_j = \Gamma_j/\Gamma'_D$. Here Δ_j denotes the line shift, related to the line-shifting coefficient by $\delta_j = \Delta_j/p$, and x'_j is the frequency relative to the shifted line center.

In the case of SRS, the lineshape [which depends on Im $\chi^{(3)}$ or Re $W(z)$] is a Voigt profile, which is the convolution on a Lorentzian of a Gaussian:

$$V(x', y) = \frac{y}{\pi} \int_{-\infty}^{+\infty} \frac{dt\, e^{-t^2}}{y^2 + (x' - t)^2} \tag{7.49}$$

For certain collisional systems, such as light molecules like H_2 and D_2, inelastic velocity-changing collisions induce a reduction of the Doppler width. This phenomenon can be explained in the following way. Because of the Heisenberg uncertainty principle, $\Delta x \, \Delta p \geq \hbar$, the instantaneous velocity of the molecule at a given position in space cannot be exactly determined. So only the average velocity can be obtained for displacement of the molecule greater than $\hbar/\Delta p$. Assuming that the absorption of a photon of momentum h/λ induces an uncertainty of the same magnitude in the molecular momentum, the velocity in the direction of observation is averaged over displacements of $\sim\lambda/2\pi$. If there are many collisions during the time corresponding to this distance, the mean velocity tends to zero and the Doppler width disappears. For intermediate cases, there is a reduction of the Doppler width.

Galatry has proposed a model that takes into account Dicke narrowing through a new parameter η related to the optical diffusion coefficient D by $\eta = kT/2\pi mD\Gamma'_D$.[59] Then in equation (7.51) $W(z)$ is substituted by the complex Galatry function

$$G(z, \eta) = \frac{1}{\sqrt{\pi}} \int_0^\infty dt\, \exp\left[-iz^* t + \frac{1}{2\eta^2}(1 - \eta t - e^{-\eta t}) \right] \tag{7.50}$$

We can note that the Galatry profile leads to the Voigt profile when $\eta \to 0$ and leads to the Lorentz profile when y becomes large.

7.5.1.3. Medium-Pressure Region

The medium-pressure region corresponds to low to high pressures, where pressure broadening and shifting dominate and where Doppler broadening has disappeared owing to Dicke narrowing. For the same collisional systems, where the $Q(J)$ lines are relatively close to each other, a new effect appears. Indeed, for simple diatomic molecules like N_2, O_2, and CO, the widths and frequency separation of the $Q(J)$ lines become similar for a certain pressure, leading to a partially unresolved Q branch. For pressures below this limit the Q-branch shape is given by a simple Lorentzian function, and a summation over all transitions with significant intensities is effected. For pressures above this limit, the Raman Q branch shows a small collapse or "nonadditivity" effect. This marks the onset of the phenomenon of "motional narrowing," which results in line interferences caused by collision-induced changes in rotational state. In other words, the broadening of the individual Q lines mixes the states associated with these different lines. For moderate overlap, the line coupling can be treated within first-order perturbation theory and the Lorentzian function is substituted by the Rosenkranz profile,[60] whose normalized expression for the SRS isotropic fundamental Q branch ($v = 0 \rightarrow v = 1$) is the following:

$$I(\omega) = \frac{kN}{\pi} \left| <0 \mid \bar{\alpha} \mid 1> \right|^2 \sum_j \Delta \rho_j \frac{\Gamma_j - (\omega - \omega_j) Y_j}{(\omega - \omega_j)^2 + \Gamma_j^2} \qquad (7.51)$$

where $\omega = \omega_1 - \omega_2$, Y_j are the line-coupling parameters characterizing the coupling of the J lines with all the other lines, and where the J dependence of α has been disregarded. N is the number density of perturbers, and k is a constant of proportionality that follows from equations (7.24) and (7.29).

7.5.1.4. High-pressure Region

As discussed earlier, at medium pressure the J lines start to overlap; this phenomenon increases at higher pressures, and finally the J lines collapse into a single band whose width decreases with the density. Furthermore, the Rosenkranz profile breaks down and a more complicated expression must be used[61,62]:

$$I(\omega) = \frac{kN}{\pi} | <0 \mid \bar{\alpha} \mid 1> |^2 \, \text{Re} \sum_{jj'} \Delta \rho_j \left(G^{-1}(\omega) \right)_{j'j} \qquad (7.52)$$

where

$$\left(G(\omega) \right)_{j'j} = (\omega - \omega_{j'} + N \delta_{j'} - i N \gamma_{j'}) \delta_{j'j}$$
$$+ N W_{j'j} (1 - \delta_{j'j}) \qquad (7.53)$$

In equations (7.52) and (7.53), $\delta_{j'j}$ is the Kronecker symbol and $W_{j'j}$ is a matrix element of the rotational relaxation rate matrix W for the $j \rightarrow j'$ transition in both 0 and 1 vibrational states.

Since the vibrational energy transfers are negligible, we have the following sum rule:

$$- \sum_{j' \neq j} W_{j'j} = \gamma_j - \gamma_v \qquad (7.54)$$

where γ_v is the pure vibrational phase relaxation. Furthermore, for most molecular collisions, the line-shifting coefficients depend only weakly on the rotational quantum number, such that $\delta_j \approx \delta_v$, and the imaginary part of W can be neglected. We can also note that to maintain thermal equilibrium, the detailed balance principle should apply.[63]:

$$\rho_j W_{j'j} = \rho_{j'} W_{jj'} \tag{7.55}$$

So at high pressure, when $Q(J)$ lines overlap, calculation of the SRS and CARS Q branch requires knowledge of $N(N - 1)/2$ matrix elements when NJ lines are taken into account. This problem can usually be solved if physical assumptions are made on the dependence of $W_{j'j}$ as a function of the rotational quantum numbers.

Several models have been used up to now to calculate all the relaxation matrix elements from the knowledge of the line-broadening coefficients γ_j using equation (7.54). We will describe here the one most used in technical applications such as CARS thermometry. This model is called the modified exponential gap law (MEGL) and is expressed as follows[64]:

$$W_{j'j} = \alpha \left\{ \frac{1 - e^{-m}}{1 - e^{-mT/T_0}} \left(\frac{T_0}{T}\right)^{1/2} \right\} \left(\frac{1 + 1.5\, E_j/kT\Delta}{1 + 1.5\, E_j/kT}\right)^2$$

$$\exp\left[-\beta \mid \Delta E_{j'j} \mid /kT \right] \tag{7.56}$$

where the parameters are α, m, Δ, and β. Here T_0 is a reference temperature and E_j is the rotational energy. Let us remark that the term in braces is sometimes replaced by $(T/T_0)^\nu$. The MEGL parameters are fitted to experimental line-broadening coefficients using equation (7.54) and then applied to determine all relaxation matrix elements whatever the temperature. The spectral band shape may then be obtained using equations (7.52) and (7.53) but may involve expensive computer time because of the requirement for matrix inversion at each frequency ω. So it is better to diagonalize the problem and work on the eigenvectors and eigenvalues.[64]

7.5.2. Collisional Effects

The ability of SRS to measure accurate linewidths and lineshifts has been extensively demonstrated. Important information on these physical parameters can usually be obtained by varying the temperature. There are a large number of experimental Raman linewidths but only a few lineshift measurements. This can be easily explained since the lineshifts are in most cases very small (one order of magnitude smaller than the linewidths). Moreover, determination of the Raman lineshifts is also important because they result from different collisional mechanisms from those responsible for linewidths.

7.5.2.1. Dicke Narrowing Among the various physical effects contributing to the linewidth, Dicke narrowing plays an important role for small molecules like H_2. The density dependence of the experimental linewidths has been studied by Bischel and Dyer at three temperatures, as shown in Figure 7.20.[65] A reduction of the Doppler width clearly appears.

7.5.2.2. Nonadditivity Effect We have seen that, when the pressure leads to significant line overlap, the Raman profile is not a simple superposition of Voigt line shapes due to interferences between the *J* lines. A good example given by Lavorel and co-workers is the determination of collisional line-broadening coefficients over a wide temperature range in the fundamental SRS Q branch of N_2.[66] Figure 7.21 shows a comparison between least-squares Rosenkranz and Voigt profile fit to SRS data at $T = 295$ K and $p = 965$ torr.

7.5.2.3. Motional Narrowing Starting from a large and accurate set of line-broadening coefficients (see the following section), the full relaxation matrix can be calculated using a relaxation model such as the MEGL model described in equation (7.56). The relaxation matrix is then introduced in equations (7.52) and (7.53), and the Raman profile can be calculated for any density and temperature. To test the capability of the fitting law to give a good lineshape, the simulation is compared to Raman spectra recorded at high density. Such a comparison is given for nitrogen

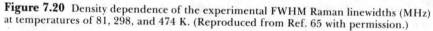

Figure 7.20 Density dependence of the experimental FWHM Raman linewidths (MHz) at temperatures of 81, 298, and 474 K. (Reproduced from Ref. 65 with permission.)

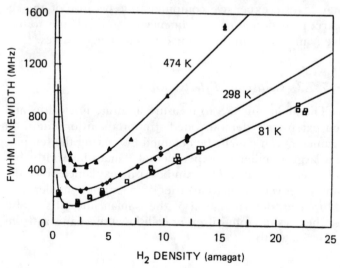

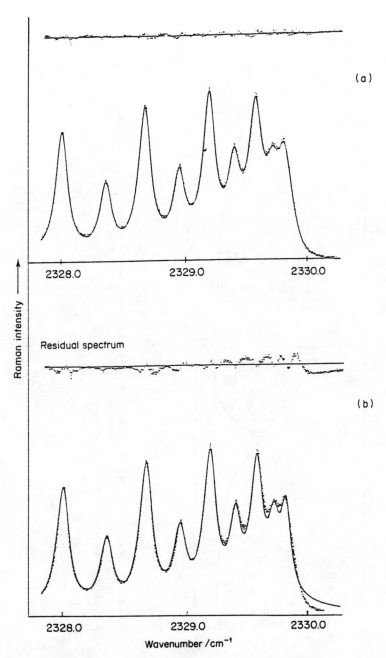

Figure 7.21 Least-squares (a) Rosenkranz and (b) Voigt profile fit (solid line) to the experimental spectrum (points) for the band head of self-perturbed N_2 molecule at $T = 295$ K and $p = 968$ torr.

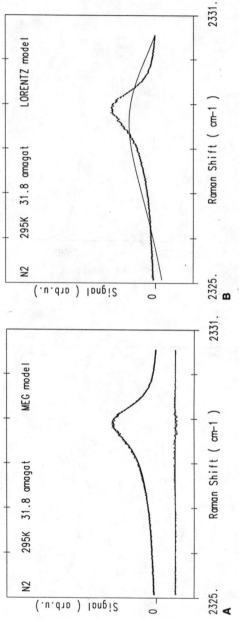

Figure 7.22 SRS spectra of N_2 at 295 K and 31.8 amagat: (a) Comparison between experimental spectrum and calculated spectrum taking into account motional narrowing. (b) Calculated spectrum not including the motional narrowing.

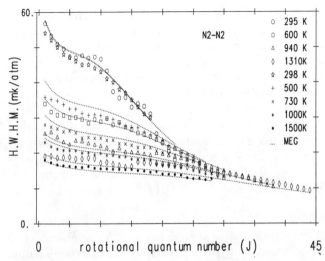

Figure 7.23 Line-broadening coefficients of N_2 calculated from MEGL model. The experimental values are indicated with various symbols according to the temperature.

at 31.8 amagat[1] and 295 K, as shown in Figure 7.22a. Figure 7.22b shows the Raman profile obtained by neglecting the effects of motional narrowing. The necessity of including this effect is evident.

7.5.3. Determination of Raman Line-Broadening Coefficients

The study of a collapsed Q branch at high density requires knowledge of the line-broadening coefficients at various temperatures to determine the full relaxation matrix W for arbitrary temperature. Consequently, many studies, particularly SRS studies, have been performed aimed at determining a complete set of $\gamma_j(T)$ coefficients. Among them one can mention the example of pure nitrogen, for which the SRS $\gamma_j(T)$ have been obtained by several groups.[66-68] (See Figure 7-23.)

7.5.4. Determination of Raman Line-Shifting Coefficients

7.5.4.1. Measurement at Low Density   The measurement of frequency shift with change in density (or pressure) is of great interest for calculating high-density spectra and modeling rotational transfer rates. Indeed, the accounting for these shifts in equations (7.52) and (7.53) allows

---

[1]The density of a gas in amagat units is the ratio of the molar volume at STP to the actual volume at a given temperature and pressure. For H_2 the density at 1 atm and 295 K corresponds to 0.925 amagat.

for a more critical test of the relaxation model.[63] Let us recall that changing off-diagonal W-matrix elements has the effect of shifting Q-branch lines toward the most probable quantum number looking like an overall frequency shift. However, line-frequency shifts depend only weakly on the rotational quantum number J. Therefore, there is a strong correlation between the transfer rates and the frequency shift. To reduce this correlation significantly, direct measurements of lineshifts are performed at low densities from a resolved Q-branch structure where the off-diagonal relaxation matrix elements give negligible effect. Such measurements have been realized for pure nitrogen by SRS for $J = 6$ up to $J = 22$ at room temperature and in the density range 0.02–0.80 amagat.[69] An example of the density dependence of the frequency shift of the Q(14) line is shown in Figure 7.24.

7.5.4.2. Measurement at High Density   Another means of reducing the correlation between off-diagonal relaxation matrix elements and frequency shift is to work at high density, where collisional narrowing is complete (and not very sensitive to the relaxation model) and consequently pure vibrational dephasing mechanisms are predominant. Such a situation can be found in the ν_1 band of CO_2, where the line-coupling

Figure 7.24 Lineshift measurements of the frequency of the Q_{14} line of N_2 as a function of density.

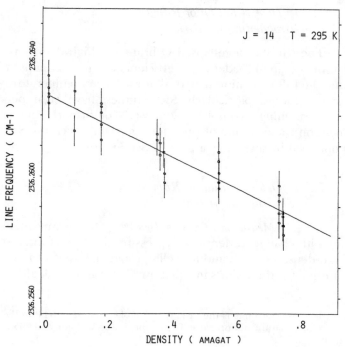

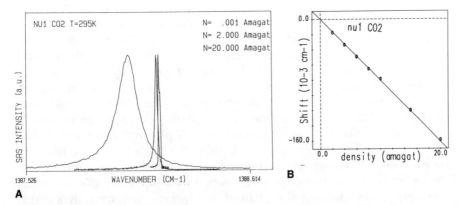

Figure 7.25 Illustration of the line shifting and line broadening on the Q branch of CO_2.

mechanism is very efficient due to the very small frequency separation between J lines.[70] An example is given in Figure 7.25a showing the overall frequency shift of the full Q branch as a function of the density. It is also interesting to see the broadening of the band due to vibrational dephasing. The collisional shift deduced from these spectra is shown in Figure 7.25b and exhibits a good linear dependence on density.

7.5.5. CARS Thermometry and Concentration Measurements

CARS is extensively used for nonintrusive measurements of temperature and concentrations in combustion media, as detailed in recent reviews.[71,72] A broadband CARS device is described in Section 7.3.2.3 is generally used. These investigations concern not only real turbulent combustion systems, but also stationary flames that allow testing of the reproducibility and accuracy of diagnostic methods. Collinear phsaematching often leads to poor spatial resolution because the CARS signal undergoes an integrative growth process. So a BOXCARS geometry is preferable, leading to a spatial resolution of about 1 mm. Numerous instantaneous spectra are generally collected and each spectrum is individually processed. The temperature deduced from lineshape study is generally compared with temperature measured by thermocouples or IR pyrometry in the temperature range of the instrument.

The noise in broadband single-pulse CARS spectroscopy limits the accuracy to around 5% to 10%. This noise has two origins: mode noise in the broadband dye laser and shot noise in the multichannel detector. In fact, flame temperature fluctuations are usually of the order of 20%; therefore, instrumental noise is not a major limitation. With a scanning CARS spectrometer, the accuracy of temperature measurement is better, typically by 3% in the range 1,500–2,000 K.

Molecular nitrogen is very often the molecule probed in CARS thermometry, which explains why this molecule has been the focus of much of the work. However, the temperature can also be deduced from the spectral shape of other species present in the combustion like O_2, CO_2, and H_2O. The coherent Raman techniques normally have one limitation when compared to spontaneous Raman scattering, because only one molecular species is probed at a time. Simultaneous information on all species present in combustion could be of great importance. A new technique called *dual broadband CARS* has been introduced using two broadband dye lasers in conjunction with a narrowband pump laser.[73] Each dye laser is mixed with the pump laser via a two-color wave-mixing simultaneously, allowing measurement of the vibrational spectra of two species. Moreover, by a three-color wave-mixing (nondegenerate CARS), the resonances that are within the frequency difference range of the two broadband sources can be probed. For example, the vibrational spectra of CO_2, N_2, and H_2O in a C_2H_4 air-flame have been simultaneously recorded, as shown in Figure 7.26.[74] Farrow and Lucht have also developed a three-laser CARS setup for simultaneous high-resolution multispecies investigations.[75] In this device, two narrow-bandwidth pump lasers are used: One is a 532-nm frequency-doubled Nd–YAG laser and the second is a narrowband dye laser. The CARS signal from the two species are generated at nearly the same frequency and thus detected using a single intensified linear diode array.

The importance of modeling is illustrated in Figure 7.27. Neglecting motional narrowing effects leads to a temperature error of 290 K. The motional narrowing is not uniquely a collisional effect, as shown in Section 7.5.2, and that explains the need for improvement in modeling to determine temperatures with higher accuracy. Since in a high-pressure environment, vibrational CARS spectra are strongly affected by motional narrowing, rotational CARS spectra could be easier to analyze, since the individual Raman lines do not overlap.

In addition to temperature measurements, the CARS technique provides information on the fluctuating properties of turbulent combustion. Concentration measurements are more difficult to perform than temperature ones because the absolute intensity is required. Temperature measurements are only based on the shape of the spectrum, with an arbitrary coefficient of intensity fitted in the least-squares analysis of the CARS spectrum. The main problem lies within the overlap of two laser beams; for absolute intensity measurements, a reference spectrum must be recorded simultaneously. Conventional CARS offers a detection sensitivity of about 1,000 ppm, and improvements of about one order of magnitude can be obtained using cancellation of the nonresonant background. For very-low-concentration measurements, resonance CARS, laser-induced fluorescence, or degenerate four-wave mixing seem to be preferable.

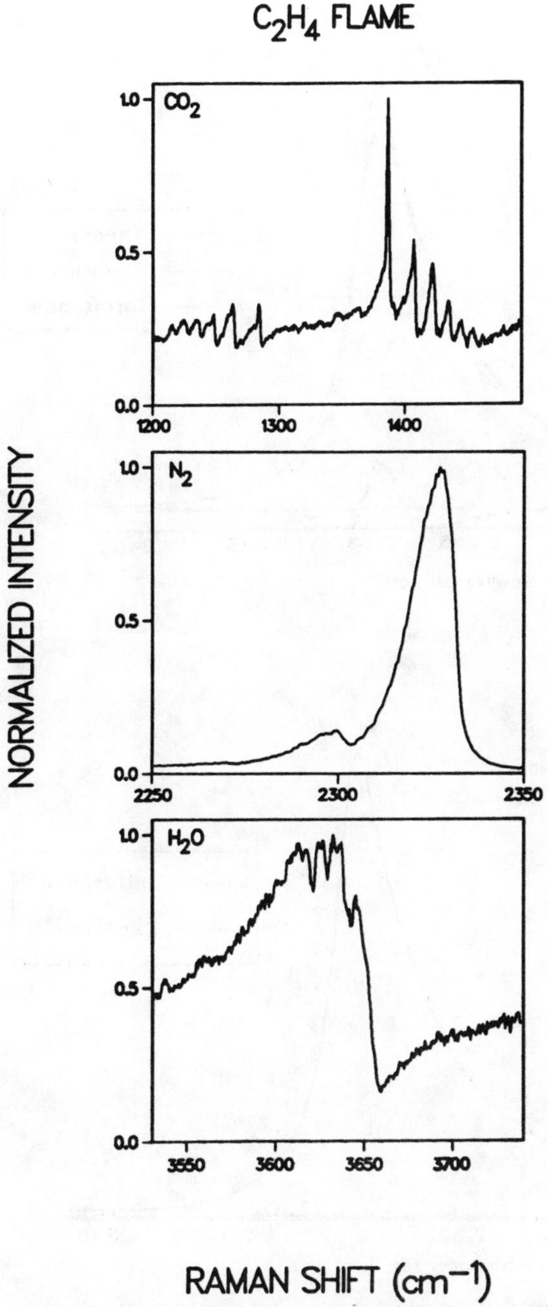

Figure 7.26 Simultaneously generated dual broadband CARS spectra of CO_2, N_2, and H_2O in the postflame zone of a premixed C_2H_4–air flame. (Reproduced from Ref. 74 with permission.)

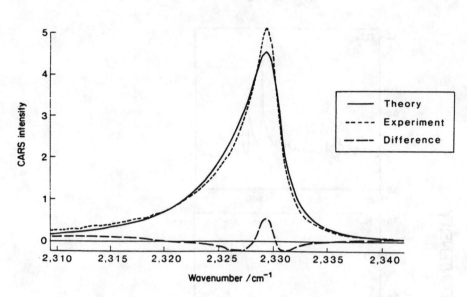

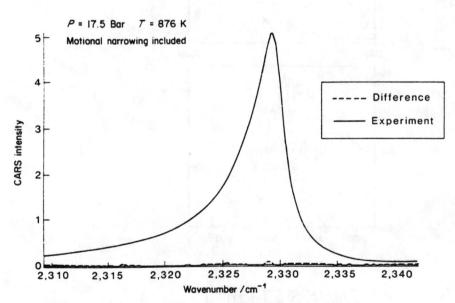

Figure 7.27 Precombustion spectrum of nitrogen in an internal combustion engine: (a) fitted using an isolated lines CARS model ($T = 587$ K); (b) motional narrowing included ($T = 876$ K). (Reproduced from Ref. 71 with permission.)

7.5.6. Raman Velocimetry

Coherent Raman techniques (SRS[76] and CARS[77]) have been used as direct methods for measuring high-speed molecular flows. Experiments on N_2 are of particular interest for measurements in wind tunnels.[78] The principle of the method is based on the Doppler shift of the molecular resonance due to molecular motion; with the velocity **v**, the Doppler shift in angular frequency is $\Delta\omega = (\mathbf{k}_1 - \mathbf{k}_2)\cdot\mathbf{v}$, where $\mathbf{k}_1$ and $\mathbf{k}_2$ are the wave vectors of the pump and probe beams. Velocity measurements are performed by comparing the Raman spectrum from flowing molecular species to a stationary spectrum. The accuracy of the method depends on the calibration method. An absolute frequency reference is obtained in a simple way by recording the Doppler-free saturated absorption spectrum of I_2 during the scan of the dye. This allows comparison of Raman spectra recorded at different times overlapping the I_2 spectra. The precision of these measurements is better than 5%.[78] In wind tunnel applications the pressure does not exceed 1 atm. At higher pressures, the technique is less accurate because the broadened linewidth makes determination of the Doppler shift more difficult. In this case, another nonlinear technique, called *stimulated Rayleigh-Brillouin spectroscopy,* seems preferable.[79]

7.6. Conclusion

In recent years, a great number of nonlinear Raman techniques have been developed and utilized in widely different fields of physics, chemistry, and biology. In this chapter, we have discussed the theory and practice of these techniques, except for picosecond Raman spectroscopy, which is discussed in Chapter 10. We have placed particular emphasis on SRS and CARS, which are extensively used for quantitative studies. Other six-wave mixing processes, including CSARS, have been reviewed briefly. We have shown that they offer new possibilities for investigation but remain experimentally difficult.

The applications we have discussed essentially concern gaseous systems because the advantages of coherent techniques over spontaneous Raman spectroscopy are more spectacular here than in condensed-phase studies. If CARS is extensively used today in combustion diagnostics methods, various applications providing information on frequency, line-broadening, and line-shifting measurements establish SRS as the real standard for high-resolution Raman studies.

REFERENCES

1. P. Regnier and J.-P. E. Taran, *Appl. Phys. Lett.* **23**, 240 (1973).
2. G. L. Eesley, *Coherent Raman Spectroscopy* (Pergamon, Oxford, 1981).

3. M. D. Levenson, *Introduction to Nonlinear Laser Spectroscopy* (Academic Press, New York, 1982).

4. Y. R. Shen, *The Principles of Nonlinear Optics* (Wiley, New York, 1984).

5. P. D. Maker and R. W. Terhune, *Phys. Rev. A* **137**, 801 (1965).

6. P. A. Franken, A. E. Hill, C. W. Peters, G. Weinreich, *Phys. Rev. Lett.* **7**, 118 (1961).

7. M. Bass, P. A. Franken, J. F. Ward, and G. Weinreich, *Phys. Rev. Lett.* **9**, 446 (1962).

8. J. F. Ward, *Phys. Rev.* **143**, 569 (1966).

9. C. Flytzanis, in *Quantum Electronics: A Treatise*, Vol. I, Part A, H. Rabin and C. L. Tang, eds. (Academic Press, New York, 1975), p. 9.

10. E. J. Woodbury and W. K. Ng, *Proc. IRE* **50**, 2367 (1962).

11. D. L. Andrews, *Lasers in Chemistry* (Springer-Verlag, Berlin, 1986).

12. P. Esherick and A. Owyoung, in *Advances in Infrared and Raman Spectroscopy*, Vol. 9, R. J. H. Clark and R. E. Hester, eds. (Heyden, London, 1982), p. 130.

13. H. W. Schrötter, H. Frunder, H. Berger, J. P. Boquillon, B. Lavorel, and G. Millot, in *Advances in Nonlinear Spectroscopy*, Vol. 15, R. J. H. Clark and R. E. Hester, eds. (Wiley, London, 1988), p. 97.

14. R. W. Terhune, P. D. Maker, and C. M. Savage, *Phys. Rev. Lett.* **14**, 681 (1965).

15. S. Yatsiv, M. Rokni, and S. Barak, *IEEE J. Quantum Electron.* **QE-4**, 900 (1968).

16. H. Berger, *Opt. Commun.* **25**, 179 (1978).

17. D. Debarre, M. Lefebvre, and M. Péalat, *Opt. Commun.* **69**, 362 (1989).

18. Y. Prior, *IEEE J. Quantum Electron.* **QE-20**, 37 (1984).

19. M. A. Yuratich and D. C. Hanna, *Mol. Phys.* **33**, 671 (1977).

20. A. C. Eckbreth, *Appl. Phys. Lett.* **32**, 421 (1978).

21. J. A. Shirley, R. J. Hall, and A. C. Eckbreth, *Opt. Lett.* **5**, 380 (1980).

22. J. Laane and W. Kiefer, *J. Raman Spectrosc.* **9**, 353 (1980).

23. J. P. Boquillon, Y. Ouazzany, and R. Chaux, *J. Appl. Phys.* **62**, 23 (1987).

24. H. Frunder, L. Matziol, H. Finsterhölzl, A. Beckmann, and H. W. Schrötter, *J. Raman Spectrosc.* **17**, 143 (1986).

25. E. K. Gustafson, Ph.D. thesis, Stanford University, 1983.

26. A. Owyoung, in *Laser Spectroscopy*, Part IV, Vol. 21, H. Walther and K. W. Rothe, eds. (Springer, Berlin, 1979), p. 175.

27. J. W. Nibler and G. A. Pubanz, in *Advances in Nonlinear Spectroscopy*, Vol. 15, R. J. H. Clark and R. E. Hester, eds. (Wiley, London, 1988), p. 1.

28. L. D. Ziegler and J. L. Roeber, *Chem. Phys. Lett.* **136**, 377 (1987).

29. R. Brakel and F. W. Schneider, in *Advances in Nonlinear Spectroscopy*, Vol. 15, R. J. H. Clark and R. E. Hester, eds. (Wiley, London, 1988).

30. L. A. Rahn, L. J. Zych, and P. L. Mattern, *Opt. Commun.* **30**, 249 (1979).

31. V. Wilke and W. Schmidt, *Appl. Phys.* **18**, 177 (1979).

32. P. Esherick, A. Owyoung, and J. Pliva, *J. Chem. Phys.* **83**, 3311 (1985).

33. P. Esherick and A. Owyoung, post-deadline paper, Conference on Lasers and Electro-optics, May 17–20, 1983, Baltimore.

34. J. J. Barett and M. J. Berry, *Appl. Phys. Lett.* **34**, 144 (1979).

35. J. J. Barett, in *Chemical Applications of Nonlinear Raman Spectroscopy*, A. B. Harvey, ed. (Academic Press, New York, 1981), p. 89.

36. C. Milan, M. Pullicino, G. Roussel, and J. Moret-Bailly, *J. Opt. (Paris)* **15**, 31 (1984).

37. R. Chaux, C. Milan, G. Millot, B. Lavorel, R. Saint-Loup, and J. Moret-Bailly, *J. Opt. (Paris)* **19**, 3 (1988).

38. A. Fischer, R. Kullmer, and W. Demtröder, *Opt. Commun.* **39**, 277 (1981).

39. M. Rotger, R. Chaux, H. Berger, and J. Moret-Bailly, *J. Opt. (Paris)* **21**, 193 (1990).

40. A. Owyoung, in *Chemical Applications of Nonlinear Raman Spectroscopy*, A. B. Harvey, ed. (Academic, New York, 1981), p. 281.

41. P. Esherick and A. Owyoung, in *Advances in Infrared and Raman Spectroscopy*, Vol. 9, R. J. H. Clark and R. E. Hester, ed. (Heyden, London, 1983).

42. W. R. Trutna and R. L. Byer, *Appl. Opt.* **19**, 301 (1980).

43. R. Saint-Loup, B. Lavorel, G. Millot, C. Wenger, and H. Berger, *J. Raman Spectrosc.* **21**, 77 (1990).

44. D. Reuter, D. E. Jennings, and J. W. Brault, *J. Mol. Spectrosc.* **115**, 294 (1986).

45. J. M. Hollas, *High Resolution Spectroscopy* (Butterworths, London, 1982).

46. T. R. Gilson, I. R. Beattie, J. D. Black, D. A. Greenhalgh, and S. N. Jenny, *J. Raman Spectrosc.* **9**, 361, (1980).

47. B. Lavorel, G. Millot, M. Lefebvre, and M. Péalat, *J. Raman Spectrosc.* **19**, 375 (1988).

48. B. Lavorel, R. Chaux, R. Saint-Loup, and H. Berger, *Opt. Commun.* **62**, 25 (1987).

49. J. L. Dunham, *Phys. Rev.* **41**, 721 (1932).

50. G. Millot, B. Lavorel, R. Chaux, R. Saint-Loup, G. Pierre, H. Berger, J. I. Steinfeld, and B. Foy, *J. Mol. Spectrosc.* **127**, 156 (1988).

51. G. Herzberg, *Molecular Spectra and Molecular Structure*, Vol. 2 (Van Nostrand Reinhold, New York, 1945).

52. J. P. Champion, *Can. J. Phys.* **55**, 1802 (1977).

53. J. P. Champion and G. Pierre, *J. Mol. Spectrosc.* **79**, 255 (1980).

54. B. Massabiaux, G. Gousset, M. Lefebvre, and M. Péalat, *J. Phys. (Paris)* **48**, 1939 (1987).

55. P. L. Holt, K. E. McCurdy, R. B. Weisman, J. S. Adams, and P. S. Engel, *J. Chem. Phys.* **81**, 3349 (1984).

56. P. L. Holt, K. E. McCurdy, J. S. Adams, K. A. Burton, R. B. Weisman, and P. S. Engel, *J. Am. Chem. Soc.* **107**, 2180 (1985).

57. J. Humlicek, *J. Quantum Spectrosc. Radiat. Transfer* **21**, 309 (1979).

58. R. H. Dicke, *Phys. Rev.* **89**, 472 (1953).

59. L. Galatry, *Phys. Rev.* **122**, 1218 (1961).

60. P. W. Rosenkranz, *IEEE Trans. Antennas Propag.* **23**, 498 (1975).

61. V. Fano, *Phys. Rev.* **131**, 259 (1963).

62. A. Ben Reuven, *Phys. Rev.* **145**, 7 (1966).

63. B. Lavorel, G. Millot, J. Bonamy, and D. Robert, *Chem. Phys.* **115**, 69 (1987).

64. M. L. Koszykowski, R. L. Farrow, and R. E. Palmer, *Opt. Lett.* **10,** 478 (1985).

65. W. K. Bischel and M. J. Dyer, *Phys. Rev.* **A33,** 3113 (1986).

66. B. Lavorel, G. Millot, R. Saint-Loup, C. Wenger, H. Berger, J. P. Sala, J. Bonamy, and D. Robert, *J. Phys. (Paris)* **47,** 417 (1986).

67. G. J. Rosasco, W. Lempert, W. S. Hurst, and A. Fein, *Chem. Phys. Lett.* **97,** 435 (1983).

68. L. A. Rahn and R. E. Palmer, *J. Opt. Soc. Am.* **B3,** 1164 (1986).

69. B. Lavorel, R. Chaux, R. Saint-Loup, and H. Berger, *Opt. Commun.* **62,** 25 (1987).

70. B. Lavorel, G. Millot, R. Saint-Loup, H. Berger, L. Bonamy, J. Bonamy, and D. Robert, *J. Chem. Phys.,* **93,** 2176 (1990).

71. D. A. Greenhalgh, in *Advances in Nonlinear Spectroscopy,* R. J. H. Clark and R. E. Hester, eds. (Wiley, London, 1988), p. 193.

72. R. J. Hall and A. C. Eckbreth, in *Laser Applications,* Vol. 5, J. F. Ready and R. K. Erf, eds. (Academic, New York, 1984), p. 213.

73. A. C. Eckbreth and T. J. Anderson, *Appl. Opt.* **24,** 2731 (1985).

74. A. C. Eckbreth and T. J. Anderson, *Appl. Opt.* **25,** 1534 (1986).

75. R. L. Farrow and R. P. Lutch, Proceedings of the Office of Basic Energy Sciences Contractors Review Meeting, Oakland, 1987.

76. G. C. Herring, W. M. Fairbank, Jr., and C. Y. She, *IEEE J. Quantum Electron.* **QE-17,** 1975 (1981).

77. E. Gustafson, J. M. McDaniel, and R. L. Byer, *IEEE J. Quantum Electron* **QE-17,** 2258 (1982).

78. G. C. Herring, S. A. Lee, and C. Y. She, *Opt. Lett.* **8,** 214 (1983).

79. G. C. Herring, H. Moosmüller, S. A. Lee, and C. Y. She, *Opt. Lett.* **8,** 603 (1983).

8 Multiphoton Absorption Spectroscopy

L. Goodman and J. Philis

8.1. Introduction

The possibility that a molecule might simultaneously absorb more than one photon was foreseen as early as 1929 (many years before lasers were even thought of) by the future Nobel laureate Maria Göppert-Mayer.[1] However, it was not until 1961 that Kaiser and Garrett carried out their famous experiment in which absorption of red photons from an intense ruby laser pulse induced ultraviolet fluorescence from Eu^{2+} doped into CaF_2 crystals.[2] The subsequent development of the tunable dye laser, particularly by Hansch in the 1970s, served as the impetus behind much of the activity that has taken place in visible–UV multiphoton electronic spectroscopy over the last two decades.

Particularly rapid growth has occurred in connection with molecular spectra excited by the simultaneous absorption of two photons. An appreciation of the power of two-photon spectroscopy in probing excited states can be obtained by contrasting the gerade⇔gerade ($g{\Leftrightarrow}g$) parity selection rule for a two-photon transition in a centrosymmetric molecule with the gerade⇔ungerade ($g{\Leftrightarrow}u$) rule for traditional single-photon electronic absorption.[3] This difference between two-photon and single-photon excitation is analogous to the difference between Raman and infrared spectroscopies.

Such ingenious technological developments as multiphoton ionization,[4] fluorescence excitation,[3] and thermal lensing[5] for the detection of elec-

319

tronic two-photon transitions have allowed this spectroscopy to become widespread. Refinements such as Doppler-free,[6] ion-dip, stimulated-emission-pumping,[7,8] and multiphoton ionization mass[9,10] spectroscopies are leading to the discovery of new vibronic (i.e., vibration-electronic) states, which are inaccessible by single-photon spectroscopy. The multiphoton ionization technique in particular, developed independently by Dalby[11] and Johnson[12,13] in 1975, has made great contributions to the spectroscopy of molecular Rydberg states. Use of supersonic jets increases the amount of spectroscopic data by allowing band resolution seldom possible with static cells.

A comprehensive discussion of multiphoton spectroscopic principles and techniques is found in the monograph by Lin and co-workers.[14] Demtröder's book on laser spectroscopy is also recommended.[15] Excellent comprehensive articles on multiphoton selection and polarization rules are by McClain and Harris[16] and Nascimento.[17] Friedrich gives a very readable comparison of the three different descriptions of multiphoton transitions that are usually employed: phenomenological rate equations, macroscopic electric susceptibilities, and quantum mechanical electric field perturbations at the molecular level.[18] In this chapter we are concerned solely with the latter description, since it evokes multiphoton spectroscopy at the level of molecular wavefunctions. Detailed reviews for specific areas include multiphoton spectroscopy of benzene and its isotopic and chemical derivatives,[19,20] aromatic molecules in general,[21] and even complex biological molecules.[22] Friedrich and McClain have reviewed the literature through 1980.[23] This is only a partial list; other basic references and review articles are referred to in the preceding reference. We shall virtually restrict this chapter to vapor-and liquid-phase two-photon spectroscopy (with emphasis on the gas phase), since the extensiveness of multiphoton spectroscopy of the solid state requires separate consideration.

Early two-photon experiments were restricted to materials of high particle density (i.e., solid and liquid samples). The improved bandwidth of the fast-repetition N_2-pumped Hansch-design dye laser compared to the broadband slow-repetition ruby-pumped system allowed lower-particle-density gas-phase experiments. In 1974 two-photon gas-phase spectroscopic studies were reported on three molecules: benzene,[27] nitric oxide,[26] and molecular iodine.[27] These were the first of many two-photon spectroscopic studies in the gas phase.

8.2. Principles

8.2.1. Physical Ideas

The simultaneous absorption of several photons requires no stationary intermediate state(s) to match the energy of the photons (Figure 8.1a). An example of such a transition is two-photon excitation of normal fluorescence, which as we shall see makes possible the display

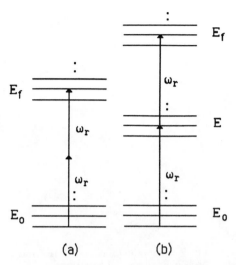

Figure 8.1 The two-photon absorption processes for a molecule: (a) nonresonant, (b) resonant. (Reproduced with permission from Ref. 14.)

and measurement of two-photon absorption bands. In this process an excited state (perhaps inaccessible by a single-photon process) can be reached by simultaneous absorption of two photons whose energies add up to the required amount. This weak higher-order process can be understood on the basis of classical electromagnetic waves. Linear (single-photon) absorption can be described by using first-order perturbation theory and discarding the $\mathbf{A} \cdot \mathbf{A}$ term in the radiation field Hamiltonian[28]:

$$H = \left(\frac{ie\hbar}{mc}\right)\mathbf{A} \cdot \nabla + \left(\frac{e^2}{2mc^2}\right)\mathbf{A} \cdot \mathbf{A} \tag{8.1}$$

Linear excitation involves a resonance transition between two definite and observable energy levels, $E_{\text{initial}} = E_0$ and $E_{\text{final}} = E_f$ of the molecule. Extending this approach to the very high photon fluxes found in pulsed laser beams requires second- or higher-order perturbation theory, since the electric field strength at the peak of the pulse is very high. If the same electric dipole approximation is made for the $\mathbf{A} \cdot \nabla$ term that is traditionally made in linear spectroscopy, a resonance appears at one-half the photon frequency corresponding to the two-photon absorption process (more generally, $\hbar\omega_1 + \hbar\omega_2 = E_f - E_0$). The two-photon (i.e., second-order) absorption strength is proportional to $|\mathbf{E}|^4$, where $\mathbf{E}$ is the electric vector, compared to the quadratic dependence found for single-photon absorption. Because light intensity varies with $|\mathbf{E}|^2$, the rate of light absorption in a transition involving the simultaneous absorption of two photons is proportional to the square of the light intensity, rather than to the linear dependence for single-photon spectroscopy (Figure 8.2). The $\mathbf{A} \cdot \mathbf{A}$ term in the radiation

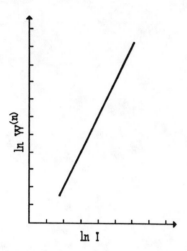

Figure 8.2 The formal intensity law (slope = n) for multiphoton processes. Here I and $W^{(n)}$ denote the laser intensity and the transition probability of an n-photon process, respectively. (Reproduced with permission from Ref. 14.)

field Hamiltonian (8.1) can still be neglected, since it is ineffective in two-photon absorption.[16]

For two-photon absorption, the absorptivity[29] δ is defined by

$$\delta = \frac{\Delta I/I_0}{cF_1 d} \tag{8.2}$$

where c is the sample concentration (molecules cm^{-3}), F_1 is the laser flux density (photons $cm^{-2} s^{-1}$), and d is the sample length (cm) along the propagation direction. Thus, the units of δ are cm^4 s molecule^{-1} photon^{-1}. One such unit, 1 cm^4 molecule^{-1} photon$^{-1} \equiv 10^{50}$ Göppert-Mayers (GM).

A feature of two-photon absorption not possible in traditional single-photon experiments involves excitation by two counterpropagating beams. When a two-photon transition is excited in this way, the Doppler shift inherent in experiments performed at laboratory temperature cancels. The consequence is that all the molecules, regardless of their velocity components, absorb at $2\hbar\omega$ without the usual single-photon Doppler broadening, even at room temperature. Moreover, because the absorption strength is collapsed into a single narrow line, detection of weak two-photon bands is facilitated.

8.2.2. Formation of the Two-Photon Transition Tensor

The quantum mechanical amplitude for a two-photon transition from the initial (ground) state, $|0\rangle$, to the final (excited) state, $|f\rangle$, involves a second-rank *tensor* formed from the product of two transition moment integrals, $\mathbf{M}_{0n} = \langle 0|\mu|n\rangle$ and $\mathbf{M}_{nf} = \langle n|\mu|f\rangle$. (The quantity

$\mu = \Sigma_i q \mathbf{r}_i$ is the sum of all charge-position products in the molecule). The analogous expression for a single-photon transition, M_{0f}, depends only on the transition dipole moment *vector*, $\mathbf{M}_{0f}$. The intermediate states $|n>$ in the two-photon amplitude are all the eigenstates of the molecule. The second-order perturbation theory expression for the two-photon transition amplitude involves the following, in which $\mathbf{\varepsilon}_1$ and $\mathbf{\varepsilon}_2$ are the polarization vectors of the absorbed photons:

$$S_{0f} = \Sigma_n \left[\frac{<0|\mu\cdot\varepsilon_1|n> <n|\mu\cdot\varepsilon_2|f>}{\hbar\omega_1 - E_{n0}} + \frac{<0|\mu\cdot\varepsilon_2|n> <n|\mu\cdot\varepsilon_1|f>}{\hbar\omega_2 - E_{n0}} \right] \quad (8.3)$$

The energy denominators in equation (8.3) involving energy gaps $E_{n0} = E_n - E_0$ and photon frequencies ω_1 and ω_2 give greater importance to low-lying intermediate states of Figure 8.1a. However, no physical participation should be attributed to the intermediate states in this case; they appear only because perturbation theory is used to describe the transition. If the laser frequency approaches resonance with a real intermediate state, as in Figure 8.1b, a drastic increase in two-photon absorption takes place. This resonance enhancement alters the vibronic structure from the nonresonant case shown in Figure 8.1a. Only in this latter case, where the vibronic structure is largely determined by Franck–Condon factors (involving solely the initial and final states), can two-photon band intensities be interpreted in a parallel fashion to single-photon ones. In the near-resonant intermediate state case of Figure 8.1b the vibronic structure of the intermediate state also plays an important role.

The two-photon transition probability $\delta^{(2)}$ is proportional to $I^2|S_{0f}|^2$ (where I is the intensity of the laser), and consequently is a fourth-rank tensor. The necessity for high-power laser pulses can be appreciated from a typical two-photon transition absorptivity: only $\sim 10^{-1}$ GM (cross section: 10^{-38} cm^2) compared to $\sim 10^{-17}$ cm^2 for a single-photon one. It is preferable to work in molecule coordinates x, y, z because it is easier to apply symmetry considerations. The tensor then becomes (α, $\beta = x, y, z$)

$$S^{\alpha\beta}_{0f} = \Sigma_n \left[\frac{<0|\mu_\alpha|n> <n|\mu_\beta|f>}{\hbar\omega_1 - E_{n0}} + \frac{<0|\mu_\beta|n> <n|\mu_\alpha|f>}{\hbar\omega_2 - E_{n0}} \right] \quad (8.4)$$

By decomposing equation (8.4) into its irreducible tensor components (a second-rank tensor has nine elements), further clarification can be achieved as the transition amplitude can be broken into three terms:

$$S_{\alpha\beta} = S^0_{\alpha\beta} + S^s_{\alpha\beta} + S^a_{\alpha\beta} \quad (8.5)$$

The first, S^0, is a diagonal isotropic tensor defined by

$$S^0_{\alpha\beta} = \frac{1}{3} \Sigma_n \frac{<f|\mu_\alpha|n><n|\mu_\alpha|0>[2E_{n0} - (\hbar\omega_1 + \hbar\omega_2)]}{(E_{n0} - \hbar\omega_1)(E_{n0} - \hbar\omega_2)} \delta_{\alpha\beta} \quad (8.6)$$

and allowing transitions only between states of the same symmetry; S^s is a symmetric (anisotropic) tensor term:

$$S^s_{\alpha\beta} = \frac{1}{2}\Sigma_n \frac{[<f|\mu_\alpha|n><n|\mu_\beta|0> + <f|\mu_\beta|n><n|\mu_\alpha|0>]}{(E_{n0} - \hbar\omega_1)(E_{n0} - \hbar\omega_2)} \times [2E_{n0} - (\hbar\omega_1 + \hbar\omega_2)]$$ (8.7)

that follows selection rules appropriate for electric quadrupole transitions. The final term, S^a, is an antisymmetric tensor term:

$$S^a = \frac{1}{2}\Sigma_n \frac{[<f|\mu_\alpha|n><n|\mu_\beta|0> - <f|\mu_\beta|n><n|\mu_\alpha|0>](\hbar\omega_1 - \hbar\omega_2)}{(E_{n0} - \hbar\omega_1)(E_{n0} - \hbar\omega_2)}$$

(8.8)

that transforms like an axial vector and follows the selection rules appropriate for magnetic dipole transitions. Note that this last term contains the difference in frequencies of the two photons absorbed as a multiplicative factor and is thus zero in an experiment that uses a single laser beam (i.e., where both exciting photons have the same frequency and polarization).

8.2.3. Two-Photon Selection Rules

8.2.3.1. *Vibronic Rules*   The $g \Leftrightarrow g$ two-photon selection rule is easily understood from the $g \Leftrightarrow u$ requirement for each transition moment in the tensor (8.4). Vibronic selection rules are implicitly contained in equations (8.6)–(8.8), if states $|0>$ or $|f>$ are vibronic (i.e., products of electronic and vibrational wavefunctions).

A simple way to show that the selection rules for two-photon transitions are the same as the well-known vibrational Raman selection rules is to consider the numerators of equation (8.4). The sum is over all states $|n>$ of a complete set; therefore, it contains all symmetries. Because we are only interested in symmetry selection rules, we identify $\Sigma_n<0|\mu_\alpha|n> <n|\mu_\beta|f>$ with $<0|\mu_\alpha\mu_\beta|f>$ and $\Sigma_n<0|\mu_\beta|n><n|\mu_\alpha|f>$ with $<0|\mu_\beta\mu_\alpha|f>$. Closure over $|n><n|$ in equation (8.4) is an improper action because n is present in the denominators, but the symmetry arguments are still valid because the denominators are scalar quantities. Now $\mu_\alpha\mu_\beta$ are the components of the tensor operator, which yields Raman selection rules. Raman scattering involves vibrational functions; for the case of two-photon absorption, electronic functions are involved. Therefore, two-photon absorption excites transition densities that contain components transforming as the quadratic products x^2, y^2, z^2, xy, yz, zx.

An interesting case arises when both of the symmetric tensor expressions, (8.6) and (8.7), vanish. It is not enough for the tensor in the brackets of equation (8.8) to be nonzero for the antisymmetric tensor term to contribute to a nonzero transition amplitude. The transition remains forbidden for two *identical* photons, leading to the term *identity-forbidden*.[16,30] By carrying out a two-beam experiment the forbidden character can be removed, the two-photon absorptivity increasing approximately as $(\omega_1 - \omega_2)^2$; see equation (8.8).

An example of an identity-forbidden transition can be found in the $A_{1g} \leftrightarrow B_{2u}$ band system of benzene, found near 260 nm in single-photon spectroscopy. This transition has been very well studied in one-, two-, and three-photon spectroscopies, but the b_{1u} (ν_{12})[31] fundamental vibrational frequency, thought to be near 1,000 cm^{-1}, remains unobserved. The b_{1u} vibration falls into the identity-forbidden category, because it generates a magnetic-dipole allowed (i.e., R_z) A_{2g} vibronic state $(B_{2u} \otimes b_{1u} = A_{2g})$. Attempts to observe the ν_{12} vibrational band in the $A_{1g} \leftrightarrow B_{2u}$ transition by a two-color experiment have been made.[32] However, vibronic coupling appears to be too weak to allow the ν_{12} vibration to be observed with the small (3,500 cm^{-1}) frequency gap employed.[33] Recent experiments with a 9,000 cm^{-1} gap and computer summing do not show ν_{12} activity either.[139]

A bona fide observation of an identity-forbidden transition was not obtained until 1984. The feasibility of observing this type of forbidden character through two-beam experiments was demonstrated by Garetz and Kittrell[34] on the $1^1\Sigma^- \leftrightarrow X^1\Sigma^+$ transition in CO. In their studies P and R rotational branches in the (7–0) band were observed using 266- and 294-nm laser beams, corresponding to a 3,580 cm^{-1} difference.

Two-photon absorptivity determined by the symmetric tensor terms (8.6) and (8.7) arises from the nonzero elements. The relevant tensor patterns, determined by molecular point groups and state symmetries $|0\rangle$ and $|f\rangle$, have been comprehensively discussed for all point groups.[14,16,17,35] Two examples, the tensors for transitions within C_{2v} and C_{2h} point groups, are given in Tables 8.1 and 8.2, respectively. In the C_{2v} case, two-photon transitions to all four (A_1, A_2, B_1, B_2) states from a ground A_1 state are allowed. The point group C_{2h} also has four symmetry species—that is, A_g, B_g, A_u, and B_u. But because of the parity rule, transitions from the totally symmetric A_g state are two-photon allowed only to A_g and B_g. Vibronic coupling may induce transitions to the two-photon forbidden pure electronic B_u and A_u states. For B_u, the symmetries of these modes are determined by

$$A_g \otimes \begin{bmatrix} A_g \\ B_g \end{bmatrix} \otimes (B_u \otimes \Gamma_{\text{vib}}) \supset A_g$$

Table 8.1 Two-Photon Absorption Tensors for C_{2v} Symmetry (For Identical Photons the Elements $S_{\alpha\beta} = S_{\beta\alpha}$)

$$S(A_1 \rightarrow A_1) = \begin{pmatrix} S_{xx} & 0 & 0 \\ 0 & S_{yy} & 0 \\ 0 & 0 & S_{zz} \end{pmatrix} \qquad S(A_1 \rightarrow A_2) = \begin{pmatrix} 0 & S_{xy} & 0 \\ S_{yx} & 0 & 0 \\ 0 & 0 & 0 \end{pmatrix}$$

$$S(A_1 \rightarrow B_1) = \begin{pmatrix} 0 & 0 & S_{xz} \\ 0 & 0 & 0 \\ S_{zx} & 0 & 0 \end{pmatrix} \qquad S(A_1 \rightarrow B_2) = \begin{pmatrix} 0 & 0 & 0 \\ 0 & 0 & S_{yz} \\ 0 & S_{zy} & 0 \end{pmatrix}$$

Table 8.2 Two-Photon Absorption Tensors for C_{2h} Symmetry Allowed Transitions (For Identical Photons the Elements $S_{\alpha\beta} = S_{\beta\alpha}$)

$$S(A_g \rightarrow A_g) = \begin{pmatrix} S_{xx} & S_{xy} & 0 \\ S_{yx} & S_{yy} & 0 \\ 0 & 0 & S_{xx} \end{pmatrix} \qquad\qquad S(A_g \rightarrow B_g) = \begin{pmatrix} 0 & 0 & S_{xz} \\ 0 & 0 & S_{yz} \\ S_{zx} & S_{zy} & 0 \end{pmatrix}$$

Thus, vibrations of a_u and b_u symmetry are possible, inducing modes in an $A_g \leftrightarrow B_u$ two-photon transition. In the final analysis two-photon transitions between electronic levels of different parity are due to distortion of molecular symmetry by asymmetric vibrations. Because, in single-photon spectra, transitions to B_u and A_u states are allowed and to A_g and B_g states forbidden, the C_{2h} example provides an instance of the general complementarity that exists between the two spectra.

    8.2.3.2. Rotational Rules The mathematics of the rotational intensity problem, which involves the dependence of tensor elements on rotational quantum numbers, is beyond the introductory scope of this chapter. Here we give an overview of the main features and conclusions that determine overall band shapes, rather than develop the theory for intensities and polarizations of individual lines in the rotational branches. Comprehensive discussions can be found in the classic works by McClain and Harris,[16] Metz et al.,[35] and Lin et al.[14]

  Band shape is determined by rotational transition energies and intensities. Rotational energy depends on two quantum numbers: that pertaining to the total angular momentum, J; and that pertaining to the angular momentum component along the molecular symmetry axis, K. It follows that $K \leq J$. The two-photon selection rules are

$$\Delta J = 0, \pm 1, \pm 2 \quad \text{and} \quad \Delta K = 0, \pm 1, \pm 2. \tag{8.9}$$

Consequently, a two-photon band contour has five branches determined by the value of ΔJ:

Branch	O	P	Q	R	S
ΔJ	-2	-1	0	1	2

and each branch in a symmetric-rotor or near symmetric-rotor molecule has one or more (a maximum of five) values of ΔK. Lowercase or capital letters o, p, q, r, s, written as left superscripts, stand for ΔK values. For example, ^{P}Q designates $\Delta J = 0$ and $\Delta K = -1$. These rules lead to quite different rotational intensity distributions from the more restricted P, Q, R branch possibilities arising from the $\Delta J = 0, \pm 1$ selection rule of single-photon spectroscopy.

  The values of ΔK for the active lines that appear in any particular band depend on the vibronic symmetry of the initial and final states. The possible ΔJ values are mainly determined by the relative polarization of the two absorbed photons. The consequence is that different rotational branches of the same vibronic band may show different polarization behavior. This behavior is strikingly different from single-photon spectroscopy, where for isotropic media the absorbing light polarization has no effect on intensity distribution. We discuss this point further in the next section.

There are additional subtleties regarding the rotational structure of two-photon spectra. These are usually expressed by developing tensor elements, S_{jk}, in terms of spherical coordinates $j = 0, 1, 2$ and $k = j, j - 1, ...,$ $-j$, with S_{00} designating a zero-rank tensor element, S_{1k} a first-rank element, and so on. For identical photons both S_{00} and S_{2k} are important; S_{1k} appears only when $\omega_1 \neq \omega_2$. If the element $S_{jk} \neq 0$, then rotational intensity features with $\Delta K = k$ and with all ΔJ from $-j$ to j are present.

An example is the excitation of an $A_1 \rightarrow B_2$ transition in a C_{2v} molecule by two identical photons. In this case, only the spherical elements S_{2+1} and S_{2-1} are nonzero and thus the selection rules $\Delta K = \pm 1, \Delta J = -2, -1, 0, 1, 2$, lead to 10 possible rotational features: ^PO, ^rO, ^PP, ^rP, ^PQ, ^rQ, ^PR, ^rR, ^PS, and ^rS.

8.2.4. Polarization

A well-known result of single-photon spectroscopy involving isotropic media is that absorption is essentially independent of the exciting light polarization, whether it be linear, circular, or elliptical. In contrast, averaging over molecular orientations does not cause the polarization dependence of two-photon spectroscopy to vanish. This polarization dependence is one of the most salient features of multiphoton spectroscopy. The very thorough discussions of orientational averaging by McClain and colleagues[16,36,37] and by Lin et al.[14] show that two-photon absorptivity $<\delta^{(2)}>$ (abbreviated to $<\delta>$) may be conveniently expressed in terms of three tensor invariants, δ_F, δ_G, and δ_H:

$$<\delta> = F\delta_F + G\delta_G + H\delta_H \tag{8.10}$$

These invariants are molecular parameters defined through

$$\delta_F = \text{const} \times \Sigma_\alpha \Sigma_\beta S_{\alpha\alpha}(S_{\beta\beta})^* \tag{8.11a}$$

$$\delta_G = \text{const} \times \Sigma_\alpha \Sigma_\beta S_{\alpha\beta}(S_{\alpha\beta})^* \tag{8.11b}$$

$$\delta_H = \text{const} \times \Sigma_\alpha \Sigma_\beta S_{\alpha\beta}(S_{\beta\alpha})^* \tag{8.11c}$$

Here $S_{\alpha\beta}$ ($\alpha, \beta = x, y, z$) are the cartesian tensor elements represented by equation (8.3) and the asterisk indicates complex conjugate. Coefficients F, G, and H depend only on the polarization vectors ε_1 and ε_2.

Since many two-photon experiments are performed with a single laser beam, the polarization behavior obtained with identical photons (i.e., photons having the same energy and polarization) is valuable. The two photons may be

1. Linearly polarized ($\uparrow\uparrow$);
2. Circularly polarized ($\curvearrowright$).

The identical photon condition requires the tensor to be symmetric, that is, $S_{\alpha\beta} = S_{\beta\alpha}$, and thus $\delta_G = \delta_H$. Substitution of F, G, and H in equation (8.10) gives

$$<\delta_{\uparrow\uparrow}> = 2\delta_F + 2\delta_G + 2\delta_H = 2\delta_F + 4\delta_G \tag{8.12}$$

$$<\delta_{\curvearrowright}> = -2\delta_F + 3\delta_G + 3\delta_H = -2\delta_F + 6\delta_G \tag{8.13}$$

Measurement of a two-photon spectrum using linear and circular polarizations enables determination of δ_F and δ_G. In practice, the ratio of peak intensities Ω under circular and linear polarization excitations is often easily measurable. From (8.12) and (8.13) it is given by

$$\Omega \equiv <\delta_{\curvearrowright}>/<\delta_{\uparrow\uparrow}> = \frac{(3\delta_G - \delta_F)}{2\delta_G + \delta_F} \tag{8.14}$$

The polarization parameter Ω has a range $0 \leq \Omega \leq 3/2$, and because of sensitivity to the value of δ_F, ($-\delta_F$ in the numerator and $+\delta_F$ in the denominator), it is valuable in determining transition symmetry. Equation (8.11a) shows that δ_F is the absolute square of the tensor's trace, $\delta_F = |S_{xx} + S_{yy} + S_{zz}|^2 \geq 0$. For transitions from a totally symmetric ground state to a nontotally symmetric excited state, $\delta_F = 0$, because in this case the diagonal tensor elements vanish. For a totally symmetric excited state, $\delta_F > 0$. From (8.14) it follows that

$\Omega < 3/2$ for transitions between totally symmetric states

$\Omega = 3\delta_G/2\delta_G = 3/2$ for transitions between a totally symmetric

ground state and nontotally symmetric excited states

More explicit polarization rules can be obtained by detailed inspection of tensor patterns. The patterns depend solely on the symmetries of the initial and final states in the transition and lead to[17]

$\Omega < 1$ for states of the same symmetry   (8.15a)

$\Omega = 3/2$ for states of different symmetry   (8.15b)

$\Omega = 0$ and 24/31 for $A \rightarrow A$ and $E \rightarrow E$ transitions,   (8.15c)
respectively, in T, T_d, T_h, O, and O_h point groups

The above preceding discussion shows how a simple two-photon experiment with a single laser beam can help identify a transition to a totally symmetric electronic or vibronic excited state. Thus, in the C_{2h} examples discussed in Section 8.2.3.2, polarization behavior can differentiate an $A_g \rightarrow B_g$ transition ($\Omega = 3/2$) from an $A_g \rightarrow A_g$ one ($\Omega < 1$). In a C_{2v} molecule (Table 8.1) an $A_1 \rightarrow A_1$ transition can be distinguished from $A_1 \rightarrow B_1$, B_2, or

A_2. An active vibration of a_2 symmetry can be identified in an $A_1 \rightarrow A_2$ electronic transition by suppression of the appropriate vibronic band in circularly polarized light ($A_2 \otimes a_2 = A_1$).

Another example is provided by the $A_1 \rightarrow B_2$ electronic band system of C_{2v} symmetry monosubstituted benzenes excited by green photons (the transition of $A_{1g} \rightarrow B_{2u}$ parentage in the parent molecule benzene). The band system has an electronically allowed component and an electronically forbidden but vibrationally induced component. The forbidden part is induced by a b_2 vibrational mode[19] and thus (8.15a) applies. The b_2 vibrational bands are strongly active in linearly polarized light and can easily be identified by their loss of activity under circularly polarized excitation. Many examples of measured values of Ω are provided in Ref. 17.

There are important polarization effects in the rotational lines of two-photon spectra that are of direct use in the assignment of vibronic bands. These effects are in contrast to the lack of polarization effects in the rotational structure of one-photon spectra. For a single laser experiment

$$<\delta_{\uparrow\uparrow}> = \frac{1}{3}M_0R_0 + \frac{2}{15}M_2R_2 \tag{8.16}$$

$$<\delta_{\uparrow\downarrow}> = \frac{M_2R_2}{5} \tag{8.17}$$

where M_j and R_j ($j = 0, 1, 2$) are molecular and rotational factors.[14,35] The former, representing the transition moment and dependent on the transition symmetry, is most clearly evaluated from Wigner $3j$ symbols. The rotational factors R_j depend on ΔK and ΔJ. For identical photons the non-vanishing rotational factors are R_0 for $\Delta K = 0$ in Q branches, and R_2 for $\Delta K = 0, \pm 1, \pm 2$ in all branches. It is convenient to take the reciprocal of the polarization parameter,

$$(\Omega)^{-1} = \frac{2}{3} + \frac{5}{3}\left(\frac{M_0R_0}{M_2R_2}\right) \tag{8.18}$$

From (8.18) it follows that the following cases are possible:

1. $M_0 = 0$: $\Omega = 3/2$, independent of the rotational quantum numbers J, K. All rotational lines have the same polarization effect, and thus no change is incurred in the rotational envelope by changing from linear to circular polarization. This case holds for electronic transitions between states of different symmetry.

2. $M_0 \neq 0$ and $R_0 = 0$: $\Omega = 3/2$ for all rotational lines of the O, P, R, and S branches of a totally symmetric transition.

3. $M_0 \neq 0$ and $R_0 \neq 0$: $\Omega < 3/2$. This applies to the Q branch of all totally symmetric transitions. Each rotational line has a different rotational factor and consequently a different value of Ω. The deviation from 3/2 depends on the relative values of (M_0R_0) and (M_2R_2).

Case 3 shows that measurement of Ω distinguishes the Q branches of totally symmetric bands from nontotally symmetric transitions. An example is the analysis by Wunsch et al. of the $A_{1g} \rightarrow B_{2u}$ spectra of benzene at moderate rotational resolution in linearly and circularly polarized light.[38]

We have limited our discussion to two-photon spectra. Polarization effects in three-photon transitions are discussed by Andrews et al.[39] and Friedrich.[40]

8.3. Detecting Multiphoton Absorption

Multiphoton absorption experiments can be categorized as involving one or two laser beams. In a two-laser-beam experiment the photons have different frequencies and one laser is usually fixed at a desired frequency, the other being tunable. Although the recent trend to two-color experiments offers advantages and covers a wide area of applications, the basic detection methods and principles for multiphoton absorption can be presented for the single-laser-beam case. The fate of a molecule after multiphoton excitation is relevant to the detection method to be chosen. Two-photon or higher-order absorption can be revealed by

1. Direct measurement of the absorbed fraction of light;

2. Detection of fluorescence (or phosphorescence) emission from the n-photon excited level (or from another level populated after the primary multiphoton excitation);

3. Measurement of ions produced by further absorption of m-photons from the n-photon prepared excited state;

4. Heat effects generated from nonradiative decay of the excited molecules;

5. Depopulation of the excited state (using a second laser) by stimulated emission or absorption to a higher state. This approach has been applied to cases where the first excitation is single-photon absorption.

8.3.1. Direct Measurement of Two-Photon Absorption

Direct measurement of two-photon absorption is extremely difficult. To obtain two-photon absorptivity by this method requires measurement of a miniscule diminution of light intensity. An approach that makes use of the I^2 dependence of two-photon excitation involves a strong laser beam I_1 (pump) and a weaker one, I_0 (probe). Two-photon absorption is induced only by the combined intensity of I_1 and I_0. With I_0 off, no absorption occurs. The decrease, ΔI, in I_0 intensity incurred by transmission through the sample is measured as the wavelength of either I_1 or I_0 is scanned.

The advantage of this approach is that even though the change in the light intensity, ΔI, is very small for two-photon absorption, I_0 itself is small. Nevertheless the small two-photon probability makes the experiment difficult. A typical detection lower bound is $\Delta I/I_0 > 0.1\%$. It is clear that the sample has to be solid or liquid to provide large numbers of absorbing molecules and it should be completely transparent to each light source.

Hopfield and colleagues[41] carried out the first two-photon absorption measurements on crystalline KI in 1963. The experiment used ultraviolet light from a xenon arc lamp as I_0 and a ruby laser (15 J per pulse, $\lambda = 694.8$ nm) as I_1. The beams were overlapped by internal reflection in the KI crystal. Hopfield et al. found that at a particular xenon lamp wavelength, transmitted ultraviolet intensity was modulated by the presence of ruby laser pulses. The intensity dip was interpreted as being due to simultaneous absorption of two photons: one photon from the Xe source and another from the ruby laser. The ultraviolet transmission was observed to decrease by ~0.2%, and to be proportional to both laser and ultraviolet intensity (i.e., following the I^2 law of Figure 8.2).

Liquid-phase organic molecules provide another area where direct two-photon absorptivity measurements have been successfully carried out. Swofford and McClain's 1975 study of two-photon absorptivity[29] is noteworthy. A tunable pulse laser was used as I_1 and a fixed frequency ($\lambda = 476.2$ nm) continuous krypton ion laser as I_0; 1-chloronaphthalene was the test molecule. The measured dip in two-photon signal represented 8% absorption of the probe beam, yielding an accurate value for δ of 1.3 GM at 35,400 cm^{-1}. For diphenylbutadiene,[29] an absorption was found near the onset of single-photon absorption, at 28,259 cm^{-1} with $\delta = 1.0$ GM. The strong two-photon absorption was attributed to a g-parity lowest (and previously undetected) excited state and led to a paradigm shift in our understanding of the sequence of polyene excited states.[42]

The direct-absolute multiphoton absorption method requires extreme stability of light sources and careful control of beam spatial uniformity. Formidable problems, which include dye region alignment ambiguities, radiation-induced heating effects, probe beam attenuation caused by stimulated Raman scattering, and detector and accompanying electronics nonlinearities explain why this method is little used.

8.3.2. Fluorescence and Phosphorescence Excitation

The fluorescence excitation method is well known from traditional single-photon spectroscopy. It makes use of the general fluorescence property of the lowest-energy excited state (S_1) in polyatomic molecules. The fluorescence quantum yield, Φ_F, is typically 0.01–0.8 and the sensitivity of the fluorescence excitation method depends on the magnitude of Φ_F. It is possible to detect a two- or three-photon $S_1 \leftarrow S_0$ transition using dye laser pulses with peak power above 5×10^9 W m^{-2}. Fluorescence will be to shorter wavelengths than for two- or three-photon absorption, $S_1 \leftarrow S_0$, but to longer wavelengths than for single-photon absorption. The intensity of the undispersed fluorescence, scanned as a function of laser wavelength, gives the n-photon fluorescence excitation spectrum. Fluorescence is usually detected by a high-gain UV-sensitive photomultiplier with appropriate glass color filters blocking scattered light from the exciting laser beam. In its simplest form (beyond displaying photomultiplier output directly on an oscilloscope), a triggered boxcar integrator or photon counting device is used to process the signal.[44]

A small fraction of the unfocused exciting laser beam is reflected onto a photodiode allowing the normalized excitation spectrum to be obtained from the ratio: fluorescence signal/(photodiode signal)^n with $n = 2$ for two-photon excitation. The reflected signal can also be used to trigger the detection electronics. This approach to obtaining a multiphoton absorption spectrum assumes constancy of fluorescence quantum yield over the scanned region, an assumption not always satisfied. An additional problem occurring for solution or crystalline phases is that the sample may undergo single-photon reabsorption of the emission. Both of these side effects may distort the excitation spectrum.

Attention must be paid to the optical efficiency of the fluorescence cell. A multilens system or an ellipsoidal reflector (Figure 8.3) can optimize collection of fluorescence by the photomultiplier. For two-photon excitation the photomultiplier signal I_F can be expressed through the practical intensity equation

$$I_F = \frac{1}{2} G \Phi_F \delta I^2 c d \tag{8.19}$$

Here G is a geometrical factor that gives the fraction of emission reaching the photomultiplier, c is the concentration, and d is the excited sample depth seen by the photomultiplier. The factor $\frac{1}{2}$ arises because absorption of two photons is required for each emitted photon. A comparison of the absorptivities, δ_i between two compounds can be made using equation (8.19).

The early Hochstrasser–Wessel–Sung two-photon experiments on gas- and crystalline-phase benzene[24,25] and more recent two- and three-photon measurements (see Section 8.4.1) have used the fluorescence excitation method. A nitrogen-pumped dye laser with green region, 490–530 nm, output for two-photon excitation (red region, 540–570 nm, for three-photon) populated C_6H_6 vibronic levels of the $S_1(B_{2u})$ state. The $A_{1g} \leftarrow B_{2u}$ fluorescence, occurring well to the blue at 250–310 nm, is easily separated from the exciting radiation.

Carbon monoxide and atomic xenon[45] provide some other examples of three-photon-excited fluorescence. Simultaneous absorption of three visible photons ($\lambda \approx 443.2$ nm) populates the (2–0) level of CO in the A^1P

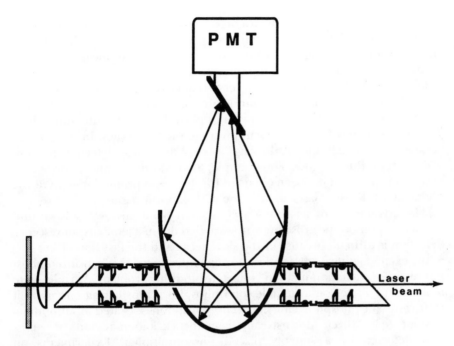

Figure 8.3 Fluorescence excitation detection of two-photon absorption. PMT is the photomultiplier tube.

state. For Xe, the three-photon $6s(^3P_1) \leftarrow 5p(^1S_0)$ transition is achieved with $\lambda = 440.76$ nm. In this case fluorescence is in the vacuum ultraviolet region ($\lambda \approx 147$ nm) and could be detected by a solar-blind (insensitive to visible light) photomultiplier.

The I^2 dependence inherent in two-photon processes suggests that for two-photon excitation the fluorescence signal can be enhanced by tight focusing of the laser beam. This is true only for three-photon and higher-order processes, however. The reason for this can be seen from the decrease in photon flux with focal spot area increase, πr^2 (i.e., $I^2 \propto r^{-4}$) combined with the r^4 dependence of the number of molecules, ΔN, as the focal region increases (e.g., for a cylindrical focal volume $\Delta V = b\pi r^2$ with $b = 8\pi r^2/\lambda$). In effect, a two-photon signal should be relatively invariant to the focusing condition. Furthermore, tight focusing may lead to saturation and/or induce depopulation of the resonant state through alternative channels, such as ionization or fragmentation, which compete with the fluorescence channel. All these reasons lead to a long focus lens (e.g., $f \approx 50$ cm) as the choice for two-photon fluorescence excitation.

Polarization information is readily achieved using the fluorescence excitation method. The usual procedure is to linearly or circularly polarize the excitation beam using a combination of Glan polarizer and quarter-wave ($\lambda/4$) retardation plates (helpful details are found in Refs. 14 and 15).

Alternatively, an appropriately oriented Fresnel rhomb may be substituted for the $\lambda/4$ plate. There are a number of refinements, such as Hohlneicher and colleagues' rotating polarization device, designed to measure $I_{F\uparrow\uparrow}$ and $I_F \uparrow\uparrow\hspace{-0.3em}\searrow$ (hence $\delta_{\uparrow\uparrow}$ and $\delta \hspace{-0.3em}\searrow$) in rapid sequence.[46]

Many molecules undergo fast intersystem crossing into the triplet manifold ($S_1 \rightarrow T_1$). For these compounds phosphorescence excitation is possible. Examples of molecules having high phosphorescence quantum yield, Φ_P, include aldehydes, ketones, and nitrogen-heterocyclics. In these cases, because Φ_F is small and Φ_P is large, phosphorescence detection is more suitable than fluorescence. For example, the two-photon phosphorescence excitation spectrum has been obtained for $^1(\pi^* \leftarrow n)$ promotion in pyrazine crystal at 1.6 K using laser light in the 680–425-nm region.[47]

The advantages of jet beams (see Section 8.3.9) are well known and much two-photon spectroscopy is now carried out in a jet. Phosphorescence detection is difficult because of the slow decay and the fast flow (1 km s^{-1}) of the excited jetborne molecules. Ito et al. introduced the "sensitized phosphorescence excitation method" to get around this problem. A triplet-state molecule impinges on a suitable solid phosphor, transferring energy to it with resultant phosphor emission. The phosphor is located at an optimal position (e.g., 10 cm downstream from the excitation region), its emission detected by a suitably placed photomultiplier. Experiments on halobenzenes[49] have shown that spectral features also arise from generation of sensitized emission by collision of the ions produced by multiphoton ionization with the phosphor. Thus, part of the sensitized phosphorescence is due to "ion-induced sensitized phosphorescence."

8.3.3. Resonance-Enhanced Multiphoton Ionization (REMPI or MPI)

Because of its high sensitivity and simplicity, resonance-enhanced multiphoton ionization (REMPI) is widely used to obtain multiphoton spectra; in many cases it is the method of choice. The REMPI principle is that a *real* intermediate resonance state, E_f, by drastically increasing ion yield, engenders a multiphoton ionization spectrum, $E_0 \rightarrow E_f$. The high collection efficiency possible for ions makes it the most sensitive detection method.

The very high peak photon flux ($>10^{27}$ photons cm^{-2} s^{-1}) at the focal spot (beam waist diameter, 0.1 mm) of a focused* pulsed laser beam induces N-photon ionization. The ionization can be simultaneous or sequential; in any case, there is the requirement that the energy of N-photons exceed the ionization potential. The N-order ionization cross section, δ_N, is

---

*The diameter of the focal spot at the diffraction limit of a perfectly coherent Gaussian laser beam (diameter D and wavelength λ) is $D_0 = f\lambda/\pi D$, where f is the lens focal length. For a multimode laser beam having divergence θ (in radians), $D_0 \approx f\theta$.

low but the probability that a bound electron will reach the ionization continuum increases greatly with increasing photon flux. At a laser flux I (photon cm^{-2} s^{-1}), the transition probability for *nonresonant* N-photon ionization is

$$W_N = \delta_N I^N \tag{8.20}$$

where the units of δ_N are cm^{2N} s^{N-1} photon^{-1}. By tuning the laser, the energy sum of two, three, or more photons may come into resonance (i.e., $nh\omega = E_f - E_0$) with a molecular state, E_f. In this case, the ionization cross section (and consequently the ion yield) is enhanced by 10^4–10^6 provided the simultaneous n-photon electronic transition $E_f \leftarrow E_0$ is allowed.

With visible laser light (photon energy equivalent to $\sim$2.5 eV), four photons are needed to exceed the ionization potential of a typical molecule. An extension of equation (8.3) gives the cross section, δ_4, for *resonant* four-photon ionization:

$$\delta_4 \propto \Sigma_i \Sigma_j \Sigma_k \left| \frac{<C|\mu_\alpha|k><k|\mu_\beta|j><j|\mu_\beta i><i|\mu_\alpha|0>}{(E_{k0} - 3\hbar\omega_L)(E_{j0} - 2\hbar\omega_L)(E_{i0} - \hbar\omega_L)} \right|^2 \tag{8.21}$$

where C is the final ionization continuum state and the indices i, j, k denote the first, second, and third virtual intermediate resonance states formed from real bound and continuum levels. The summation is over all molecular states. In general, the N-order transition rate consists of a product of N electric dipole matrix elements and $(N - 1)$ sums, running over all the intermediate resonance states of the molecule.

An important feature of equation (8.21) is that δ_4 is parametric to the photon energy, $h\omega_L$. Three possible resonant terms are present in the four-photon process represented by equation (8.21). Resonances may appear at fundamental, doubled, or tripled laser frequencies. The existence of a real state at $\sim$5.0 eV causes the energy denominator, $(E_f - E_0 - 2\hbar\omega_L)$, to vanish (since in our example $\hbar\omega_L \sim 2.5$ eV) and δ_4 drastically increases. The result: a two-photon resonant, four-photon ionization spectrum, symbolized as $2R + 2$ or $[2 + 2]$. A $3R + 1$ (i.e., three-photon resonant), four-photon ionization may appear if the real state, E_f, is at $\sim$7.5 eV.

An expression for the ion current can be found (at an elementary and phenomenological level) by solving kinetic equations for molecular-level populations participating in the entire process.[14,50–52] The rate equations, analogous to chemical rate equations, provide a simple view of the intensity dependence of multiphoton absorption. For the typical case shown in Figure 8.4, the MPI current is given by

$$(\text{MPI})_{cur} = \frac{N_0 \delta_2^{(r)} \delta_2^{(i)} I^4}{\delta_2^{(i)} I^2 + \gamma} \quad \text{for } 2R + 2 \tag{8.22a}$$

The quantity $\delta_2^{(r)} I_2$ is the two-photon pumping rate from E_0 to E_f, while $\delta_2^{(i)} I^2$ is the two-photon ionization rate from E_f to the ionization continuum.

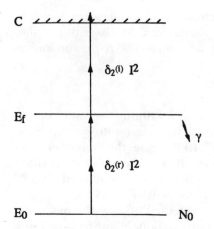

Figure 8.4 Four-photon ionization involving a resonance to E_f. Cross sections $\delta_2^{(i)}$ and $\delta_2^{(i)}$ are for two-photon resonant absorption $E_0 \rightarrow E_f$ and $E_f \rightarrow C$ ionization steps, respectively. A decay channel from E_f with rate γ is also shown. N_0 is the ground-state population.

When added to the E_f decay rate, this latter quantity gives the total depopulation rate of the resonant state, E_f. Typical molecular values for absorptivities are $\delta_2(r) = 10^{-50}$ cm^4 s molecule^{-1} photon^{-1} and $\delta_2^{(i)} = 10^{-40}$ cm^4 s molecule^{-1} photon^{-1}. If the last step, $E_f \rightarrow C$, is a single-photon process, the MPI current is

$$(MPI)_{cur} = \frac{N_0 \delta_2^{(r)} \delta_1^{(i)} I^3}{\delta_1^{(i)} I + \gamma} \qquad \text{for } 2R + 1 \tag{8.22b}$$

In this case a typical value of $\delta_1^{(i)}$ is 10^{-19}–10^{-20} cm^2 s molecule^{-1} photon^{-1}.

Equations (8.22a) and (8.22b) contain two main assumptions: ground- and resonant-level populations remain constant over the pulse duration, and laser linewidth exceeds the resonance width. Consequent to these assumptions:

1. The laser-power-dependent denominator, $(\delta_{N-n}^{(i)} I^{N-n} + \gamma)$, implies that n-photon resonant, N-photon ionization signals do not necessarily obey the I^N law of Figure 8.2 expected in nonresonant N-photon processes.

2. At the limit of low intensities or small absorptivity ($\gamma \gg d_{N-n}^{(i)} I^{N-n}$) the overall ionization probability, $\delta^{(N)} \propto \delta_n^{(r)} \delta_{N-n}^{(i)} I^N$.

3. For moderate intensities, $\delta_{N-n}^{(i)} I^{N-n} \gg \gamma$; normally $\delta_{N-n}^{(i)} I^{N-n}) \gg \delta_n^{(r)} I^n$. Saturation of the ionization step then leads to $\delta^{(N)} \propto \delta_n^{(r)} I^n$ with the consequence that an n-photon resonantly enhanced MPI spectrum faithfully reflects the n-photon cross section of the $E_f \leftarrow E_0$ transition. Tuning the laser generates an $nR + [N - n]$ REMPI spectrum, including polarization behavior appropriate for this resonance transition. Thus, for $2R + 1$ the MPI current intensity (power) dependence is ideally quadratic; for $3R + 1$, ideally cubic.

4. Radiationless transitions and photochemical processes, in general, increase the dissipative (relaxation) rate, γ. If γ greatly exceeds the ionization rate, $\delta_{N-n}^{(i)} I^{N-n}$, the molecular electronic state E_f will not be observable in MPI spectra. This frequently applies to higher valence states, because for these states the promoted electron is in an antibonding orbital. There is evidence that molecular Rydberg states, although lying in the VUV region, are more photochemically stable (i.e., have lower relaxation rates) than high-energy valence states, and thus MPI spectroscopy has a propensity for Rydberg states. The detection of Rydberg states is a *major* contribution of MPI spectroscopy; these transitions are difficult to unravel from the stronger valence transitions in other spectroscopies. Resourceful use of multiphoton selection rules, complementary to one-photon, have allowed detection of $3s$ and other g symmetry Rydberg states in benzene, for example.[12,13,53,54] Many other examples are in the literature.[14,23,50,51]

5. The polarization behavior is reflected in Ω, the circular/linear polarized-light signal ratio discussed in Section 8.2.4. Polarization effects are seen in REMPI, although the ionization step $E_f \to C$ may be complicating. The polarization dependence is clear and secure when every molecule reaching the resonant state is ionized. (See item 3.) Then γ in equation (8.22a) or (8.22b) can be omitted, with the following result:

$$\Omega = \frac{\delta_2^{(r)}{}_{\circlearrowright}}{\delta_2^{(r)}{}_{\uparrow\uparrow}} \tag{8.23}$$

In practice, this means that in an MPI experiment, the ratio Ω must be measured at various laser intensities.

These conclusions are idealized. In practice, focused laser beam geometric effects in the photon–molecule interaction region can severely alter intensity dependence laws. The net result of nonisotropic high-power laser flux in the interaction region is that a $\frac{3}{2}$ power law may actually govern the MPI signal.[14,52,55] In a focused beam with a "dog bone" geometry, the flux may exceed critical saturation magnitude at focus, dropping off as distance-squared from the focal point. Increasing the laser flux expands the ionization region, but no more signal is created from the already saturated volume.

A typical REMPI experimental setup is shown in Figure 8.5. Because the laser beam represents ionizing radiation, an ionization chamber or proportional counter, customary in nuclear physics, is used. A typical ionization cell has two parallel platinum electrodes (e.g., 2×2 cm, separated by 1 or 2 cm) in a metal or quartz cylinder. An alternative arrangement is a thin wire surrounded by a second cylindrical electrode. The laser beam, entering the cell through a quartz window, is focused between the parallel plates with an $f = 3$–30 cm converging lens; the choice of focal length depends on laser power. One electrode is connected to a current (or charge) ampli-

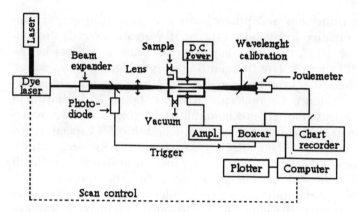

Figure 8.5 REMPI experimental scheme.

fier, and the other is biased to a 50–100-V positive or negative potential. Sample vapor, usually at 0.1–20 torr, is introduced into the evacuated chamber. Electron and positive-ion clouds are created with each laser pulse (repetition rate 5–50 Hz). The ionization current, formed by the oppositely moving charged particles, corresponds to a current of $\sim 10^{-11}$ A, measurable by a sensitive DC amplifier, picoammeter, vibrating-reed electrometer, or preamplified boxcar integrator. Very good electrical isolation between electrodes and chamber is required to minimize leakage currents. The chamber and its components must also have high mechanical stability (e.g., elimination of vibrations from vacuum rotary pumps). Quartz chambers require shielding by a Faraday cage (a metallic box).

The form of MPI signals is better than those created by α, β, or γ radiation because recombination and diffusion of the charged particles are minimal. A very useful feature is the time coincidence afforded by simultaneously triggering the laser and detection unit. This feature virtually eliminates background interference due to sources that are random in time and allows efficient detection by a boxcar integrator. Proportional counting (Figure 8.6) is a technique that can be used to intensify weak MPI signals. The counter contains an "amplifying gas"—for example, P-10 gas (90% Ar + 10% CH_4). Cascades of collisionally produced ions induced by the high accelerating potential between the electrode allow charge amplification by a factor of up to $\sim 10^3$. The detection limit of this technique may be as few as ~ 10 ions.

In closing, we note that the no-free-lunch principle applies. Unlike the pure two-photon spectra obtained by fluorescence detection, MPI sometimes leads to overlapping two- and three-photon spectra. For the intense laser pulses required to produce MPI, line-broadening effects,[14,57] AC Stark shifts,[14,58] and above-threshold ionization[59,60] may also occur. For an account of the considerable progress in understanding MPI processes in molecules afforded by mass analysis of photoionization products, see Chapter 9.

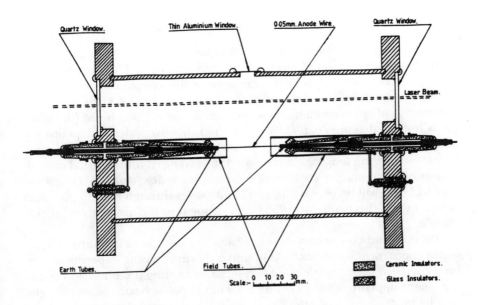

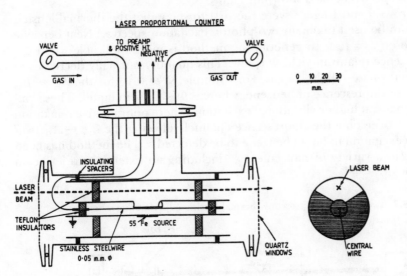

Figure 8.6 Proportional counters. (Reproduced with permission from Ref. 56.)

8.3.4. Thermal Lensing (Thermal Blooming)

A thermal lens is produced in an absorbing sample by nonradiative conversion of electronic energy into heat. The phenomenon is simple: Local temperature increases cause the sample to function as a negative lens. Dissipated heat from the excited molecules leads to a temperature gradient across a diameter section of the exciting beam (the sample is hottest at beam center). Because the refractive index n depends on temperature, the resulting spatial variation of n is equivalent to forming a concave-diverging lens within the liquid (or even gas) for the usual case of a negative temperature coefficient of refractive index, dn/dT. A Gaussian-shaped beam will diverge or "bloom" as it passes through the sample. The power of the thermal lens ($1/f$) is proportional to the absorptivity of the sample.

Albrecht and co-workers demonstrated the usefulness of thermal lensing in spectroscopic studies.[61] Although a thermal-lensing experiment can be carried out with a single dye laser, separate heating (exciting) and probing beams improve sensitivity. The low-energy probe beam senses changes in n by measurement of intensity fluctuations detected through a pinhole as the wavelength of the lens-producing laser is scanned. A dual-beam thermal lens spectrometer is shown in Figure 8.7.

Twarowski and Kliger[62] were the first to show that the thermal lensing effect can be used to obtain two-photon excitation spectra.[5] Neat benzene was chosen as a test. In general, the method is most applicable when the fluorescence quantum yield Φ is low. This quantity normally decreases in the liquid phase relative to gas. For example, $\Phi \approx 0.18$ for the $A_{1g} \leftarrow B_{2u}$ gas-phase fluorescence of benzene, whereas $\Phi < 0.07$ in liquid. These values decrease at higher vibrational excitation, where nonradiative relaxation channels overcome the fluorescence channel, allowing the $B_{2u} \leftarrow A_{1g}$ two-photon excitation in liquid benzene to be detected.[62] The method has been applied to a variety of molecules,[63,64] including an extension to jet beam spectroscopy.[65]

Figure 8.7 Two-photon spectroscopy thermal lensing experimental scheme.

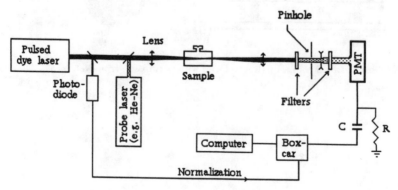

8.3.5. Photoacoustic (Optoacoustic) Detection

Heat generated by nonradiative decay of excited molecules also produces thermal expansion and consequently pressure fluctuation in the sample. The photoacoustic method is based on this effect. Detection of the induced pressure (acoustic) wave as the pulsed (or chopped) light beam is wavelength-scanned gives information on molecular excitation. Only the absorbed light energy converted into heat via radiationless decay contributes to the photoacoustic signal. Again, detection of the $B_{2u} \leftarrow A_{1g}$ two-photon excitation in liquid benzene provides an example.[66,67] It is possible to estimate two-photon absorptivities, δ, by this approach. For the C_6H_6 $B_{2u} \leftarrow A_{1g}$ vibronic principal coupling band, 14_0^1 δ is estimated as 6.6×10^{-2} GM (with an uncertainty factor of 3) by the photoacoustic method.

Good acoustic coupling between the sample cell and the detector is a requisite. Proper design of the photoacoustic cell (i.e. shape and dimensions), good choice of transducer (microphone) and gated amplification can yield a highly sensitive spectrometer.[66–69] The actual observable is the microphone signal magnitude, which measures the number of excited molecules. A typical cell transversal time for the pressure wave is $\sim 10^{-3}$ s.

Photoacoustic spectroscopy has provided the first direct detection of nonradiative decay from molecular Rydberg states.[70] Wilder and Finley recorded two-photon photoacoustic spectra of CH_3I and CD_3I for the first s-Rydberg transition.[70] Their paper provides a complete description of a photoacoustic spectrometer. Time-resolved two-color photoacoustic spectroscopy[71,72] is another area of application. These studies demonstrate the sensitivity of the photoacoustic method for monitoring stimulated-emission pumping processes.

8.3.6. Counterpropagating Beams: Doppler-free Two-Photon Spectroscopy

The Doppler effect is a source of rotational–vibrational line broadening in molecular spectra. This broadening is also a serious problem in analyzing fine-structure components, hyperfine components, Zeeman splittings, and so on, in atomic optical spectroscopy. As mentioned in Section 8.1, two-photon (and in principle, n-photon) spectroscopy has the ability to eliminate Doppler broadening. In Doppler-free two-photon optical spectroscopy the resolution at a resonance line, $v_0/\Delta v$, is $>> 10^6$, and both individual molecular rotational–vibrational transitions and atomic hyperfine structure are resolvable.

The Doppler effect is due to isotropic thermal (Maxwellian) motion of particles. Suppose that light of frequency v_0 is propagating along the x-axis. The particle "sees" this light at a Doppler-shifted frequency, $v_0(1 \pm v_x/c)$ (the $\pm$ signs depending on relative particle–light direction). Absorption occurs at $E_f - E_0 = \hbar \nu_0$; the spectral envelope reflecting the velocity distribution will be a broad Gaussian band centered at ν_0 with FWHM[73]:

$$\Delta \nu_D = 7.16 \times 10^{-7} \left(\frac{T}{A}\right)^{1/2} \nu_0 \tag{8.25}$$

In equation (8.25), A stands for the relative atomic or molecular mass and T, the absolute temperature. The quantity $\Delta \nu_D$ represents the broadening of the resonance line, ν_0; at room temperature, $\nu_0/\Delta\nu$ varies from 10^5 for light molecules to 10^6 for heavy ones.

The principle behind Doppler-free two-photon spectroscopy is the use of two counterpropagating beams, the molecule undergoing two-photon excitation by simultaneously absorbing one photon from each beam. Cancellation of the linear Doppler shift takes place because $\hbar\nu_0(1 + v_x/c) + \hbar\nu_0(1 - v_x/c) = 2\hbar\nu_0$ when the light frequency, ν_0, fulfills the two-photon quantization condition, $E_f - E_0 = 2\hbar\nu_0$. The essential point is that the Doppler effect is eliminated for all molecules, notwithstanding their velocity. The result is a very narrow two-photon absorption band, which represents a large signal spike. The two-photon resonance is usually monitored by subsequent fluorescence, the spectral linewidth being determined by the transverse decay rate of the excited state. Figure 8.8 illustrates a two-photon counterpropagating-beam experiment. Possible interfering effects arise from simultaneous absorption of two photons from either beam, causing a Doppler-broadened background. The extension to three photons requires three coplanar beams each set apart by 120°, so that the total momentum, $\Sigma\hbar\mathbf{k}_i = 0$ where $\mathbf{k}_i$ are the wave vectors. In practice, this is a difficult experiment, rarely carried out.

Doppler-free two-photon excitation can be accomplished by reflecting the incident beam from a single laser back into coincidence with itself by means of a spherical reflector. The reflector must be carefully aligned to ensure overlap of the beams. Both beams are focused, the incident beam by a lens and the reflected beam by the spherical reflector. It goes without

Figure 8.8 Doppler-free two-photon absorption involving two counterpropagating beams. The Doppler-free fluorescence peak is shown superimposed on the shaded Doppler-broadened (half-bandwidth $\Delta\omega D$) background.

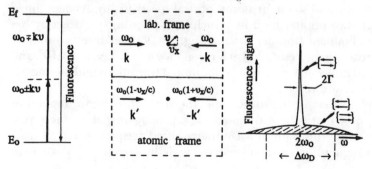

saying that the bandwidth of the exciting laser has to be much smaller than the Doppler width. Good sources for experimental details are Refs. 14, 15, and 74–79. A CW single-mode dye laser pumped by an Ar or Kr ion laser is useful for Doppler-free experiments because typical bandwidths are <1 GHz (0.03 cm^{-1}). Amplification of such a narrowband CW laser is also possible. The polarization of the counterpropagating beams can be chosen so that the Doppler-broadened background (see Figure 8.8) completely disappears. This effect has been demonstrated for the Na $3s \rightarrow 5s$ transition.[79] The power of the counterpropagating-beam method can be seen by the improvement in precision of fundamental atomic constants obtained by Doppler-free two-photon spectroscopic measurements on atomic hydrogen.[80]

8.3.7. Two-Beam Multiphoton Spectroscopy

The use of two pulsed laser beams has already been mentioned. Here we describe recent developments. These methods are most applicable to supersonic jet-cooled molecules; at room temperature many rotational and low-frequency vibrational levels are populated leading to diffuse spectra. Supersonic jet spectroscopy will be discussed in Section 8.3.9.

Two-beam molecular multiphoton spectroscopy has been reviewed by Ito et al.[81], a pictorial representation is shown in Figure 8.9. A real intermediate level A, usually the first singlet [S_1] state, is involved. Multiphoton ionization or fluorescence (phosphorescence) excitation methods are applicable. In most cases the two beams must be temporarily and spatially matched. The beams (frequencies ν_1 and ν_2) are generated from two dye lasers activated by a single pump laser (output split by a beam splitter) or two individual lasers. The first case is more appropriate to short delays (1–10 nsec) between the beams, realized by choosing different beam path lengths. The second is required for long (~ 100 nsec) delays. This variable-delay-time feature allows the evolution and relaxation of the intermediate state to be studied and consequently the mechanism of the multiphoton process to be understood. Delay times can be measured using a fast response photodiode–oscilloscope combination at the sample position.

8.3.7.1. Double Resonance 8.3.7.1.1. TWO-COLOR MPI SPECTROSCOPY.

The principle is illustrated in Figure 8.9a. Population of the origin or vibronic level of an excited molecular state A is achieved by laser I_1, of fixed frequency, ν_1. A second frequency-tunable laser (I_2, ν_2), further excites the molecule to a Rydberg state, R. Ions formed by absorption of an additional photon (from I_1 or I_2), allow measurement of ion current as a function of dye laser frequency ν_2, to yield the $A \rightarrow R$ absorption spectrum. A prime example of this method is the work of Williamson and Compton on molecular iodine.[82] A feature of two-color spectroscopy is that the spectrum may be simplified over that obtained in a single-color exper-

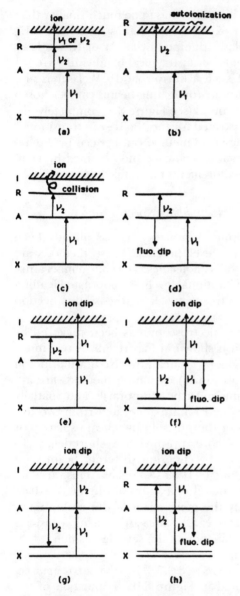

Figure 8.9 Two-beam double resonance spectroscopies. (Reproduced with permission from Ref. 81.)

iment, because of Franck–Condon and/or symmetry factors introduced by starting from state A rather than from the ground state. Williamson and Compton's I_2 experiments exemplify this feature. The power of the method is shown by a study of (1,4)-difluorobenzene Rydberg spectra.[83] At least six *s* and *d* Rydberg series were found.

An example of two-photon excitation to state A is provided by NO, the $A^2\Sigma^+$ state lying 5.48 eV above the $X^2\Pi$ ground state. Ebata et al.[84] observed the resonance-enhanced MPI spectrum to a Rydberg state from a single rovibronic level in the $A^2\Sigma^+$ state, populated via two-photon resonance obtained with laser I_1.

8.3.7.1.2. TWO-COLOR IONIZATION THRESHOLD SPECTROSCOPY.

A useful application of two-color, resonance ionization spectroscopy is the determination of adiabatic ionization potentials (IP_0) and of vibrational frequencies of the molecular ion ground state. The experimental accuracy is much greater (typically ~ 1 cm^{-1}) compared to 100 cm^{-1} for classical photoelectron spectroscopy. Pumping the molecule to state A (laser I_1, ν_1) and tuning the frequency of laser I_2, as in Figure 8.9a, directly ionizes the molecule as soon as ($\nu_1 + \nu_2$) exceeds the ionization continuum threshold; an ion current is then observed. The active vibrations revealed by this approach allow information on the relative geometrical structures of the molecule in state A and in the ion ground state to be obtained. Similar potential surfaces favor a sharp rise in the ion current for $\Delta\nu_i = 0$ transitions when any of the vibrational levels of state A are pumped.

The many applications of threshold spectra include Smalley et al.'s[85] study of benzene and naphthalene, and Parker and El-Sayed's[86] measurement of IP and lifetime of the lowest excited Rydberg state of DABCO, using as the first step, two-photon excitation to the Rydberg state. In the latter study the two-beam MPI technique was used to bypass interfering dissociative states. Another aromatic molecule example is the jet-cooled pyrazine study of Goto et al.,[87] reporting two-color MPI spectra obtained via various vibrational levels of the S_1 ($n\pi^*$) state.

Another interesting area of application is in studying clusters. Ito and colleagues[88,89] measured ionization potentials of Ar, H_2O, CH_3CN, $CHCl_3$, and CCl_4 van der Waals complexes of fluorobenzene, and hydrogen-bonded complexes of phenol with benzene, CH_3OH, and dioxane in a supersonic jet. The 200–2,000 cm^{-1} reduction of ionization potential (IP) in the van der Waals complexes was found to be closely related to the solvent molecule polarizability, regardless of whether the solvent is polar or not.[88] A noteworthy conclusion is that the first hydrogen-bonding solvent molecule attached to phenol induces a great reduction in the IP, but the second solvent molecule tends to reverse the reduction. Clusters of phenol and 7-azaindole have also been studied,[89] using two-color threshold photoionization spectroscopy. Hager et al. determined the IP of indole clusters containing polar and nonpolar solvent species to investigate the importance of charge-induced dipole and charge-dipole attractive forces in the binding of the ion-neutral clusters.[90] Indole's importance is traced to its relationship to protein photochemistry and its unusual condensed-phase solvation and photoionization dynamics.[90]

The existence of rotational isomers can be established and individual rotational isomer IPs measured by two-color MPI spectroscopy. An example is the study by Oikawa et al.[91] of several disubstituted benzenes such as fluorophenols, (1,4)-dimethoxybenzene, and so on. The difference in IP between cis and trans isomers can be small (1 cm^{-1} for 2-fluorophenol) or very substantial (261 cm^{-1} for 3-fluorophenol).

8.3.7.1.3. TWO-COLOR MPI ASSISTED BY OTHER PHENOMENA. Neutral atom and molecule discrete bound states existing above the first ionization threshold are designated autoionizing states. A textbook example of such a state is doubly excited $(2s2p)$ ^1P helium, lying ~36 eV above the He+ ground state. The name of these neutral atomic (molecular) high-lying states stems from their ability to yield ions; that is, the atom or molecule is autoionized because these states are embedded in the continuum. Autoionizing states play a role in multiphoton ionization as shown in Figure 8.9b. In general, the assisting states are Rydberg, not valence.[85,86,92,93] A completely different situation is presented in Figure 8.9c. The high-lying state R is not an autoionizing state. However, when in such a highly excited Rydberg state, the molecule can gain the necessary energy to overcome IP$_0$ by collisions. This mechanism is most important if the cross section for the transition $R \rightarrow$ ion is very small and $(\hbar\nu_1 + \hbar\nu_2) < $ IP$_0$.[94]

Another means for obtaining ionization when $(\hbar\nu_1 + \hbar\nu_2) < $ IP$_0$ involves thermal assist. The radiation (thermal photon) produced from high-temperature ovens or hot surfaces, used to produce atomic vapors, is intense enough to effectively contribute to photoionization in atomic MPI experiments.[95]

8.3.7.2. Double-Resonance Dip Spectroscopy

Fluorescence intensity or ionization current can be modulated by a second (probe) laser, I_2, which depopulates the resonant state. A signal dip results from the population loss.

8.3.7.2.1. FLUORESCENCE (PHOSPHORESCENCE) DIP SPECTROSCOPY. Fluorescence dip spectroscopy, illustrated in Figure 8.9d, detects the $A \rightarrow R$ transition by observing $A \rightarrow X$ fluorescence. The origin or (ro)vibronic level of the intermediate state A is populated by a fixed-frequency (ν_1) laser, I_1. Radiation from a tunable dye laser (I_2) is introduced; if ν_2 is resonant to $A \rightarrow R$, the resultant decrease in number of A-state molecules causes the fluorescence signal to decrease. The $A \rightarrow R$ absorption spectrum is obtained by measuring the intensity dip while scanning ν_2. High Rydberg states of NO[84] and highly excited states of benzene[96] have been observed using the fluorescence dip technique. One of the benzene states is attributed to the long-elusive valence $^1E_{2g}$ state. Glyoxal provides an example of phosphorescence dip spectroscopy.[97]

8.3.7.2.2. IONIZATION DIP SPECTROSCOPY. Cooper et al.[98] and Murakami et al.[99] introduced this useful spectroscopic technique depicted in Figure 8.9e. The ion current due to [1 + 1] REMPI is first observed using a fixed-frequency laser, I_1. The constant ion current is proportional to the state A population. If the second tunable laser (I_2, v_2) induces the transition $A \rightarrow R$, the current decreases. The dependence of the observed ion dip signal on v_2 reflects the $A \rightarrow R$ absorption. An example is the detection by Kakinuma et al.[96] of $3d_1$, $3d_2$, and $4d_2$ benzene Rydberg states.

8.3.7.2.3. STIMULATED-EMISSION PUMPING. Stimulated-emission pumping, shown in Figure 8.9f, makes use of stimulated emission to a vibrational level of the ground molecular state to depopulate level A. Stimulated emission can occur when the ground-state vibrational frequency is equal to the difference in the two laser frequencies, ($v_1 - v_2$). Stimulated-emission fluorescence dips were first observed by Kittrel et al.[100] This technique can reveal relaxation processes among ground-state vibrational levels,[101] as exemplified by recent studies on (1,4)-difluorobenzene[102] and on glyoxal.[103]

A requisite of the stimulated-emission pumping approach is close temporal and spatial matching of the two laser beams. Ito and colleagues[104] showed how this requirement may be somewhat relaxed (Figure 8.9g). Photoionization is produced by combined laser beams ($v_1 + v_2$) with intensities I_1 and I_2 low enough so that ions are not produced by each laser alone. The ion current is measured with both lasers on as v_2 is scanned. A current dip appears if level A is depopulated by stimulated emission. Since $(\hbar v_1 + \hbar v_2) > \mathrm{IP}_0$, the observed ground-state vibrations are restricted to $\hbar v < (2\hbar v_1 - \mathrm{IP}_0)$. Ground-state vibrational levels of jet-cooled 3-fluorotoluene and aniline have been obtained using this approach.[104]

8.3.7.2.4. POPULATION-LABELING SPECTROSCOPY. The population-labeling approach, shown in Figure 8.9h, starts from a vibrational level of the ground molecular state. A fixed-frequency beam, v_1, resonant to state A, is produced by laser I_1 and the subsequent fluorescence or photoionization current is measured. A frequency-tunable laser, I_2 (v_2), in resonance to a higher R state, is now introduced and depopulates the ground-state vibronic level, producing a consequent dip in the original fluorescence intensity or ion current. This method, first applied in 1976 for Na_2 molecules,[105] is effective for the definite assignment of $X \rightarrow R$ transitions by labeling the common vibronic state in X.

8.3.8. Three- and Four-Wave Mixing

Two-photon resonances can be detected in four-wave sum- or difference-frequency mixing experiments. Only a brief outline of these nonlinear phenomena will be given as the theory is long and the notation rather complex[106-107]; the subject is discussed in more detail in Chapter 7.

An electromagnetic wave $\mathbf{E}(\mathbf{r}, t)$ induces polarization in a medium; that is, $\mathbf{P} = f(\mathbf{E})$. In a weak field, $\mathbf{P}(\mathbf{r}, t)$ is linear in $\mathbf{E}(\mathbf{r}, t)$ so that $\mathbf{P}_i = \varepsilon_0 \Sigma_j \chi_{ij} \mathbf{E}_j$ in SI units. In a laser beam the electric field is so strong that additional nonlinear terms become significant.[2,108-110]

$$\mathbf{P} = \varepsilon_0 (\chi^{(1)} \mathbf{E} + \frac{1}{2} \chi^{(2)}) : \mathbf{E} \cdot \mathbf{E} + \frac{1}{4} X^{(3)} \mathbf{E} \cdot \mathbf{E} \cdot \mathbf{E} + \cdots) \qquad (8.26)$$

Here $\chi^{(1)}(\omega)$ is the well-known first-order electric susceptibility (polarizability) and $\chi^{(2)}(-\omega_i, \omega_j \pm \omega_k)$, $\chi^{(3)}(-\omega_i, \omega_j, \omega_k, \omega_m)$ are the second- and third-order susceptibilities; $X^{(n)}$ is a tensor that has 3^{n+1} elements.

The quadratic term $\chi^{(2)}$ is responsible for sum- and difference-frequency generation, $\omega_i = \omega_j \pm \omega_k$, caused by the two laser frequencies ω_j, ω_k. This term vanishes for systems with a center of symmetry, including optically active liquids, but efficiently occurs in anisotropic media such as birefringent crystals. A common application is frequency doubling[108,109] of a visible laser. Nonlinear crystals used for second harmonic generation must have large $\chi^{(2)}$, good transparency to both incident ω_j and generated $\omega_i = 2\omega_j$ radiation and must be resistant to damage under intense laser illumination.

For isotropic media (atomic vapors and molecular liquids and gases) the first nonlinear term in the polarization expansion, equation (8.26), is the $\chi^{(3)}$ one. This term is responsible for so-called four-wave mixing and it describes the interaction between four waves; three input and one output. Third-harmonic generation ($\omega_i = 3\omega_j$) provides an example. Combining input beams to generate new frequencies is an important technique for the production of coherent radiation in the ultraviolet and vacuum ultraviolet regions.[106,111,112] The input frequencies are denoted by plus signs and the output by minus signs. Thus, sum-frequency generation; input ω_1, ω_2, and ω_3, is denoted as $\chi^{(3)}(-\omega_4; \omega_1, \omega_2, \omega_3)$, that is, $\omega_4 = \omega_1 + \omega_2 + \omega_3$.

Because $\chi^{(3)}$ is of high order we might expect insignificant nonlinear optical effects due to this term, particularly for low-density vapor systems. However, when the optical frequencies are close to an atomic or molecular resonance, $\chi^{(3)}$ is greatly enhanced. Two-photon resonances in the third-order susceptibility have been demonstrated by Hochstrasser et al.[113] for nitric oxide and sulfur dioxide gases. In these experiments two pulsed laser beams (ω_1, ω_2) were focused simultaneously in the same region with ω_2 fixed and ω_1 scanned. Two monochromators selected the generated frequency $\omega_3 = 2\omega_1 - \omega_2$. The change in intensity of ω_3 while scanning ω_1 represents the two-photon molecular resonances (occurring at $2\omega_1$) observed in this three-wave mixing experiment. The signal intensity is proportional to $I_1^2(\omega_1) I_2(\omega_2)$ and to N^2 where N is the density of the gaseous sample. A feature of three-wave mixing is the invariance of the spectrum to the value of ω_2. McClain et al.[114,115] used CW lasers to measure two-photon absorptivities of organic molecules via three-wave mixing. An example of four-wave sum- and difference-frequency mixing is provided by studies on carbon monoxide, carried out in both static cell experiments[116] and a pulsed supersonic jet.[117]

8.3.9. Supersonic Jets: A Clean Spectroscopic Medium

Laser spectroscopy combined with supersonic beams opens a broad highway for molecular spectroscopic studies.[118] Highly resolved jet-cooled absorption spectra, without the rotational congestion accompanying gas-phase room-temperature spectra, can be obtained using simple detection techniques, such as fluorescence excitation or multiphoton ionization. Another unique feature is that jets allow spectra of free radicals to be analyzed. The dispersed fluorescence from a particular excited (ro)vibronic level yields ground-state vibrational levels with a resolution determined by the monochromator bandwidth used to disperse the fluorescence light. The vibrational structure of an ion's ground state can be explored using two-color multiphoton ionization. Internal rotation–torsional vibrations, rotational conformers and isomers, van der Waals and hydrogen-bonded complexes, and chemical intermediates can all be effectively studied in a supersonic jet.

Supersonic beams, developed by Kantrowitz and Grey[119] in 1951, were not applied to molecular spectroscopy until after 1973.[120-122] The key to supersonic jet formation is a small pinhole nozzle (diameter D) that separates a high pressure (P_0) gas chamber, from a low pressure (P) constantly evacuated one. The outcome is the supersonic jet profile shown in Figure 8.10. The mean free path at pressure P_0 and temperature T_0 is l_0. If $D >> l_0$, there are many collisions between particles (i.e., atoms or molecules) as they are driven out of the nozzle. The exciting molecules now have a forward-directed flow; that is, thermal energy has been transferred into translational energy. The enthalpy H_0 associated with random particle motion is

Figure 8.10 Supersonic jet formation by expansion of a gas at pressure P_0 (temperature T_0) into a vacuum chamber ($P << P_0$). At distance x_i the temperature $T_i << T_0$.

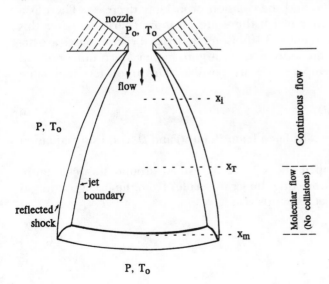

P, T_0

converted into directed mass flow downstream from the pinhole, and this conversion continues until no more collisions occur.

As the distance from the nozzle increases, the molecules' velocity distribution width narrows. Energy conservation requires

$$H_0 = H_i + \frac{1}{2}mv_i^2 \tag{8.27}$$

where H_0, H_i are the molar enthalpies in chamber P_0 and at distance x_i after the nozzle, respectively; m is the molar mass. Thermodynamical concepts (e.g., $H_i = C_p T_i$, $C_p - C_v = R$, $\gamma = C_p/C_v$, adiabatic expansion law, and so on) yield

$$\frac{T_i}{T_0} = \left(\frac{P_i}{P_0}\right)^{\gamma-1/\gamma} = \left(\frac{\rho_i}{\rho_0}\right)^{\gamma-1} = [1 + \frac{1}{2}(\gamma - 1)M_i^2]^{-1} \tag{8.28}$$

for a point x_i downstream from the nozzle. At this distance, T_i represents a low Maxwell–Boltzmann temperature, corresponding to a narrow velocity distribution. The pressure at this point in the expansion has become P_i and the density ρ_i. For a monoatomic gas $\gamma = \frac{5}{3}$. M_i is the Mach number, defined as the ratio of the flow velocity v_i to the speed of sound; a_i, at temperature T_i; that is,

$$M_i = \frac{v_i}{a_i} = \frac{v_i}{(\gamma RT_i/m^2)} = A\left(\frac{x_i}{D}\right)^{\gamma-1} \tag{8.29}$$

where A is a constant that depends on γ ($A = 3.26$ for monoatomic gases). Typical parameter values $D = 0.01–0.1$ cm, $P_0 = 30$ atm to 100 torr, $P = 1–10^{-6}$ torr, yield values of $M_i >> 1$, explaining the term *supersonic*. Actually, even for $M_i \to \infty$, the directed speed v_i of the jet particles is only ~1.3 times the most probable speed of particles in a bulb.

As x_i increases, T_i, ρ_i, and the collision probability decrease. There is a terminal position x_T after which there are no more collisions. After this point, the flow is free (Figure 8.10). Commonly used carrier gases are argon and helium. For the more massive argon, higher Mach numbers and consequently lower temperatures are possible. The argon terminal value Mach number is

$$M_T = 133(P_0 D)^{0.4} \tag{8.30}$$

The value of x_T can be calculated from P_0 (atm) and D (cm) using equation (8.29).

A stable barrel-shaped shock wave is formed around the jet (Figure 8.10), the boundary protecting the jet molecules from the background gas of the chamber. The jet terminus is at

$$x_m = 0.67D\left(\frac{P_0}{P}\right)^{1/2} \tag{8.31}$$

The optimal position for excitation is close to x_T. Very low temperatures can be achieved at this position, provided the pumps maintain $x_m > x_T$. Equations (8.28) and (8.29) show that the temperature T_i is independent of the backing pressure P_0 and decreases as x_i increases. However, the terminal Mach number M_T depends on the product $(P_0 D)$. For argon the terminal temperature T_T is

$$T_T = \frac{T_0}{1 + 5896(P_0 D)^{0.8}} \tag{8.32}$$

The product $P_0 D$ is related to the binary collision probability. The freedom to increase P_0 and/or D is limited by the pump speed, because the volume of expanding gas is proportional to $P_0 D^2$ and its maximum value cannot overcome the pump throughput (displacement) N. With this condition, that is, $N = $ constant $P_0 D^2$, equation (8.30) gives working relationships $M_T \propto (P_0 N)^{0.2}$ and $M_T \propto (N/D)^{0.4}$. Therefore, very low temperatures can be achieved by increasing P_0 and decreasing D near the maximum-volume limiting case of the expanding gas $= N$. The disadvantage is that the signal decreases as D decreases.

Pulsed valves are ubiquitous in jet apparatus since they reduce sample consumption and gas throughput load on the vacuum system. They are ideally interfaced to a pulsed laser by synchronizing the valve and laser pulses. Typically, 10–15 cm diffusion–rotary pumps, isolated by a liquid nitrogen trap, are required to maintain the necessary $<10^{-5}$-torr vacuum. At this pressure electron multipliers such as a channeltron can function to measure ion current from MPI detection. For fluorescence detection, where the photomultiplier is outside the vacuum chamber, a less-stringent vacuum is required.

The sample is usually seeded at 0.1% to 1% in He or Ar carrier gas. Rapid rotational–translational equilibration makes possible rotational temperatures of just a few degrees Kelvin. The degree of vibrational cooling depends on the form of the vibrational mode as well as on properties of the carrier gas. Argon is more efficient than helium because of the velocity slip effect; fewer collisions with heavier atoms are required to enhance the sample forward flow. The translational, rotational, and vibrational temperatures T_t, T_r, and T_v achieved in a jet are ordered $T_t < T_r < T_v$.

An additional important process that takes place in supersonic jets is complex (cluster) formation. The formation of these species may be a problem (e.g., obscuring portions of the spectrum) or interesting and useful. A time-of-flight or quadrupole mass spectrometer in conjunction with MPI is needed. Cluster formation requires, at minimum, three-body collisions; the collision rate (and thus cluster formation) is proportional to $P_0^2 D$. The released heat of condensation increases the temperature of the jet. Helium thus gives lower terminal temperatures than argon because of the reduced rate of condensation. Because each particle in the jet has a similar velocity, a complex once formed remains alive, whereas in a static cell it dissociates immediately through the high-kinetic-energy collisions that occur in a random velocity sample.

8.4. Examples

The examples cited in this section illustrate the principles and techniques discussed previously. In particular, we use as illustrations two-photon spectra arising from the benzene $^1B_{2u} \leftarrow {}^1A_{1g}$ excitation, well studied by traditional ultraviolet spectroscopy; multiphoton Rydberg-state jet spectroscopy of acetone; three-photon spectroscopy of Cl_2, and two-beam spectroscopy of copper ions in crystals.

8.4.1. Benzene $^1B_{2u} \leftarrow {}^1A_{1g}$ Two-Photon Spectrum

The C_6H_6 $^1B_{2u} \leftarrow {}^1A_{1g}$ TP spectrum is parity forbidden; that is, no origin band is observed. The results of analyses of well-resolved spectra obtained in the vapor phase by different techniques,[38,123,124] as well as in crystals at low temperatures,[24,25] indicate that the forbidden character is induced by four vibrations (Figure 8.12), all of u parity: b_{2u} modes ν_{14} and ν_{15}, e_{1u} ν_{18}, and (weakly) e_{2u} ν_{17}. An overview of the fundamental region of the hot-band fluorescence excitation spectrum is shown in Figure 8.11. Detection of the fluorescence with a cell similar to that shown in Figure 8.3 reveals intense Q and much weaker R and S branches associated with the dominant two-photon activity of the b_{2u} modes, as well as the rotational

Figure 8.11 Hot-band fundamental region in the two-photon $^1B_{2u} \leftarrow {}^1A_{1g}$ fluorescence excitation spectrum of benzene vapor. The upper scale shows the laser wavelength, and the lower scale the two-photon energy in wavenumbers. (Reproduced from Ref. 124.)

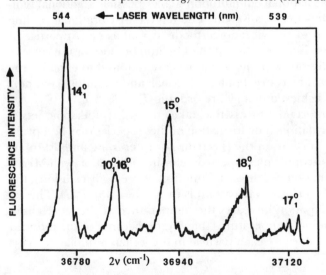

envelopes of the other less active modes. It is noteworthy that these details are realizable despite the small Boltzmann factor (<0.007) for the >1,000 cm^{-1} b_{2u} and e_{1u} ground-state modes.

8.4.2. Isotopic Labeling Effects on Benzene Two-Photon Spectra

There is an important difference between the active modes appearing in TP and OP (one-photon) spectra of the $^1B_{2u} \leftarrow ^1A_{1g}$ benzene transition: the prominent TP vibronic coupling mode ν_{18} possesses substantial hydrogen character; on the other hand, the only OP coupling mode, ν_6 (shown in Figure 8.12), is primarily skeletal. Thus, the way is paved for deuterium isotope effects to appear in the TP spectrum that are not possible in OP.

The first 1,100 cm^{-1} of the cold-band region of the TP C_6H_6 $^1B_{2u} \leftarrow ^1A_{1g}$ jet spectrum is shown in Figure 8.13, showing the active mode ν_{18}. In contrast, Figure 8.14 shows the same region for the similarly obtained spectrum for C_6H_5D demonstrating the activity of an additional b_2 symmetry mode, ν_{9b}, and totally symmetric ν_1 and ν_{12} in addition to the now split ν_{18}. The isotopic perturbation affects vibrational kinetic energy because of the change in mass, but has no effect on the potential energy. The result, illustrated by the spectrum in Figure 8.14, is that the normal vibrations of the

Figure 8.12 Zero-point displacement diagrams for some ground-state normal vibrational modes active in the one- and two-photon spectra of benzene. (Reproduced from Ref. 125.)

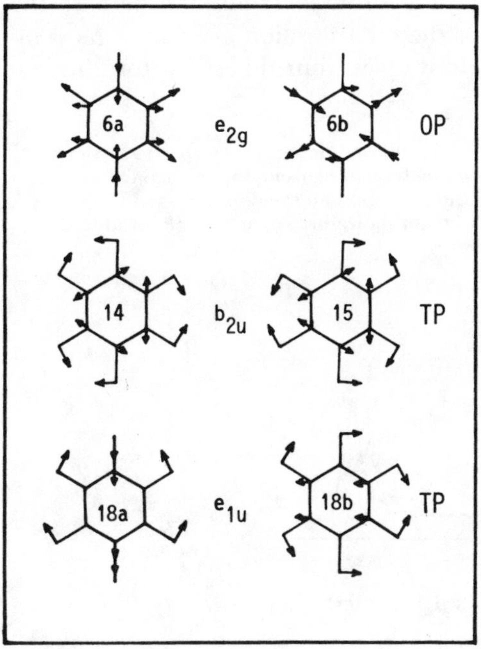

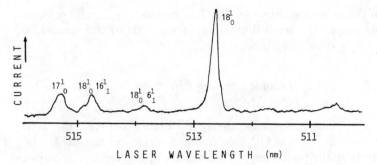

Figure 8.13 Two-photon resonant [2 + 1] multiphoton ionization $^1B_{2_u} \leftarrow {}^1A_{1_g}$ spectrum of jet-cooled benzene vapor, showing the fundamental cold-band region $700–1,100 \text{ cm}^{-1}$ above the missing origin.

unsubstituted molecule become scrambled in the labeled molecule, as well as undergoing frequency shifts.[126] An important outcome is that TP spectra have provided stringent tests for the benzene-excited-state force field, which were not previously possible.

8.4.3. Polysubstitution Effects on Benzene Two-Photon Spectra

A striking example of the complementary behavior of OP and TP spectra is revealed by substituted benzene spectra. For example, the $^1B_{2u} \leftarrow {}^1A_{1g}$ TP spectra of a 1,4-homosubstituted benzene such as (1,4)-dimethyl benzene reverts to a forbidden transition because the $g \leftrightarrow g$ rule is not satisfied, whereas the OP transition intensity is most intense among the polysubstituted derivatives (four times that for a monosubstituted benzene).[127,128]

Figure 8.14 Two-photon resonance [2 + 1] multiphoton ionization spectrum, showing the fundamental cold-band region of monodeuterated benzene vapor obtained in a jet. This spectrum should be compared with the corresponding region ($700–1,100 \text{ cm}^{-1}$ above the missing origin) for the C_6H_6 spectrum shown in Figure 8.13. (Reproduced from Ref. 126.)

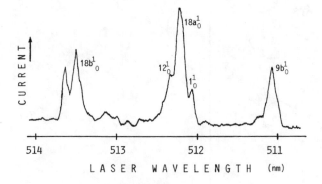

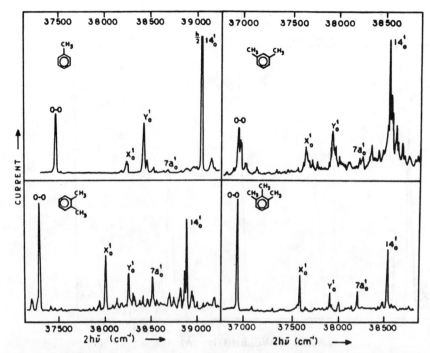

Figure 8.15 Normalized two-photon resonant (2 + 1) multiphoton ionization spectra for several methylbenzenes obtained in a supersonic jet; *h* represents the peak height of the 14_0^1 band in toluene. (Reproduced from Ref. 128.)

A particularly interesting case is provided by the difference between OP and TP spectra in 1,2,3-trimethylbenzene. Despite being symmetry and parity allowed, the $B_2 \leftarrow A_1$ OP transition is perturbation forbidden. In contrast, as is illustrated by the 0-0 band intensities in [2 + 2] MPI jet spectra in Figure 8.15, the TP transition for this molecule is the most intense of any orientation.[128]

8.4.4. The a_2 Torsional Mode in Acetone

Two-photon Rydberg state spectra of acetone, CH_3COCH_3, illustrates the power of two-photon jet spectroscopy to reveal important information for a molecule that has been well studied by traditional spectroscopic methods. Acetone has not lived up to its potential as a benchmark molecule for obtaining an understanding of the torsional interaction in coupled methyl tops, because only one methyl torsional vibration is active in its infrared and Raman spectra. The infrared- and Raman-inactive a_2 torsional mode has historically remained unmeasured. This lacuna has prevented the calibration of methyl–methyl interaction potential models.

An approach to this problem that successfully reveals the missing ground-state a_2 torsional fundamental is multiphoton ionization spectros-

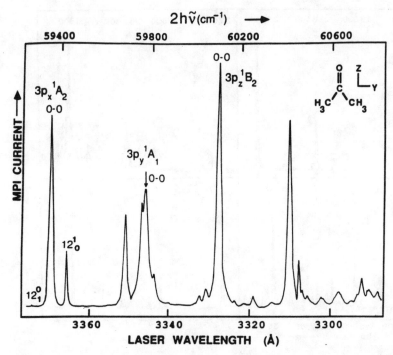

Figure 8.16 Rydberg $3p \leftarrow n$ two-photon $(2 + 1)$ multiphoton ionization spectrum of jet-cooled acetone seeded into 3.5 atm He. (Reproduced from Ref. 129.)

copy involving an acetone Rydberg transition.[129] The [2 + 1] MPI spectrum of jet-cooled acetone in the 338–328 nm (59,000–61,000 cm^{-1}, one-photon) region is shown in Figure 8.16. The three origin bands and associated vibronic activity of the $3p \leftarrow n$ Rydberg transitions are clearly resolved.

Figure 8.17a shows the $A_2 \leftarrow A_1$ $3p_x \leftarrow n$ origin region in linearly polarized light. The result of raising the vibrational temperature by changing jet expansion parameters is shown in Figure 8.17b. Figure 8.17c demonstrates that both of these bands, but not the A_2 origin, are strongly attenuated in circularly polarized light, securing their A_1 vibronic symmetry, which in turn requires a_2 symmetry vibrations. Thus, two-photon jet spectroscopy provides the missing acetone ground-state a_2 torsional fundamental frequency, 77 ± 2 cm^{-1}.[129] The two-photon determined a_2 torsional frequency (along with the infrared determined b_2 one) has served as impetus for development of accurate torsional potential barrier models.[130,131]

8.4.5. Multiphoton Ionization Spectra of Cl_2

MPI spectroscopy of Cl_2 provides examples of three- and two-photon excitations.[132–135] Li, Wu, and Johnson[132] studied [3 + 1] spectra corresponding to the $2^3\Pi_u$ and $2^1\Pi_u \leftarrow X^1\Sigma_g^+$ Rydberg transitions

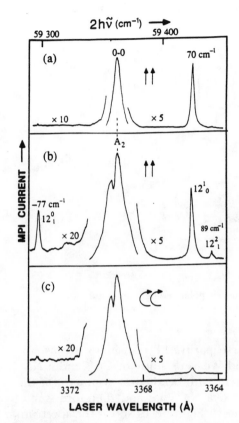

Figure 8.17 Two-photon REMPI spectrum of the $3p_x \leftarrow n$ ($^1A_2 \leftarrow {}^1A_1$) Rydberg region of acetone (see Figure 8.16) seen under excitation and cooling conditions of (a) linearly polarized light and 2.3 atm Ar, (b) linearly polarized light and 0 atm Ar, (c) circularly polarized light and 0 atm Ar. (Reproduced from Ref. 129.)

(Figure 8.18) with emphasis on pressure effects and third harmonic generation. The spectra show N and O branches consistent with the three-photon selection rule $\Delta J = \pm 3, \pm 2, \pm 1$, and 0. Pressure effects on the spectra give information on how Cl_2 photodissociates. Additional sharp bands were detected using a linearly polarized laser when the pressure was reduced to 0.2 torr, coinciding with [3 + 2] and [4 + 1] resonances in atomic chlorine. That chlorine atoms were formed in both the $^2P_{3/2}$ and $^2P_{1/2}$ ground states demonstrates that the dissociation channel is via the $1^3\Pi_u$ triplet repulsive state. Only the $^2P_{3/2}$ state would result from a $1^1\Pi_u$ channel.

8.4.6. Two-Photon Spectroscopy of Impurity Ions in Crystals

Two-photon spectroscopy is especially appropriate for studying the electronic spectra of transition metal and rare earth ions in crystals.[136] These ions often appear at centers of symmetry, and since inner-

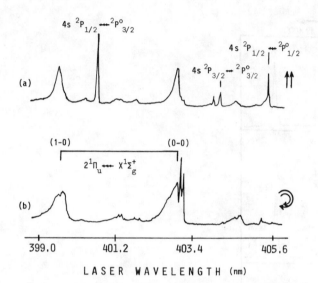

Figure 8.18 MPI spectra of chlorine: (a) linearly polarized light at 0.2 torr; (b) circularly polarized light at all pressures (<1 torr). (Reproduced with permission from Ref. 132.)

shell (e.g., $d \to d$) transitions preserve parity, TP transitions are usually allowed. An example is McClure and Weaver's study of Cu^+ and a Na^+ site in NaF. The spectra are consistent with octahedral symmetry at the Na^+ site. The Cu^+ ion lies exactly at a Na^+ site and therefore is surrounded by an octahedron of F^- ions.[137] Figure 8.19 shows the spectrum due to the $3d^{10} \to 3d^9\,4s$ transitions at 300 K. It illustrates the two polari-

Figure 8.19 Two-photon excitation spectrum of NaF:Cu^+ at 300 K; (a) spectrum polarized parallel to a cube axis; (b) spectrum polarized 45° to the cube axis. (Reproduced with permission from Ref. 137.)

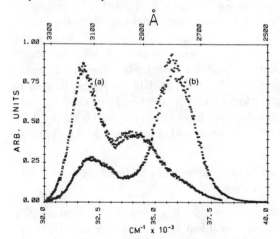

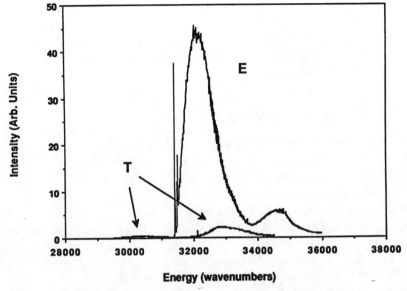

Figure 8.20 The two-photon spectrum of NaF:Cu$^+$ at 2 K using two beams: **E**, both beams polarized parallel and along a cube axis; **T**, beams polarized at right angles along cube axis. (Reproduced with permission from Ref. 138.)

zations: $^1A_{1g} \rightarrow {}^1E_g$, polarized parallel to a cube axis (a Cu—F bond) and $^1A_{1g} \rightarrow {}^1T_{2g}$, polarized at 45° to the cube axis. These spectra do not completely separate the two transitions because the $^1A_{1g} \rightarrow {}^1E_g$ spectrum still appears in the 45° polarization experiment, although reduced to one-fourth of the 0° intensity. Subtracting one-fourth of $^1A_{1g} \rightarrow {}^1E_g$ reveals not only the pure $^1A_{1g} \rightarrow {}^1T_g$ spectrum, but also a weak $^1A_{1g} \rightarrow {}^3T_{2g}$ transition underneath. This triplet state has A_{2g}, T_{2g}, and T_{1g} components. A clean separation of these components is made possible by two-beam spectroscopy. The results of Berg and McClure's polarized two-color, two-photon experiment[138] are shown in Figure 8.20. States of E_g symmetry appear when the beams are polarized parallel to each other and to a cube axis. States of T_{2g} symmetry appear when they are polarized at right angles and along orthogonal cube axes.

8.5. Conclusions

Two-photon and traditional spectra are frquently opposite in their response to the same perturbation, with the result that two-photon spectroscopy can obtain new information about excited states even of molecules that have already been very well studied by traditional one-photon procedures. This conclusion ranges over all molecular classes, from complex organics and diatomics to impurity ions in inorganic crystals, and extends to low symmetry. An important reason for doing two-photon spectroscopy is that it makes possible observation of transitions as allowed spectra instead of indirectly through perturbations (e.g., Herzberg–Teller) on forbidden spectra. The technology for obtaining the spectra as well as a few examples have been described.

REFERENCES

1. M. Göppert-Mayer, *Ann. Phys. (Leipzig)* **9**, 273 (1931).
2. W. Kaiser and C. G. B. Garrett, *Phys. Rev. Lett.* **7**, 229 (1961).
3. W. M. McClain, *Acc. Chem. Res.* **7**, 129 (1974).
4. P. M. Johnson, *Acc. Chem. Res.* **13**, 20 (1980).
5. D. S. Kliger, *Acc. Chem. Res.* **13**, 129 (1980).
6. E. Giacobino and B. Cagnac, in *Progress in Optics*, Vol. 17, E. Wolf, ed. (North-Holland, Amsterdam, 1980).
7. C. Kittrell, E. Abramson, J. L. Kinsey, S. A. McDonald, D. E. Reisner, R. W. Field, and D. H. Katayama, *J. Chem. Phys.* **75**, 2056 (1981).
8. W. D. Lawrance and A. E. W. Knight, *J. Phys. Chem.* **87**, 389 (1983).
9. R. B. Bernstein, *J. Phys. Chem.* **86**, 1178 (1982).
10. H. J. Neusser, *Int. J. Mass. Spectrom. Ion Process* **79**, 141 (1987).
11. G. Petty, C. Tai, and F. W. Dalby, *Phys. Rev. Lett.* **34**, 1207 (1975).
12. P. M. Johnson, *J. Chem. Phys.* **62**, 4562 (1975).
13. P. M. Johnson, *J. Chem. Phys.* **64**, 4143 (1976).
14. S. H. Lin, Y. Fujimura, H. J. Neusser, and E. W. Schlag, *Multiphoton Spectroscopy of Molecules* (Academic Press, New York, 1984).
15. W. Demtröder, *Laser Spectroscopy* (Springer, Berlin, 1981).
16. W. M. McClain and R. A. Harris, in *Excited States*, Vol. 3, E. C. Lim, ed. (Academic Press, New York, 1977), p. 1.
17. M. A. C. Nascimento, *Chem. Phys.* **74**, 51 (1983).
18. D. M. Friedrich, *J. Chem. Ed.* **59**, 472 (1982).
19. L. D. Ziegler and B. S. Hudson, in *Excited States*, Vol. 5, E. C. Lim, ed. (Academic Press, New York, 1982) p. 41.
20. L. Goodman and R. P. Rava, in *Advances in Chemical Physics*, Vol. 54, I. Prigogine and S. A. Rice, eds. (Wiley, New York, 1983) p. 177.
21. L. Goodman and R. P. Rava, *Acc. Chem. Res.* **17**, 250 (1984).
22. R. B. Birge, *Acc. Chem. Res.* **19**, 138 (1986).
23. D. M. Friedrich and W. M. McClain, *Ann. Rev. Phys. Chem.* **31**, 559 (1980).
24. R. M. Hochstrasser, J. E. Wessel, and H. N. Sung, *J. Chem. Phys.* **60**, 317 (1974).
25. R. M. Hochstrasser, H. N. Sung, and J. E. Wessel, *Chem. Phys. Lett.* **24**, 7 (1974).
26. R. G. Bray, R. M. Hochstrasser, and J. E. Wessel, *Chem. Phys. Lett.* **27**, 167 (1974).

27. D. L. Rousseau and P. F. Williams, *Phys. Rev. Lett.* **33**, 1368 (1974).

28. J. I. Steinfeld, *Molecules and Radiation* (MIT Press, Cambridge, MA, 1986).

29. R. Swofford and W. M. McClain, *Rev. Sci. Instrum.* **46**, 246 (1975).

30. M. Friedrich and W. M. McClain, *Chem. Phys. Lett.* **32**, 541 (1975).

31. E. B. Wilson, *Phys. Rev.* **45**, 706 (1934).

32. W. Hampf, H. J. Neusser, and E. W. Schlag, *Chem. Phys. Lett.* **46**, 406 (1977).

33. R. P. Rava, L. Goodman, and K. Krogh-Jespersen, *J. Chem. Phys.* **74**, 273 (1981).

34. B. A. Garetz and C. Kittrell, *Phys. Rev. Lett.* **53** 156 (1984).

35. F. Metz, W. E. Howard, L. Wunsch, H. J. Neusser, and E. W. Schlag, *Proc. Roy. Soc. Lond.* **A363**, 381 (1978).

36. P. R. Monson and W. M. McClain, *J. Chem. Phys.* **53**, 29 (1970).

37. W. M. McClain, *J. Chem. Phys.* **55**, 2789 (1971).

38. L. Wunsch, F. Metz, H. J. Neusser, and E. W. Schlag, *J. Chem. Phys.* **66**, 386 (1977).

39. D. L. Andrews and W. A. Ghoul, *J. Chem. Phys.* **75**, 530 (1981).

40. D. M. Friedrich, *J. Chem. Phys.* **75**, 3258 (1981).

41. J. J. Hopfield, J. M. Worlock, and K. Park, *Phys. Rev. Lett.* **11**, 414 (1963).

42. B. Hudson and B. Kohler, *Chem. Phys. Lett.* **14**, 299 (1972).

43. B. Hudson and B. Kohler, *Chem. Phys. Lett.* **23**, 139 (1973).

44. W. Demtröder, *Laser Spectroscopy* (Springer, Berlin, 1981), pp. 195–211, 223–226.

45. F. H. M. Faisal, R. Wallenstein, and H. Zacharias, *Phys. Rev. Lett.* **39**, 1138 (1977).

46. B. Dick, H. Gonska, and G. Hohlneicher, *Ber. Bunsenges. Phys. Chem.* **85**, 746 (1981).

47. P. Esherick, P. Zinsli, and M. A. El-Sayed, *Chem. Phys.* **10**, 415 (1975).

48. H. Abe, S. Kamei, N. Mikami, and M. Ito, *Chem. Phys. Lett.* **109**, 217 (1984).

49. I. Suzuka, T. Tomioka, and Y. Ito, *Chem. Phys. Lett.* **172**, 409 (1990).

50. D. H. Parker, J. O. Berg, and M. A. El-Sayed, in *Advances in Laser Chemistry*, A. H. Zewail, ed. (Springer, Berlin, 1978), pp. 320–335.

51. P. M. Johnson and C. E. Otis, *Ann. Rev. Phys. Chem.* **32**, 139 (1981).

52. D. S. Zakheim and P. M. Johnson, *Chem. Phys.* **46**, 263 (1980).

53. R. L. Whetten, K. Fu, and E. R. Grant, *J. Chem. Phys.* **79**, 2626 (1983).

54. R. L. Whetten, S. G. Grubb, C. E. Otis, A. C. Albrecht, and E. R. Grant, *J. Chem. Phys.* **82**, 1115 (1985).

55. S. Speiser and J. Jortner, *Chem. Phys. Lett.* **44**, 399 (1976).

56. M. Towrie, J. W. Cahill, K. W. D. Ledingham, C. Raine, K. M. Smith, M. H. C. Smyth, D. T. Stewart, and C. M. Houston, *J. Phys. B: Atom. Mol. Phys.* **19**, 1989 (1986).

57. C. C. Wang, J. V. James, and J.-F. Xia, *Phys. Rev. Lett.* **51**, 184 (1983).

58. P. Kruit, W. R. Garrett, J. Kimman, and M. J. van der Wiel, *J. Phys. B, Atom. Mol. Phys.* **16**, 3191 (1983).

59. P. Kruit, J. Kimman, H. G. Muller, and M. J. van der Wiel, *Phys. Rev. A* **28**, 248 (1983).

60. G. Petite, F. Fabre, P. Agostini, M. Crance, and M. Aymar, *Phys. Rev. A.* **29**, 2677 (1984).

61. R. L. Swofford, M. E. Long, and A. C. Albrecht, *Science* **191**, 183 (1976).

62. A. J. Twarowski and D. S. Klinger, *Chem. Phys.* **20**, 259 (1977).

63. J. K. Rice and R. W. Anderson, *J. Phys. Chem.* **90**, 6793 (1986).

64. P. R. Salvi, P. Foggi, R. Bini, and E. Castellucci, *Chem. Phys. Lett.* **141**, 417 (1987).

65. M. F. Hineman, R. G. Rodriguez, and J. W. Nibler, *J. Chem. Phys.* **89**, 2630 (1988).

66. A. C. Tam and C. K. N. Patel, *Nature* **280**, 304 (1979).

67. C. K. N. Patel and A. C. Tam, *Rev. Mod. Phys.* **53**, 517 (1981).

68. A. Rosencwaig, *Photoacoustics and Photoacoustic Spectroscopy* (Wiley, New York, 1980).

69. J. Davidson, J. H. Gutow, and R. N. Zare, *J. Phys. Chem.* **94**, 4069 (1990).

70. J. A. Wilder and G. L. Finley, *Rev. Sci. Instrum.* **58**, 968 (1987).

71. D. J. Moll, G. R. Parker, and A. Kupperman, *J. Chem. Phys.* **80**, 4800 (1984).

72. S. N. Thakur, D. Guo, T. Kundo, and L. Goodman, *Chem. Phys. Lett.* in press.

73. W. Demtröder, *Laser Spectroscopy* (Springer, Berlin, 1981), pp. 84–89.

74. J. A. Gelbwachs, P. F. Jones, and J. E. Wessel, *App. Phys. Lett.* **27**, 551 (1975).

75. E. Giacobino, F. Biraben, G. Grynberg, and B. Cagnac, *J. Phys. (Paris)* **38**, 623 (1977).

76. E. Riedley, R. Moder, and H. J. Neusser, *Opt. Commun.* **43**, 388 (1982).

77. V. S. Letokhov and V. P. Chebotayer, *Nonlinear Laser Spectroscopy* (Springer, Berlin, 1977), pp. 1–35, 155–183.

78. G. Grynberg and B. Cagnac, *Rep. Prog. Phys.* **40**, 791 (1977).

79. F. Biraben, B. Cagnac, and G. Grynberg, *Phys. Rev. Lett.* **32**, 643 (1974).

80. *Laser Spectroscopy*, Vol. 7. T. W. Hansch and Y. R. Shen, eds. (Springer, Berlin, 1985): (a) F. Biraben and L. Julien, pp. 67–70; (b) C. J. Foot et al., pp. 33–36.

81. M. Ito, T. Ebata and N. Mikami, *Ann. Rev. Phys. Chem.* **39**, 123 (1988).

82. A. D. Williamson and R. N. Compton, *Chem. Phys. Lett.* **62**, 295 (1979).

83. M. Fujii, T. Kakinuma, N. Mikami, and M. Ito, *Chem. Phys. Lett.* **127**, 297 (1986).

84. T. Ebata, M. Naohiko, and M. Ito, *J. Chem. Phys.* **78**, 1132 (1983).

85. M. A. Duncan, T. G. Dietz, and R. E. Smalley, *J. Chem. Phys.* **75**, 2118 (1981).

86. D. H. Parker and M. A. El-Sayed, *Chem. Phys.* **42**, 379 (1979).

87. A. Goto, M. Fujii, and M. Ito, *J. Phys. Chem.* **91**, 2268 (1987).

88. N. Gonohe, H. Abe, and N. Mikami, and M. Ito, *J. Phys. Chem.* **89**, 3642 (1985).

89. K. Fuke, H. Yoshiuchi, K. Kaya, Y. Achiba, K. Sato, and K. Kimura, *Chem. Phys. Lett.* **108**, 179 (1984).

90. J. Hager, M. Ivanco, M. A. Smith, and S. C. Wallace, *Chem. Phys.* **105**, 397 (1986).

91. A. Oikawa, H. Abe, N. Mikami, and M. Ito, *Chem. Phys. Lett.* **116**, 50 (1985).

92. J. Hager, M. A. Smith, and S. C. Wallace, *J. Chem. Phys.* **83**, 4820 (1985).

93. J. Hager, M. A. Smith, and S. C. Wallace, *J. Chem. Phys.* **84**, 6771 (1986).

94. G. J. Fisanick, T. S. Eichelberger IV, M. B. Robin, and N. A. Kuebler, *J. Phys. Chem.* **87**, 2240 (1983).

95. J. M. Gagne, A. Briand, T. Berthoud, and K. N. Piyakis, *Appl. Phys.* **B50**, 29 (1990).

96. T. Kakinuma, M. Fujii, and M. Ito, *Chem. Phys. Lett.* **140**, 427 (1987).

97. A. Goto, M. Fujii, N. Mikami, and M. Ito, *Chem. Phys. Lett.* **119**, 17 (1985).

98. D. E. Cooper, C. M. Klimcak, and J. E. Wessel, *Phys. Rev. Lett.* **46**, 324 (1981).

99. J. Murakami, K. Kaya, and M. Ito, *Chem. Phys. Lett.* **91**, 401 (1982).

100. C. Kittrel, E. Abramson, J. L. Kinsey, S. A. McDonald, D. E. Reisner, R. W. Field, and D. H. Katayama, *J. Chem. Phys.* **75**, 2056 (1981).

101. T. Suzuki, N. Mikami, and M. Ito, *J. Phys. Chem.* **90**, 6431 (1986).

102. S. H. Kable and A. E. W. Knight, *J. Chem. Phys.* **86**, 4709 (1987).

103. H. L. Kim, S. Reid, and J. D. McDonald, *Chem. Phys. Lett.* **139**, 525 (1987).

104. T. Suzuki, M. Hiroi, and M. Ito, *J. Phys. Chem.* **92**, 3774 (1988).

105. M. E. Kaminsky, R. T. Hawkins, F. V. Kowalski, and A. L. Shawlow, *Phys. Rev. Lett.* **36**, 671 (1976).

106. D. C. Hanna, M. A. Yuratich, and D. Cotter, *Nonlinear Optics of Free Atoms and Molecules* (Springer, Berlin, 1979).

107. V. R. Shen, *Rev. Mod. Phys.* **48**, 1 (1976).

108. R. M. Hochstrasser, and H. P. Trommsdorff, *Acc. Chem. Res.* **16**, 376 (1983).

109. P. A. Franken, A. E. Hill, C. W. Peters, and G. Weinreich, *Phys. Rev. Lett.* **7**, 118 (1961).

110. J. Reintjes, *Appl. Opt.* **19**, 3889 (1980).

111. R. T. Hodgson, P. P. Sorokin, and J. J. Wynne, *Phys. Rev. Lett.* **32**, 343 (1974).

112. C. R. Vidal, *Appl. Opt.* **19**, 3897 (1980).

113. R. M. Hochstrasser, G. R. Meredith, and H. P. Trommsdorff, *Chem. Phys. Lett.* **53**, 423 (1978).

114. R. J. M. Anderson, G. R. Holtom, and W. M. McClain, *J. Chem. Phys.* **66**, 3832 (1977).

115. R. J. M. Anderson, G. R. Holtom, and W. M. McClain, *J. Chem. Phys.* **70**, 4310 (1978).

116. K. Tsukiyama, M. Tsukakoshi, and T. Kasuya, *Appl. Phys.* **B50**, 23 (1990).

117. F. Merkt and T. P. Softley, *Chem. Phys. Lett.* **165**, 477 (1990).

118. D. H. Levy, *Ann. Rev. Phys. Chem.* **31**, 197 (1980).

119. A. Kantrowitz and J. Grey, *Rev. Sci. Instrum.* **22**, 328 (1951).

120. R. E. Smalley, L. Wharton, and D. H. Levy, *Acc. Chem. Res.* **10**, 139 (1977).

121. R. E. Smalley, L. Wharton, and D. H. Levy, *J. Chem. Phys.* **63**, 4977 (1975).

122. R. E. Smalley, D. H. Levy, and L. Wharton, *J. Chem. Phys.* **64**, 3266 (1976).

123. A. Sur, J. Knee, and P. Johnson, *J. Chem. Phys.* **77**, 654 (1982).

124. J. M. Berman and L. Goodman, *J. Chem. Phys.* **87**, 1479 (1987).

125. L. Goodman and R. P. Rava, *Adv. Chem. Phys.* **54,** 177 (1983).

126. R. P. Rava, J. G. Philis, K. Krogh-Jespersen, and L. Goodman, *J. Chem. Phys.* **79,** 4664 (1983).

127. A. L. Sklar, *J. Chem. Phys.* **10,** 135 (1942).

128. R. P. Rava, L. Goodman, and J. G. Philis, *J. Chem. Phys.* **77,** 4912 (1982).

129. J. G. Philis, J. M. Berman, and L. Goodman, *Chem. Phys. Lett.* **167,** 16 (1990).

130. A. G. Ozkabak, J. G. Philis, and L. Goodman, *J. Am. Chem. Soc.* **112,** 7854 (1990).

131. A. G. Ozkabak and L. Goodman, *Chem. Phys. Lett.* **176,** 19 (1991).

132. L. Li, M. Wu, and P. M. Johnson, *J. Chem. Phys.* **86,** 1131 (1987).

133. B. G. Koenders, D. M. Wieringa, K. E. Drabe, and C. A. De Lange, *Chem. Phys.* **118,** 113 (1987).

134. B. G. Koenders, D. M. Wieringa, G. J. Kuik, K. E. Drabe, and C. A. De Lange, *Chem. Phys.* **129,** 41 (1989).

135. L. Li, R. J. Lipert, H. Park, W. A. Chupka, and S. D. Colson, *J. Chem. Phys.* **88,** 4608 (1988).

136. D. S. McClure, in *Laser Techniques in Luminescence Spectroscopy*, T. Vo-Dinh and D. Eastwood, eds. (ASTM, Philadelphia, 1990), pp. 21–35.

137. D. S. McClure and S. C. Weaver, *J. Phys. Chem. Solids* **52,** 81 (1991).

138. J. Berg and D. S. McClure, *J. Chem. Phys.* **90,** 3915 (1989).

139. D. Guo, T. Kundo, S. N. Thakur, and L. Goodman, unpublished.

9 *Laser Mass Spectrometry*

K. W. D. Ledingham and R. P. Singhal

9.1. Introduction

Mass spectrometry[1] is a long-established analytical technique for measuring atomic, isotopic, and molecular masses in which the sample is normally introduced in gaseous form to the instrument operated at ultrahigh vacuum (UHV) conditions. The atoms or molecules are ionized by electron impact to form ions that are then mass analyzed by either a magnetic sector, quadrupole, or time-of-flight (TOF) spectrometer.

Magnetic sector and quadrupole mass spectrometers analyze masses singly or over a limited mass range; thus, multielement analysis takes considerably longer than a TOF analysis where all the masses are measured simultaneously. Double-focusing magnetic sector spectrometers have mass resolutions ($M/\Delta M$) as high as 10^5; a linear TOF has a typical resolution of about 10^2 although a reflectron TOF[2] can have a resolution up to a few 10^3. The lasers in laser mass spectrometry are used for two purposes. Firstly, if the source to be analyzed is a solid, lasers are utilized to gasify the target (ablation or desorption) producing ions that are subsequently mass ionized. Secondly, the neutrals produced in the ablation step can themselves be postionized by a second laser system, taking the place of electron impact ionization in standard mass spectrometry. There are a number of different arrangements of lasers and mass spectrometers, and in this chapter three principal configurations will be identified and described: laser microprobe analysis, laser postionization of neutrals (either resonant or

365

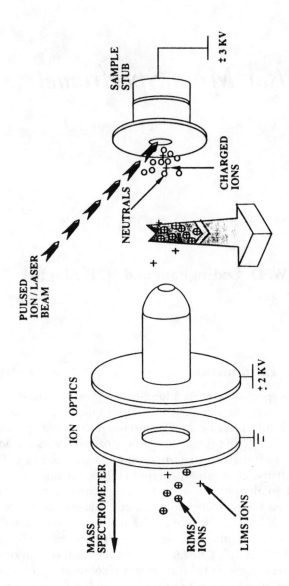

Figure 9.1 Schematic diagram for laser mass spectrometry. When a laser beam desorbs ions from the sample stub, the instrument is termed a laser microprobe. If the neutrals are ionized resonantly or nonresonantly, the process is called RIMS or SALI, respectively. RLA is the process in which a single resonant laser beam both ablates and ionizes.

nonresonant), and finally a new technique called resonant laser ablation (RLA). Acronyms abound in the field of laser mass spectrometry and with the help of Figure 9.1 a number of these will be explained (see also the Appendix). When an ion beam hits the target and sputters off ions that are subsequently mass analyzed, the procedure is called secondary ion mass spectrometry (SIMS). Although this is not a laser mass spectrometric technique, comparisons with SIMS are often made. If a pulsed laser beam strikes the target and desorbs/ablates either positive or negative ions, which are then analyzed, the instrument is a laser microprobe. This procedure is also called laser ionization mass spectrometry (LIMS) or alternatively laser-induced mass analysis (LIMA) or laser microprobe mass analysis (LAMMA).

In the sputtering or ablation process, the yield of ions is several orders of magnitude lower than that of neutrals, as estimated by the Saha equation.[3] When the neutral plume is postionized by a second laser system, the process is called surface analysis by laser ionization (SALI) if the laser wavelength is off-resonance, and resonant ionization mass spectrometry (RIMS) if the lasers are resonant. In principle, laser postionization of neutrals should always be more sensitive than LIMS if there is a large overlap between the ionizing laser beam and the ablated plume.

9.2. Laser Microprobe Analysis/Laser Ionization Mass Spectrometry (LIMS)

Before discussing in detail the laser microprobe and its application[4,5] the laser–target interaction should be described briefly. The interaction of laser light with solids has been reviewed in detail by many authors (see, for example, Refs. 6, 7, 8), and is a complicated process dependent on laser flux. Vertes et al.[9] have suggested that three different regimes of ionisation can be distinguished, assuming irradiation by a Q-switched laser:

1. *Low laser flux regime* (10^9–10^{10} W m^{-2}): Surface heating and ionization of target particles dominate, with material transfer across the solid–vacuum interface being negligible. This regime is favored for molecular-weight determinations of organic molecules with little fragmentation.

2. *Medium flux* (10^{11}–10^{12} W m^{-2}): Mass transfer is nonnegligible, with laser energy being absorbed in both the ionized vapor and the target. If organic molecules are involved, much fragmentation occurs, which might yield structural information.

3. *High flux* (10^{13}–10^{15} W m^{-2}): This regime is characterized by plasma formation and sample cratering as well as ion and neutral formation. Clustering and droplet formation also contribute to the ablation of the target. The detection of elemental species is maximized, providing fair possibilities for bulk trace analysis.

Schematic diagram of LIMA

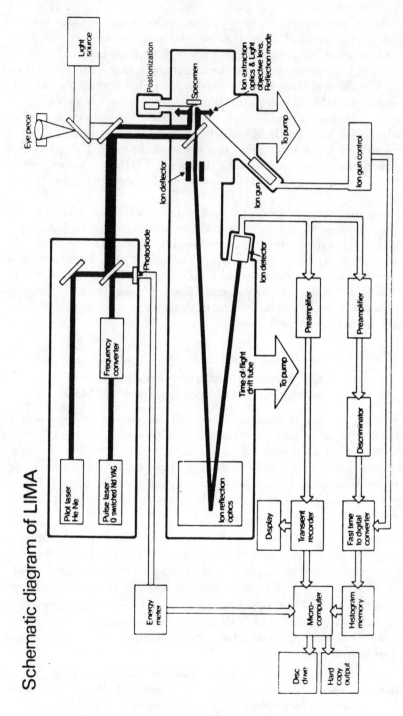

Figure 9.2 The CMS LIMA instrument. (Reproduced by permission of CMS Ltd.)

As the wavelength of the ablating laser is reduced to the UV, the potential for multiphoton ionization of neutrals increases, especially for elements and molecules with high ionization potentials. The sample must also exhibit efficient optical absorption at the ablation wavelength to enable energy coupling between the photon pulse and the solid. A diagram of one of the most popular laser microprobes is shown in Figure 9.2.

In LIMS, ablation and ionization are normally carried out using a focused Nd:YAG laser operated at fundamental wavelength (1.064 μm) or one of its harmonics, doubled (532 nm), tripled (355 nm), or quadrupled (266 nm), although for good ionization efficiency 266 nm is preferred. The laser flux can be varied between 10^{10} and 10^{15} W m^{-2} with a pulse length typically of 5 ns and a spot size between 0.5 and 5 μm when the laser is operated in TEM$_{00}$ mode. The typical volume of material consumed in this process ranges from 0.01 μm^3 at desorption threshold to about 5 μm^3 at maximum useful flux. By biasing the sample either positively or negatively with respect to a grounded extraction electrode (see Figure 9.1), positive or negative ions are extracted and transmitted through a reflectron time of flight mass spectrometer with a resolving power of a few thousand. The elemental detection limits are between 0.1 and 100 ppm and complete elemental and isotopic detection between hydrogen and uranium is possible. In Figure 9.3 aluminum-lithium powder produced by high-pressure gas atomization solidifies into small (~50-μm-diameter) spherical particles with

Figure 9.3 Analysis of impurities within micrometer-sized cell walls on individual particles of aluminum–lithium powder. (Reproduced by permission of CMS Ltd.)

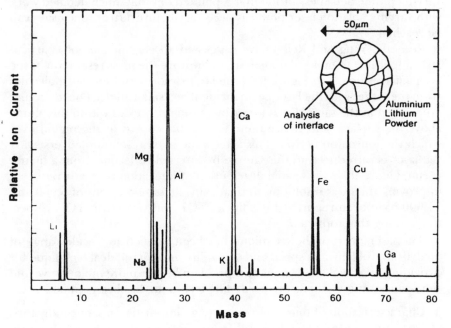

a cellular structure. LIMS has been used to determine the concentration of impurities within the micrometer-sized cell walls on individual particles.

Another recent and important application of the laser microprobe is the study of cluster formation. In a recent study a doubled Nd:YAG laser was used to ablate a graphite target.[10] The flux was 10^{12} W m^{-2}, the pulse duration 10 ns and the spot size on the target 1 mm^2. Carbon clusters, C_n, with n up to 214, were observed in the reflectron mass spectrometer (to be described later). These are shown in Figure 9.4 and display all the prominent features observed by other authors.[11,12] Clusters of low n values are highly abundant. The intensity of observed clusters falls up to $n = 31$, beyond which a significant increase in their formation is measured. In addition, for $n < 31$, clusters for all values of n are obtained, but for $n > 31$ only those with even n values are observed. The clusters are clearly visible up to $n = 214$. For still larger n values there is an indication of clusters being present but the signal-to-background ratio is not good. It is interesting to note that between $n = 32$ and $n = 214$, the general envelope of cluster formation and survival probability has a characteristic shape with a maximum around $n = 50$ and nearly linear decrease beyond that. Superimposed on the smooth envelope are clusters with $n = 60$ and 70, which are more abundant, while $n = 62$ and 72 exhibit greatly depleted strengths. Clusters with $n = 60$ (Buckminsterfullerene) and 70 have theoretically been shown to be particularly stable and are thought to have edgeless, speroidal cage structures, while the smaller clusters have linear chain or ring structures. It is argued[10] that the different clusters may be formed with varying plume densities, although it is pointed out that much detailed work with varying ablation laser power is necessary before a firm conclusion can be reached.

Recently, Yang and Reilly have suggested an elegant method, particularly applicable to molecular analysis, to improve the mass resolution of the laser microprobe.[13] In Figure 9.5a the laser desorbs and ionizes molecules from the target with the laser beam incident at a small angle. The molecules are ionized at the sample surface or by plasma processes within the plume above the surface, and hence there is a spatial spread of the region over which the ionization occurs. This affects the mass resolution adversely. If the laser is introduced to the sample by internal reflection from a quartz prism, Figure 9.5b, the spatial distribution of ionization is greatly reduced, improving the mass resolution dramatically. Mass resolutions of 3,900 and 11,000 have been reported for a linear TOF and reflectron TOF, respectively, using this approach.

The laser microprobe technique has been applied to a wide range of analytical problems. Its speed and versatility make it an ideal technique for providing an initial assessment of a problem before more quantitative and time-consuming methods are used. The technique has been applied to

1. Characterization of microelectronics (e.g., impurities in semiconductors, composition of photoresists, and failure analysis of integrated circuits);

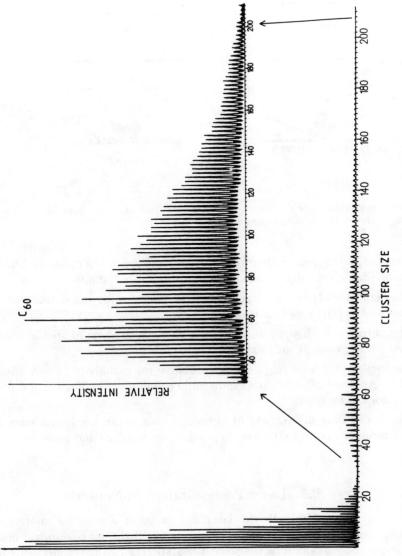

Figure 9.4 Time-of-flight spectrum of carbon clusters produced in the ablation of a graphite target using a laser beam of 10^{12} W m^{-2} flux at 532 nm.

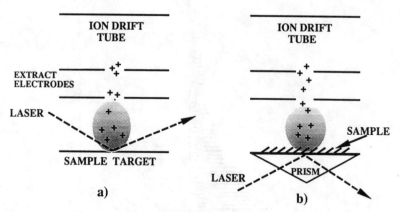

Figure 9.5 Laser-induced surface ionization with (a) external and (b) internal reflection. (Reproduced with permission from Ref. 13.)

2. Organic and inorganic chemistry (e.g., quality control of catalysts, lubricants, plasticizers, dyes, and characterization of pollutants);

3. Biomedical analysis (e.g., trace metals in tissues, cells, and dental materials and impurity inclusions in drugs);

4. Metallurgy (e.g., hydrogen in alloys and steels, analysis of special coatings for the nuclear industry);

5. Geological problems (e.g., semiquantitative microanalysis of core samples). Analyses of individual grains and crystallites in geological and mineralogical samples;

6. Depth profiling in a variety of samples with a depth resolution from a few monolayers up to 0.5 μm, depending on material and laser flux.

9.3. Laser Postionization of Neutrals

Although the laser ion probe is a sensitive analytical technique, it is necessary to understand in detail the mass removal and ionization processes before observed mass spectra can be related to the sample under study. These two processes are coupled and lead to matrix-dependent ionization, which makes quantification difficult. An alternative approach is thus to separate the ablation and ionization processes in time and space. Laser postionization of neutrals should be less troubled with matrix problems and more sensitive than LIMS, since the numbers of neutrals created in the ablation process is often very much greater than the number of ions. Although the motivation for laser postionization of neutrals has been appreciated for some time, it is only recently that practical forms of instrumentation have become available. The technique essentially involved the creation of neutral atoms from a solid sample by a laser or an ion beam that, after a delay, are ionized nonresonantly or resonantly by a second laser system. If an untuned laser system is used followed by time-of-flight mass spectrometry, the process is called SALI.[14] The emphasis of this approach is the suitable choice of a laser to provide uniform ionization for all atoms and molecules. Becker has indicated that in his laboratory the most commonly used wavelengths for inorganic analysis have been 248 and 193 nm (5.0 and 6.4 eV photon^{-1}, respectively) with a pulsewidth of ~10 ns and a focused flux of ~10^{14} W m^{-2}.[15] These were obtained from excimer lasers, although a quadrupled Nd:YAG laser is clearly a suitable alternative. At this flux, all atoms within the laser volume with ionization potentials less than 10 or 12.8 eV are ionized with the absorption of two photons. SALI provides a uniform, high detection efficiency with a high degree of quantification for a given element between different chemical matrices.

For SALI analysis of organic molecular species a single vacuum ultraviolet (VUV) photon is used. This has been shown to provide uniform detection efficiency coupled with the reduced photofragmentation that is common in multiphoton ionization. The VUV radiation is generated by a method based on the ninth harmonic of a Nd:YAG laser (118 nm). First, the fundamental wavelength (1.064 μm) is tripled (355 nm) using conventional nonlinear crystals. The 355 nm is further tripled by focusing the beam in a gas cell of Xe phase-matched with Ar to provide a beam of fluence equivalent to 10^{12} photons per pulse. Single VUV ionization is considered to be a "soft" ionization method, and furthermore, since single-photon ionization cross sections for molecules are more uniform than multiphoton cross sections, quantitative analysis is more likely. Figure 9.6 shows a spectrum of 7-methylguanine exhibiting a strong parent peak signal (165 amu). The spectrum was obtained using Ar+ sputtering followed by single-photon (118-nm) ionization.[15]

An alternative approach for laser postionization of neutrals has been developed by Zare et al.[16] Their preferred method is to desorb molecules, essentially intact, from surfaces using an infrared (CO_2) laser and after a suitable delay to ionize the neutrals produced using a quadrupled Nd:YAG laser. The ions are subsequently detected using a time-of-flight mass spectrometer. The appropriate delay (70–90 μs) between the desorbing CO_2 laser and the ionizing Nd:YAG laser pulses is chosen, so that the ionizing laser pulse intercepts as many molecules as possible. Typically, the desorbing CO_2 laser has a pulse energy of 10 mJ and the ionizing Nd:YAG laser 1 mJ slightly focused to provide an ionization flux of about 10^{10} W m^{-2}. Zare has called the process REMPI (resonance-enhanced multiphoton ionization), and although many molecules have electronic transitions corresponding to 266 nm, the present authors prefer to consider this procedure nonresonant, since a single wavelength ionizes many different molecules. REMPI has been shown to be a particularly useful technique for the analysis of nonvolatile compounds and those that decompose on heating. The most important feature of many of these spectra is the dominance of the

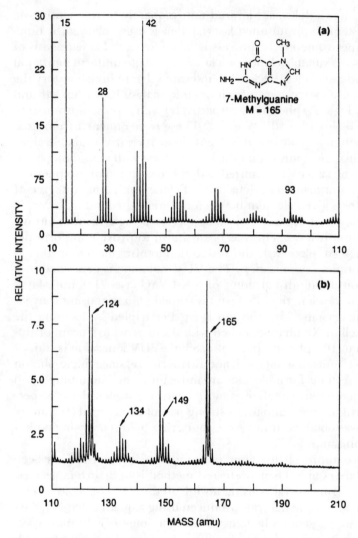

Figure 9.6 Mass spectrum of 7-methylguanine. The spectrum was accumulated over 1,000 pulses with 118-nm radiation for single-photon ionization following pulsed Ar^+ sputtering. (Reproduced by permission from Ref. 15.)

parent peak. Figure 9.7 shows an example of a cocktail of equimolar amounts of a number of PTH amino acids. The parent peaks labeled a–e are almost equal in height.[17] Hahn et al. have also demonstrated how the signal height of the parent peak varies with sample concentration and have found a linear dependence over five orders of magnitude (Figure 9.8).[18] The amount of adsorbate desorbed per laser shot ranges from nanomoles (100 monolayers) to femtomoles (10^{-4} monolayers).

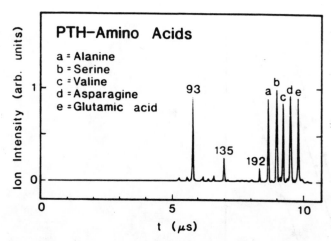

Figure 9.7 Laser desorption–multiphoton ionization TOF mass spectra of an equimolar mixture of five PTH-amino acids. (Reproduced by permission from Ref. 17.)

Off-resonance laser postionization of neutral atoms or molecules has been shown to be highly sensitive, to be reasonably quantifiable, and to have reduced matrix effects. If high selectivity is also required the ionizing lasers should be tuned to specific atomic or molecular resonances leading to processes called RIMS (resonance ionization mass spectrometry) for atoms or R2PI (resonant two-photon ionization) for molecules. In RIMS, ionization of the neutral plume is performed by tuning the postionizing laser frequency to an atomic transition. The resonant excitation process is highly probable, and even for moderate laser fluxes, almost complete ionization of atoms of a particular element may be achieved without affecting atoms of other elements. This results in several orders of magnitude higher selectivity in elemental detection than with nonresonant postionization. By using two or more tunable lasers, it is possible to increase further the level of elemental selectivity, and the possibility of single-atom detection may be contemplated.[19] Trace detection at 1 ppt[20] (1 part in 10^{12}) and detection of small numbers ($\sim 10^8$) atoms[21] has been reported. The underlying physics of RIMS is discussed in the following paragraphs.

Atoms of each element have a unique set of excited states, which may be reached by the absorption of one or more photons of the correct frequency provided certain selection rules are satisfied. An excited atom would normally decay back to the ground state with a characteristic time, the mean life of the excited state, which is typically of the order of 10 ns. However, before decaying, the atom can absorb another photon that may take the atom to a higher excited state or cause ionization. The cross section for resonant absorption of one photon is typically of the order of 10^{-15} or 10^{-16} m^2, and a fluence of 10 mJ m^{-2} is sufficient to cause saturation ex-

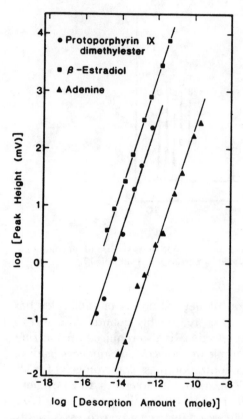

Figure 9.8 Parent ion signal for protoporphyrin IX dimethylester, β estradiol, and adenine versus amount desorbed per laser pulse. (Reproduced by permission from Ref. 18.)

citation of the atoms present in the laser beam. The cross sections for photoionization are about 10^6 times smaller and laser fluences of 1 kJ m^{-2} are required for saturation ionization of the excited atoms. The photoionization cross sections become smaller the further one goes into the continuum. For ionization, the energy of absorbed photons must be greater than the ionization potential (IP) of the atom. In RIMS, two or more photons are used and the probability of ionization is simply the product of the probabilities for the individual steps in the RIMS scheme.

As discussed earlier, transitions between bound states proceed with high probability and require modest laser fluences. At these fluences, nonresonant ionization of other atoms is very small, and very high selectivities are achieved. The photoionization step requires much higher laser fluences and affects the selectivity and sensitivity of the RIMS process. Also, to reach these higher laser fluences the output of most commercially available tunable lasers must be moderately focused, which has the disadvantage of reducing the interaction volume and hence the number of atoms that can be analyzed. Two other ionization procedures that have been suggested in order to alleviate this problem are shown in Figure 9.9.[22] In Figure 9.9b the atom is excited to a Rydberg state close to the continuum and can be ionized with 100% efficiency by the application of a pulsed field of about 10^6 V m^{-1}. The other method (c) is to ionize the atom via autoionization states. The rate-limiting step in (b) and (c) has a cross section some two orders of magnitude larger than that in (a) and requires much lower fluences to reach saturation. The laser linewidth typically used in RIMS measurements is between 0.1 and 1.0 cm^{-1}, and hence all isotopes of an element are ionized simultaneously. Separation of the various isotopes is achieved in the mass spectrometer.

Figure 9.9 (a) An electron in its ground state absorbs a photon and is raised to an excited state. The ionization by absorption of a second photon has a small cross section. (b) The atom is excited to a Rydberg state and is finally ionized by a pulsed electric field with high efficiency. (c) The final step is to an autoionization state with a large cross section.

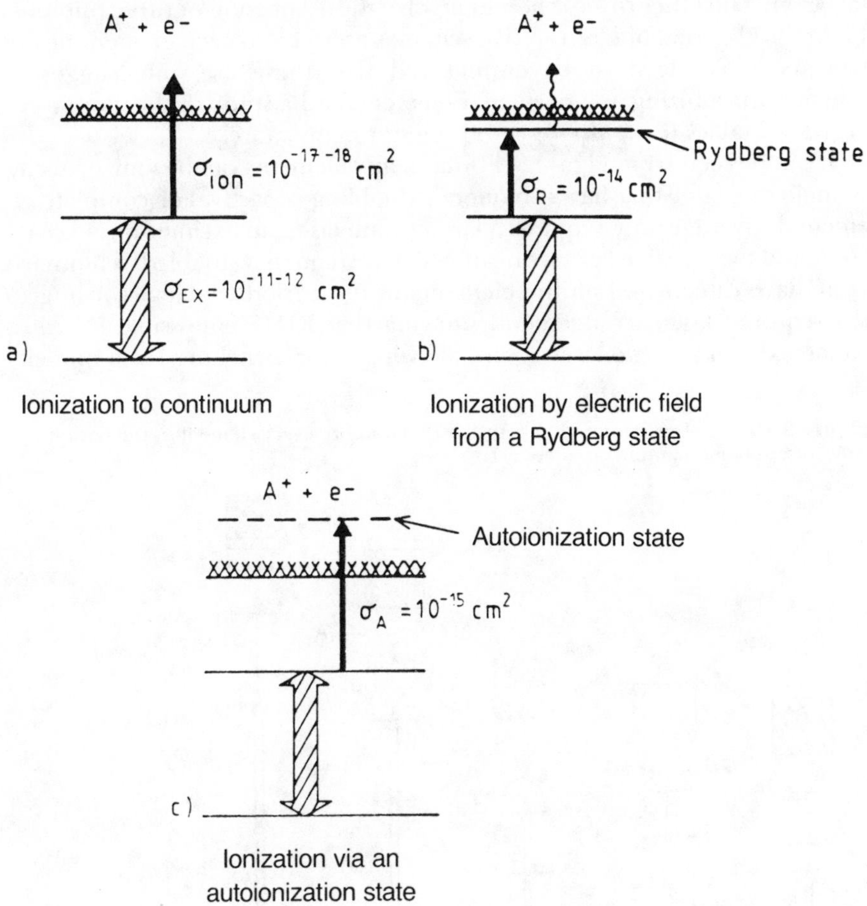

a) Ionization to continuum

b) Ionization by electric field from a Rydberg state

c) Ionization via an autoionization state

The energy required to ionize an atom varies from 3.89 eV for Cs to 24.49 eV for He, the majority of atoms having an IP less than 10 eV. The energy carried by a photon is $E_\gamma = hc/\lambda$ and is 2.48 eV for laser light of wavelength 500 nm and 4.66 eV for $\lambda = 266$ nm, the quadrupled output of a Nd:YAG laser. Atomic transitions between the ground state and low-lying excited states, with the exception of those in He and Ne, may be resonantly excited with commercially available tunable laser systems, and a variety of schemes have been proposed and applied for RIMS. Hurst and Payne have proposed five basic ionization schemes according to the relative energy positions of the intermediate states to the continuum.[23,24] These are shown in Figure 9.10. In the first scheme a level exists at an energy that is more than half the value required to ionize the atom. Hence, the atom can be ionized by two photons from the same dye laser, with the absorption of the first photon being resonant. In the second scheme the output of the tunable laser must be frequency doubled to excite the atom resonantly, which is then ionized by a photon from the much more intense fundamental beam. The other three schemes involve the absorption of three photons. Figure 9.11 shows one of the five schemes ascribed to each element of the periodic table. It must be emphasized that these are only suggested schemes for ionizing any particular element, and a study of the atomic energy level tables (e.g., Moore[25]) will suggest many more.

It can be seen from Figure 9.11 that many elements can be ionized using a single dye laser that has a frequency-doubling capacity. For complete elemental coverage, one requires a large pump laser, an excimer or Nd:YAG laser, and two tunable lasers, one of which is frequency doubled. Thonnard et al. have categorized all the elements in the periodic table according to the required laser arrangement, showing that RIMS is possible for most elements with a single laser setup, allowing rapid switching from one ele-

Figure 9.10 The five ionization schemes that can ionize every element in the periodic table except helium and neon. (See text.)

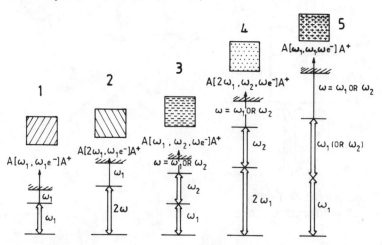

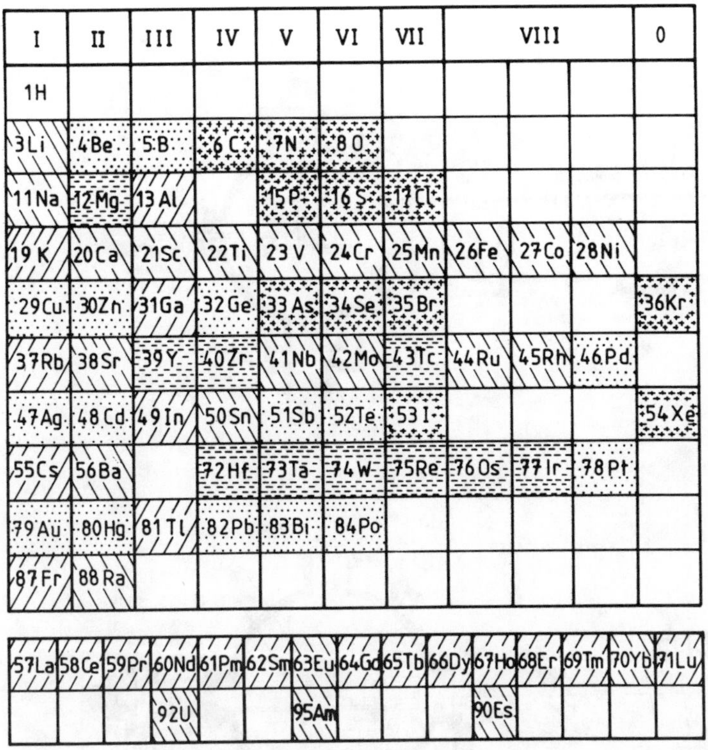

I	II	III	IV	V	VI	VII	VIII			0
1H										
3Li	4Be	5B	6C	7N	8O					
11Na	12Mg	13Al		15P	16S	17Cl				
19K	20Ca	21Sc	22Ti	23V	24Cr	25Mn	26Fe	27Co	28Ni	
29Cu	30Zn	31Ga	32Ge	33As	34Se	35Br				36Kr
37Rb	38Sr	39Y	40Zr	41Nb	42Mo	43Tc	44Ru	45Rh	46Pd	
47Ag	48Cd	49In	50Sn	51Sb	52Te	53I				54Xe
55Cs	56Ba		72Hf	73Ta	74W	75Re	76Os	77Ir	78Pt	
79Au	80Hg	81Tl	82Pb	83Bi	84Po					
87Fr	88Ra									

57La	58Ce	59Pr	60Nd	61Pm	62Sm	63Eu	64Gd	65Tb	66Dy	67Ho	68Er	69Tm	70Yb	71Lu
			92U		95Am					90Es				

Figure 9.11 Periodic table with an appropriate scheme (see Figure 9.10) for each element.

ment to another.[26] In fact, 39 elements can be ionized with a single dye combination, enabling computer-controlled element changes in a matter of seconds. A data service has been established at the National Institute of Standards and Technology USA to provide the necessary information to apply the techniques of RIMS to routine use in analytical chemistry.[27,28] This service will collect atomic data, choose appropriate resonance ionization schemes, and give pertinent operating details of successful RIMS studies. Twenty elements have been covered by this compilation so far.

In the following, the design of a commercial RIMS instrument designed at Glasgow University is described in detail (Figure 9.12).[29] The spherical sample chamber is 30 cm in diameter with 11 ports facing into the center of the chamber, the point at which the sample stub is held. The sample is mounted on an $xyz\Theta$ manipulator and can be inserted and withdrawn from the sample chamber using a rapid transfer probe. Fast sample exchanges (5 min) can be made without disruption of the main chamber pressure. The sample chamber is pumped by an oil diffusion pump fitted with a cold trap and a titanium sublimation pump, having a total pumping speed of over 800 $1s^{-1}$ and capable of maintaining a base pressure of less than 10^{-9} torr in the chamber.

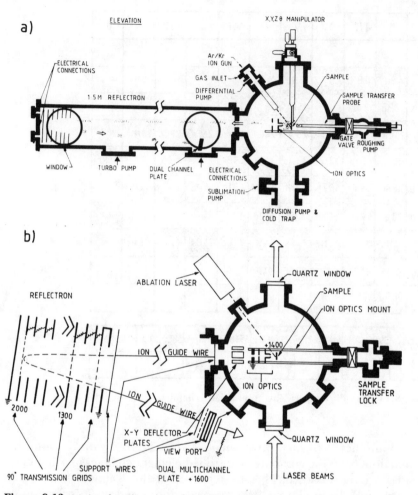

Figure 9.12 (a) An elevation view of the Glasgow Resonant Ionisation Mass Spectrometer.[29]
(b) A view of the spectrometer showing the electrostatic ion reflector in greater detail. An initial spread of ion energies is compensated by the reflectron, since high-energy ions penetrate deeper into the reflecting field and hence spend a longer time than lower-energy ions. This improves the resolution, and the guide wire increases the transmission.

The ion extract optics and the reflectron time-of-flight mass spectrometer are shown in Figure 9.12b. The sample is maintained at a voltage of about $+2,100$ V with the first extraction electrode at $+1,400$ V. Two further electrodes are included, the final one being earthed, before the ions pass through xy deflection plates and down the time-of-flight system. The reflectron time-of-flight system, with an overall drift length of 3 m, consists of a high-transmission reflectron providing energy focusing. The principal factor that limits the resolution of a conventional time-of-flight mass spectrometer is the spread of initial ion energies in the ablation process. This spread of ion energies can be compensated using a reflectron time-of-flight mass spectrometer[2] in which high-energy ions penetrate deeper into an electrostatic ion reflector and hence experience a longer flight time than ions of a lower energy. The FWHM resolution of the present system is about 1,000 for ions of about 40 amu. The ions are detected by a Galileo double-microchannel plate detector with a 0.2-ns rise time and a gain of up to 10^7, operated at about $-2,000$ V on the front plate. If the sample is maintained at a voltage greater than the reflectron voltage, ions created at the sample by the argon ion gun or ablation laser are suppressed since they are largely dumped onto the earthed back-plate of the reflectron. A further high-transmission grid operated at a voltage of about 80% of the reflectron voltage is placed in front of the microchannel plate detector to reduce the detection of lower-energy ions produced by scattering from the grids in the reflectron. A thin wire, 0.005 cm in diameter, follows the ion flight path through the flight tube, providing an electrostatic guide for the ions, increasing the transmission of the mass spectrometer. The wire, operated at about -10 V, is only important when the reflectron is run at voltages lower than the optimal 3 kV. A turbomolecular pump with a speed of 200 ls^{-1} differentially pumps the time-of-flight tube.

Ablation can be carried out using a laser or an argon ion gun. In the former case the ablation source is a Quantel Nd:YAG 585 laser that can be operated in fundamental, doubled, tripled, or quadrupled mode with pulse powers of 400, 160, 60, and 25 mJ, respectively, in a pulse of 10-ns duration and a repetition rate of 10 s^{-1}. The beam profile is nearly Gaussian. If an ion gun is used the process is called SIRIS (sputter-initiated resonant ionization spectroscopy). Here a modified Kratos minibeam 1 is used. This gun has DC ion currents up to 30 μA with beam spot diameters of between 1 and 2 mm. Ion energies up to 5 keV are available and the gun can be pulsed by applying a voltage pulse of between 0.2 and 10 μs to the grid of the ion gun. A turbomolecular pump with a speed of 80 ls^{-1} acts as a differential pumping system for the argon ion gun and during operation the partial pressure of argon in the sample chamber rises to 10^{-8} torr.

The ionization laser system consists of a Spectron SL2Q + SL3A Nd:YAG pump laser with frequency-doubling, tripling, and quadrupling facilities. This laser has a pulse repetition rate of 10 Hz and a pulse duration of 15 ns. The doubled and tripled outputs have energies of 300 and 150 mJ, respectively. These outputs are used to pump two Spectrolase 4000 dye lasers, one spanning wavelengths between 540 and 700 nm with a bandwidth of 0.1 cm^{-1} and an output of typically 20 mJ per pulse, the other covering wavelengths between 400 and 700 nm with a bandwidth of 0.1 cm^{-1} and an output of about 1 mJ per pulse. Frequency-doubling crystals extend the wavelength coverage down to between 270 and 350 nm. Uranium hollow-cathode spectra and an etalon are used for calibration of the dye lasers.

The data acquisition system measures and stores mass spectra and laser pulse energies on a pulse-to-pulse basis. A LeCroy 2261 transient recorder coupled to an IBM-PC AT forms the center of the system. Ion signals from the detector are digitized by a transient recorder that provides 640 time channels (11-bit resolution) each of 10-ns width. The transient recorder provides a "ready" pulse to activate a pulse generator that triggers the Quantel ablation laser and the Spectron Nd:YAG. The time between the ablation and dye laser pulses can be varied between 0.1 and 10 μs, although typically this is about 1 μs. A signal from a fast photodiode provides a time zero signal and activates a Stanford delay pulse generator, which provides a stop signal for the transient recorder. Signals from the uranium hollow cathode, a Fabry–Perot etalon, and a Molectron joulemeter are digitized by a LeCroy 2259 ADC. Data are then transferred to the IBM PC via a GPIB interface.

The instrument described earlier is being routinely used for RIMS analysis and for studies of the characteristics of laser ablation from solid samples. Figure 9.13 shows a gold RIMS signal at 200-ppb sensitivity,[30] while Figure 9.14 displays the $^{40}Ca^+$ ion signal as a function of ablation laser fluence.[29] The calcium ion signal is proportional to the neutral calcium atom yield in the ablation process, which is a rapidly changing function of the fluence of the ablation laser.

RIMS is a descriptor primarily associated with the detection of atoms, which usually have a simple set of energy levels. The same techniques can be used with molecules, but molecular structure is considerably more com-

Figure 9.13 RIMS signal for gold in copper at 10-ppm level, accumulated over 10^4 shots.[30]

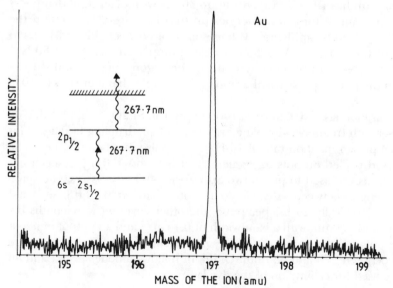

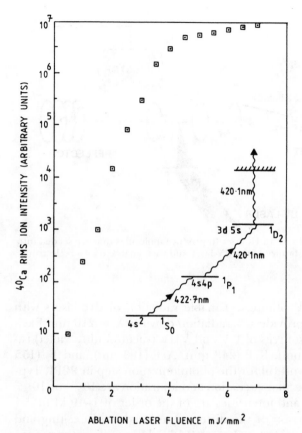

Figure 9.14 Graph of ^{40}Ca ions produced by the RIS scheme shown in the inset as a function of ablation laser fluence.[29] Above 5 kJ m^{-2} the resolution and stability of the reflectron deteriorate rapidly.

plex, since each electronic level has an associated set of vibrational and rotational levels. Because of this complexity the absorption features of molecules at room temperature are in general broad and structureless with a consequent loss in selectivity. To improve selectivity these can be "cooled down" to a few sharp peaks by introducing the molecules into supersonic beams. This can be achieved either by seeding the molecules into a light carrier gas (e.g., He or Ar at a few atmospheres pressure, expanding the gas through a narrow orifice into vacuum) or by laser desorbing the molecules into a jet of the carrier gas (Figure 9.15).[31] Rapid cooling down to a few degree Kelvin is realized by converting the energy of the internal degrees of freedom into translational energy of the carrier gas via collisions. The most common RIMS scheme for analytical purposes using molecules is resonant two-photon ionization (R2PI), in which one photon excites the molecules to an electronic state and a second photon causes molecular ionization. Since molecules normally have ionization potentials between 7 and

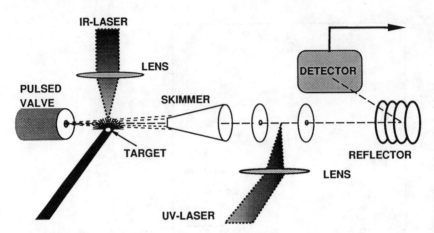

Figure 9.15 Sample irradiated by an IR laser to produce molecules that are seeded into a supersonic jet. These are postionized by a UV laser and then analyzed by a reflectron TOF mass spectrometer.[31]

13 eV, R2PI requires UV photons. Tunable radiation of dye lasers with frequency doubling can provide UV radiation down to $\lambda = 210$ nm. Excimer lasers are powerful sources of UV and VUV (vacuum ultra violet) radiation [e.g., XeCl (308 nm), KrF (248 nm), ArF (193 nm), and F_2 (155 nm)] and are particularly useful for the photoionization step in R2PI. Typical cross sections for molecular excitation and ionization are 10^{-21} or 10^{-22} m^2 and saturation fluxes and intensities are of the order of 1–10 kJ m^{-2}.

An interesting comparison of the different techniques of exciting and ionizing a molecule is provided in Figure 9.16, where aniline spectra recorded under various conditions are presented.[32] Jet-cooling sharpens the resonance features considerably. The similarity of the single-photon and the R2PI spectra demonstrates that the cross section for ionization may be wavelength-independent over the range studied. In contrast to atoms, molecules in intermediate excited states can undergo different photophysical and photochemical transformations. Photodissociation can compete with the photoexcitation and photoionization processes. For larger laser fluences, the fragmentation of the molecule is more extensive, and a range of smaller molecular and atomic products results. A mass spectrometer, however, can analyze the fragmentation products, and these can help in defining a "fingerprint" of the molecule. Marshall has measured the fragmentation pattern for toluene as a function of laser fluence, and the tendency for increasing fragmentation is obvious (Figure 9.17).[33]

A method to increase parent ion yield was demonstrated by Antonov et al. by the use of a short but much greater power laser pulse.[34] Photoionization mass spectra of benzene with 20-ns duration, 6 kJ m^{-2} (3×10^{12} W m^{-2}) and with 20 ps, 75 kJ m^{-2} (3.7×10^{15} W m^{-2}) laser pulses were obtained. Considerably lower fragmentation resulted from the picosecond pulse, even though the power was three orders of magnitude greater. The

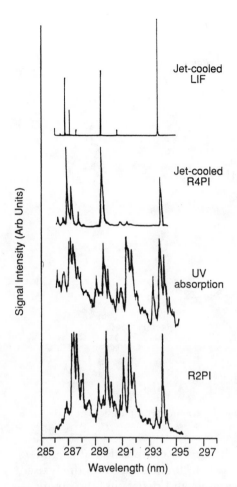

Figure 9.16 Aniline spectrum recorded by jet-cooled laser-induced fluorescence, jet-cooled R4PI (two photons to excite, two photons to ionize), single-photon UV absorption and R2PI.[32]

reason is that in the extremely strong laser field of the picosecond pulses, the rate of radiation-induced transitions is so high that molecules acquire sufficient energy to ionize before they dissociate. With the availability of tunable picosecond pulsed lasers, it is hoped that this method will be used more extensively to obtain spectra with less fragmentation of the parent molecule.

RIMS has been extensively used in a number of different fields, but in this chapter only three will be discussed in detail: profiling in semiconductors, detection of isomers and isobars, and the detection of stable isotope tracers.

Depth-resolved information about the elemental composition of semiconductor materials is extremely important in electronics technology and industry. Many optical and electronic devices use layers of GaAs and Al-

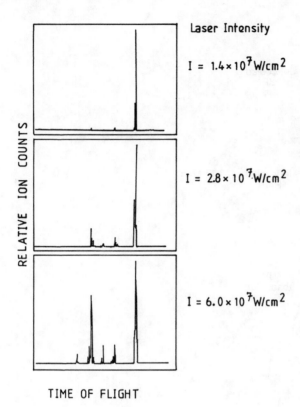

Laser Intensity

$I = 1.4 \times 10^7 \text{W/cm}^2$

$I = 2.8 \times 10^7 \text{W/cm}^2$

$I = 6.0 \times 10^7 \text{W/cm}^2$

TIME OF FLIGHT

Figure 9.17 Toluene fragmentation pattern with various laser intensities.

GaAs with large and usually abrupt changes in composition at the interfaces. SIMS is a highly developed analytical technique and is routinely used
to characterize the semiconductor materials and determine the composition of the layered structures. The basis of SIMS, as mentioned earlier, is
the detection of ions ablated from the sample surface by an energetic ion
beam. Generally, a magnetic sector or a quadrupole mass spectrometer selects the ions to be investigated. SIMS has three main drawbacks that limit
its potential: Matrix effects cause the sputtering yield to vary unpredictably
with surface composition, isobaric interferences reduce the sensitivity, and
only sputtered ions are counted, whereas the neutral atoms, several orders
of magnitude more abundant, are ignored.

RIMS is ideally suited for the characterization of III–V compound semiconductors and has already been shown to possess quantitative reliability
in the analysis of layered structures.[35] Figure 9.18 shows SIMS and RIMS
depth profiles of molecular beam epitaxy (MBE)-grown $CoSi_2$ in Si, recorded under identical sputtering conditions. The RIMS profile does not
exhibit isobaric interference effects due to $^{29}Si\, ^{30}Si$ which accounts for about
50% of the total signal in SIMS at mass 59. The sputtering yield of neutral

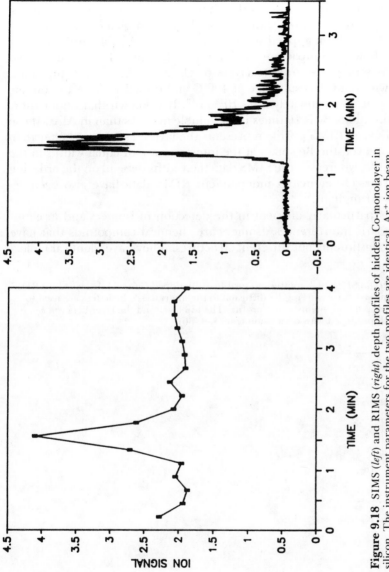

Figure 9.18 SIMS (*left*) and RIMS (*right*) depth profiles of hidden Co-monolayer in silicon. The instrument parameters for the two profiles are identical. Ar$^+$ ion beam parameters; 250-nA (continuous) energy 10 keV. The SIMS background (50% of peak signal) is presumably due to ^{29}Si^{30}Si. The RIMS signal shows essentially no background during sputtering through silicon. (Reproduced with permission from Ref. 35.)

atoms is relatively insensitive to the chemical composition of the surface and for layered structures with abrupt changes in elemental concentrations, RIMS has been applied successfully to obtain reliable profiles.[36] Figure 9.19 shows a comparison of depth-profiles for a Be-doped layered GaAs/AlAs sample. Alternative layers of GaAs and AlAs, 100 nm thick, were grown on a GaAs substrate at 400°C and doped with Be to nominal uniformity at approximately 2×10^{25} m^{-3}. It is known that under these growth conditions Be is 10 times more soluble in GaAs than in AlAs. It can be seen that the RIMS profile represents the layered structure better than SIMS. Moreover, the Be spikes at the interfaces are of equal widths in the RIMS profile, whereas in the SIMS data the widths depend on the ordering of the layers. Other matrix-independent RIMS data have also been reported by Arlinghaus.[37]

RIMS is particularly efficient in the detection of isomers and in eliminating isobaric interferences. Isomers are chemical compounds that have the same constituents but different structures. They have identical molec-

Figure 9.19 (a) SIMS (0_2^+ sputtering) profile of Be-doped GaAs/ALAs layers. (b) RIMS of Be in same structure showing Be diffusion to the interfaces. The letters are used to identify corresponding points in the profile. The top layer and the layer between *b* and *c* are GaAs. (Reproduced with permission from Ref. 36)

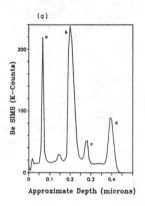

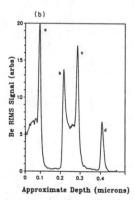

ular weights. With conventional analytical methods, isomers are extremely difficult to distinguish from each other, and generally one produces fragmentation patterns using electron impact ionization (EI) and studies the differences in these with a mass spectrometer. Only limited discrimination may be achieved in this way and EI is not useful if the isomer is in trace quantity. The isomers in general, however, have different absorption spectra. Their ionization potentials may also be different. Both of these features may be exploited in maximizing isomeric discrimination in optical ionization methods. Supersonic jet-cooling of the isomeric mixture is used to suppress some of the rotation–vibrational structure, and sharp absorption peaks result. Figure 9.20 shows R2PI spectra for four isomers of dichlorotoluene.[38] The wavelength spectra were obtained by expanding several parts per million of each compound in a 1-atm reservoir of argon into vacuum at 10^{-6} torr. There was no fragmentation in the photoionization process and only parent ions were monitored in the time-of-flight mass spectrometer. For the isomers of cresol, Tembreull and Lubman determined a discrimination level of 1:300–1:500 between any two isomers with a detection limit of at least 20 ppb in 1 atm of nitrogen. Isomers also have slightly different ionization potentials and by accurately choosing the ionization laser energy, it is possible to ionize only the lower IP isomer, leaving the high IP isomer in a neutral state.

RIMS is unique in its ability to reduce isobaric interference in analysis. Isobaric interference happens when two or more compounds with different chemical compositions have nominally the same mass. A popular example is that of CO, N_2, and CH_4 with molecular weights 27.9949, 28.0061, and 28.0313, respectively. Although a high-resolution mass spectrometer ($M/\Delta M > 5,000$) could resolve these, only the targeted compound gives a signal in RIMS, because of the ionization selectivity. This is particularly useful in analyses in the presence of organic molecules since these tend to fragment easily into a large number of smaller-weight species from atomic carbon to a sequence of $C_n H_m$ hydrocarbons. The resulting background is largely avoided in RIMS.

Radioactive tracers are used routinely in clinical and biomedical analyses. Such studies suffer from the hazards usually associated with radioactive materials. It is now possible, however, to use low-abundance stable isotopes as tracers. The detection and analysis is confidently and accurately carried out by RIMS. Generally, atomic isotope shifts are of the order of 0.01 cm^{-1}, and selective resonant excitation–ionization of the isotope requires narrow-bandwidth CW lasers. For Ba and Pb isotopes, Bushaw and co-workers have demonstrated that double-resonance RIMS (DRIMS) using two high-resolution single-frequency CW dye lasers and a CO_2 laser for photoionization is capable of achieving isotope abundance sensitivities in excess of 10^9 and detection limits of less than a femtogram.[40,41] The restriction of narrowband CW lasers for selective isotopic excitation–ionization may be circumvented by resonantly exciting molecular vibrational levels,

390

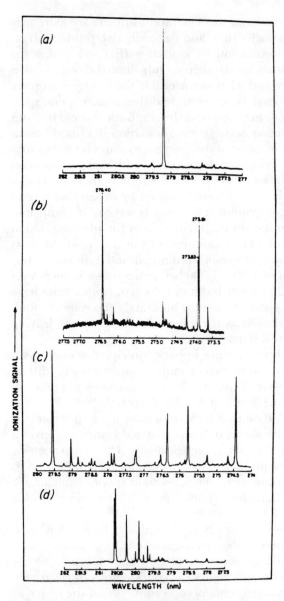

Figure 9.20 Resonant two-photon ionization spectra taken in an expansion of 1 atm back-pressure of Ar of (a) 2,4-dichlorotoluene, (b) 2,6-dichlorotoluene, (c) 2,5-dichlorotoluene, (d) 3,4-dichlorotoluene. Only the molecular ion is monitored in these spectra using a TOF mass spectrometer. (Reproduced with permission from Ref. 38.)

where the isotopic shifts are typically a few cm^{-1} and where resonances are easily selected by using tunable pulsed dye lasers. Lubman and Zare used a multicolor excitation scheme to achieve a spectroscopic selectivity of ^{127}I ^{129}I: ^{127}I$_2$ of greater than 5 × 10^3.[42] Additional selectivity from a quadrupole mass spectrometer was predicted to yield a discrimination of better than 1 part in 10^8.

Arlinghaus et al. have recently demonstrated the potential of stable isotope tagging for DNA sequencing in the human genome project.[43] Large numbers of stable isotopes can be used to label DNA, thereby multiplexing the separation process. Isotopes of iron and tin were used, and detection of subattomole (<10^{-18} mole) quantities of labeled DNA was demonstrated with good resolution. Arlinghaus et al. hope that it should be possible to detect more than 10^7 bases per day, which is orders of magnitude better than is feasible at present with conventional techniques.

9.4. Resonant Laser Ablation (RLA)

In RIMS the ionizing laser is typically located 1–2 mm above the sample surface and intercepts the plume of neutral atoms ablated from the surface. The limited geometrical overlap between the ionizing laser and the ablated plume reduces the sensitivity of RIMS. In a recent paper McLean et al. have shown that if a moderately focused tunable laser beam at grazing incidence to a GaAs target is used for both ablation and resonant ionization, then an enhancement of the ion yield of up to two orders of magnitude results.[44] The resonance widths are of the order of 0.05 nm. They call this process resonant laser ablation (RLA). RLA has also been observed with the laser directed normally onto the surface[45] and in conjunction with 50 torr of He buffer gas.[46] However, considerable broadening, by several nanometers, of the resonant signals was observed with poorer enhancements. Wang et al. and Borthwick et al. have carried out detailed experiments to understand the physical processes in RLA.[47,48] In certain circumstances this technique promises to be a very powerful analytical method in place of LIMS and RIMS.

The mass spectrometer system used for resonant ablation studies of semiconductors[44,47] is shown in Figure 9.21 and comprises an analysis chamber, an entry chamber for rapid transfer of samples, and a 1-m linear time-of-flight mass analyzer of mass resolution approximately 100. The operational pressure is typically ~10^{-8} mbar and the ion detector an electron multiplier. In the reported studies a GaAs sample was mounted on a 25-mm-diameter stainless steel stub and fixed with conductive epoxy.[44,47] The laser arrangement consisted of a Lumonics TE 860-3 XeCl (308 nm) pumping a Lumonics EPD 330 dye laser employing Rhodamine 6G. The dye laser output was frequency-doubled using a KDP crystal mounted on an Inrad autotracker. The pulse length was 5 ns and the bandwidth was

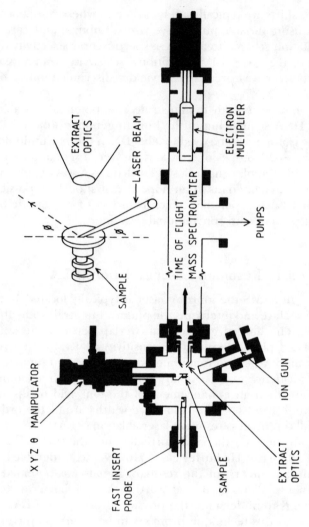

Figure 9.21 The mass spectrometer used for RLA studies of Al and Ga.[44,47] The inset shows the ablation geometry with the laser incident on the sample surface at an angle φ.

~0.01 nm. Both the fundamental and frequency-doubled laser beam components were directed at grazing incidence to the surface, and moderately focused by a quartz lens ($f = 50$ cm) to give a beam diameter of 0.5 mm. The left inset in Figure 9.22 is a partial Grotian diagram of Ga showing the states involved in the excitation and ionization processes of the experiments. The UV output was tuned to around the $4^2P_{1/2}$–$4^2D_{3/2}$ (287.5 nm) transition of Ga, and the further absorption of a fundamental photon resulted in ionization of the excited atoms. The total laser intensity onto the surface (area ~2.5 mm) of the sample was typically $< 8 \times 10^{10}$ W m^{-2}, of which the UV component was about 20%.

In an experiment using an AlGaAs sample the Al transition from the $3p$ ground state to the excited $3d$ state proceeds through the absorption of a single photon. The fine structure in the Al ground state allows two resonances, at wavelengths of 308.297 and 309.367 nm. The doublet separation

Figure 9.22 The ion signal of Ga as functions of UV laser wavelength, λ, and total laser energy E. The grazing incident angle of the laser beam onto the sample surface is $\theta = 5°$. The corresponding total ablation energy is (a) $E = 0.39$ mJ, (b) $E = 0.66$ mJ, (c) $E = 0.79$ mJ, and (d) $E = 0.98$ mJ. The resonant enhancement can clearly be seen with the resonant width increasing asymmetrically as the laser energy increases. The left inset is a partial Grotian diagram of Ga relevant to the experiments.

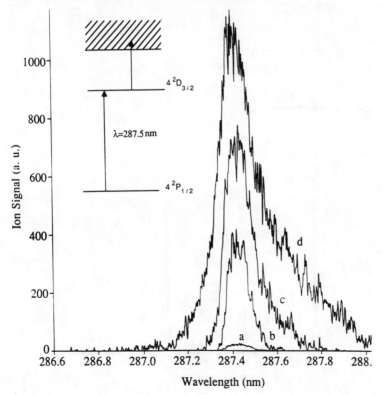

of the excited $3d$ state is too small to resolve in this experiment. The excited Al atoms can be ionized by the absorption of a red photon and mass-analyzed in the time-of-flight mass spectrometer. In Figure 9.23 the measured ion yield as a function of the laser wavelength is shown. The resonance effects are clearly seen with a signal enhancement of greater than two orders of magnitude. The resonances are less than 0.05 nm wide at half maximum height—an observation in sharp contrast with that made in Refs. 45 and 46.

Wang et al. have studied RLA of Ga in detail.[47] At relatively low pulse energies the spectral profiles had widths of about 0.1 nm and were centered at the corresponding atomic transition $4^2P_{1/2}$–$4^2D_{3/2}$ calibrated using a uranium–argon hollow-cathode lamp. The sharp resonance indicated that the neutrals were first desorbed from the surface of the sample, resonantly excited, and then subsequently ionized in the gas phase by the same laser pulse. This is a two-step process that is quite different from that observed using a nonresonant laser of high power: For example, in LIMA

Figure 9.23 The ion signal for Al as a function of the ablation laser wavelength.[44] The enhanced yield at wavelengths corresponding to excitation of the $3d$ from the ground-state doublet is clearly observed.

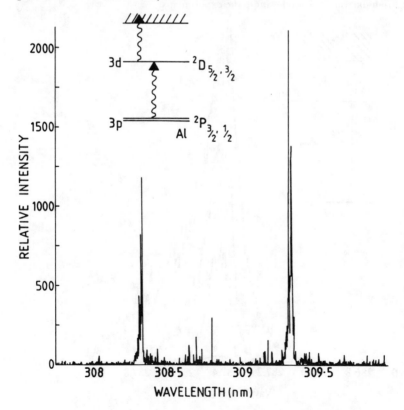

and LAMMA, ions are produced predominantly by plasma reactions, and the ablation and ionization processes cannot be entirely separated spatially and temporally. The processes here are thus very similar to the resonant postablation ionization involving two lasers. However, Wang et al. have shown in a recent paper that the ionization step may be accomplished via energy pooling collisions rather than by photoionization.[49] The high efficiency of these collisional processes, together with the complete geometric overlap of the ablation and ionizing processes, results in RLA being effected at very low laser fluences.

Although RLA has been demonstrated using pure or major samples, Borthwick et al. have recently used the technique to detect minors in stainless steel standard reference materials.[48] Figure 9.24 shows the Al signal when the laser is tuned (1) to the Al resonance at 308.2 nm and (2) off-resonance at 307.5 nm. The impurity ion signals of Na and K are relatively unaffected, but the Al ion signal is the strongest feature in the mass spectrum. In the stainless steel sample used (NIST1263A), the Al abundance is 5 parts per 1,000, and the observation corresponds to a detection limit of 10 ppm. A particular advantage of RLA is the efficient use of the ablated

Figure 9.24 RLA detection of aluminium in stainless steel sample NIST 1263A (a) on-resonance and (b) off-resonance. The constant surface contaminant Na and K off-resonance signals are clearly seen irrespective of the wavelength, while the Al signal disappears off-resonance.

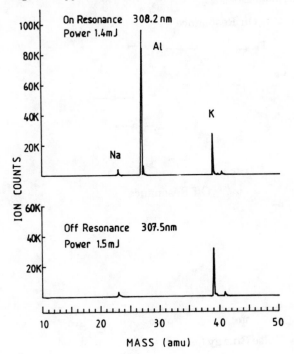

plume due to a complete overlap with the laser. This allows the use of much lower laser fluences with a corresponding reduction in the yield of background LIMA ions. It is hoped that quantitative information would be available from RLA on minors at levels down to 10 ppm.

The yield of neutrals in laser ablation of a solid surface is generally several orders of magnitude higher than the yield of ions. However, it is difficult to measure the relative yields accurately. Resonance laser ablation can provide this information directly by tuning the ablating laser off the resonance value. Figure 9.25 shows the result of such a measurement for Al and the expected large neutral to ion yield is quantified. This also shows that RLA for Al is about four orders of magnitude more sensitive than LIMS. In addition, the ablation of neutrals is expected to be affected to a lesser extent by the chemical environment of the surface, and thus RLA promises to be less matrix-dependent.

Some RLA neutrals are produced in an excited state, the relative populations being determined by the temperature at the surface during the ablation process. By tuning the laser to excite the excited-state neutrals resonantly, it should be possible to measure the surface temperature. These measurements will help in understanding the physics of the ablation process.

Figure 9.25 Power dependence of the RLA Al signal on- and off-resonance, as a function of laser pulse energy. The ratio of these two signals is a measure of the minimum neutral to ion yield in laser ablation.

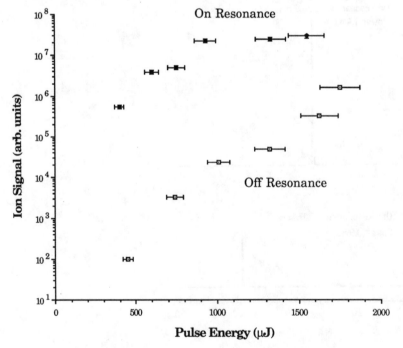

9.5. Conclusions and Future Directions

The importance of the laser microprobe (LIMS) technique is that it is a rapid, sensitive, elemental, and molecular survey analytical technique especially useful for analyzing electrically insulating samples. As a quantitative technique it has limitations, but its ability to carry out micro-area analysis via its optical microscope system will continue to make this instrument attractive.

The enormous potential for laser postionization of neutrals is just beginning to be fulfilled. In the SALI approach, using a properly chosen untuned UV laser, inorganic chemical analysis can be carried out such that the desorbed species are ionized independent of their photoionization cross sections. For organic species the ionization is relatively efficient and the fragmentation is minimal. The molecular photoionization cross sections are known to vary little for the fixed wavelengths chosen.

The importance of RIMS or SIRIS is that the procedure is highly selective as well as sensitive. Parts per million sensitivity is available for almost every element; parts per billion, for some; and parts per trillion, for a few. These limits are by no means intrinsic and often are due to lack of suitable standards. The absence of large matrix effects makes this method potentially very attractive for quantification, especially for characterizing layered structures in semiconductors. RLA has yet to be proven systematically, but potentially it is likely to be as sensitive as RIMS, yet with the advantage of a less complicated laser system.

Although RIMS has enormous potential as an ultratrace, quantitative analytic technique, instrument makers still consider its principal usage will be in research laboratories. The most often stated reason for this is that the dye laser has two disadvantages. First, for complete elemental coverage several different dyes and their solvents are required. The lasers must be emptied and refilled, and since this procedure cannot easily be computer controlled, it is difficult to use in an industrial environment. Second, the new Control of Substances Hazardous to Health (COSHH) regulations in the UK and similar legislation in Europe and in the United States will make the use of organic dyes and their solvents more restrictive. For these reasons new all-solid-state laser systems with broad tuning ranges are being sought by RIMS analysts and other laser spectroscopists to replace the dye laser. Because of the number of dyes available, complete wavelength coverage with roughly equal efficiency (within a factor of 2) from 350 to above 800 nm can be obtained from a dye laser. With efficient doubling, this range can be extended down to about 200 nm. To be equivalent to a dye system the wavelength coverage from a single all-solid-state laser is thus a demanding one. The two most promising systems are the Ti-sapphire laser and optical parametric oscillators.

In a recent paper Ledingham and Singhal have made an assessment of Ti-sapphire lasers and optical parametric oscillators (OPO) as sources of

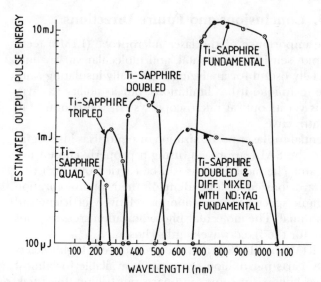

Figure 9.26 Estimated wavelength coverage of a Ti-sapphire laser with second, third, and fourth harmonics. The gap between the fundamental and second harmonic may be filled by difference mixing between Ti-sappahire doubled and the Nd:YAG fundamental pump laser. No account has been taken of absorption in the UV.[50]

variable wavelength for resonant ionization mass spectrometry.[50] Their conclusions can be summarized as follows. A Ti-Sapphire laser pumped by a TEM_{00} doubled Nd:YAG laser can reasonably give a 100-mJ, 10-ns, 10–50-Hz output variable between 660 and 1,100 nm. Doubling and mixing using BBO crystals as shown in Figure 9.26 can give total wavelength coverage between 200 and 1,100 nm at reasonable flux levels and with a spectral width of about 1 GHz. Similarly, a line-narrowed quadrupled TEM_{00} Nd:YAG-powered BBO–OPO can give complete wavelength coverage between 400 nm and 2 μm. If the output is further doubled by a BBO crystal, wavelength coverage between 200 nm and 2 μm is possible with high efficiency. At present the optical bandwidth from the OPO is rather broad, but this less desirable feature is likely to be only temporary.

Within the next few years genuinely rugged all-solid-state lasers systems will become freely available and the full potential of laser mass spectrometry will be realized.

Acknowledgments

KWDL and RPS would like to thank the members of the Glasgow LIS group who have helped over the years to promote this research: R. Jennings, A. Marshall, L. Wang, P. T. McCombes, A. P. Land, C. J. McLean, M. H. C. Smith, M. Towrie, S. L. T. Drysdale, C. M. Houston, A. Clark, I. S. Borthwick, and J. Sander.

REFERENCES

1. H. E. Duckworth, R. C. Barber, and V. S. Venkatasubramanian, *Mass Spectroscopy*, 2nd ed. (Cambridge University Press, Cambridge, 1986).

2. B. A. Mamyrin, V. I. Karataev, D. V. Schmikk, and V. A. Zagulin, *Sov. Phys.–JETP* **37,** 45 (1973).

3. N. Fürstenau, *Fresenius Z. Anal. Chem.* **308,** 201 (1981).

4. L. Moenke-Blankenburg, *Laser Microanalysis* (Wiley, New York, 1989).

5. D. M. Lubman, ed. *Lasers and Mass Spectrometry* (Oxford University Press, Oxford, 1990), p. 103.

6. L. Balazs, R. Gijbels, and A. Vertes, *Anal. Chem.* **63,** 314 (1991).

7. S. J. Hein and E. H. Piepmeier, *Trends Anal. Chem.* **7,** 137 (1988).

8. R. S. Adrain and J. Watson, *J. Phys. D: Appl. Phys.* **17,** 1915 (1984).

9. A. Vertes, P. Juhasz, L. Balazs, and R. Gijbels, *Microbeam Analysis,* P. E. Russell, ed. (San Francisco Press, San Francisco, 1989), p. 273.

10. R. P. Singhal, S. L. T. Drysdale, R. Jennings, A. P. Land, K. W. D. Ledingham, P. T. McCombes, and M. Towrie, Resonance Ionization Spectroscopy (RIS) *90, IoP Conf. Series* **114,** 185, 1991.

11. E. A. Rohlfing, D. M. Cox, and A. Kaldor, *J. Chem. Phys.* **81,** 3322 (1984).

12. A. O'Keefe, M. M. Ross, and A. P. Baronavski, *Chem. Phys. Lett.* **130,** 17 (1986).

13. M. Yang and J. P. Reilly, *Anal. Instrum.* **16**(1), 133, (1987).

14. C. H. Becker and K. T. Gillen, *Anal. Chem.* **56,** 1671 (1984).

15. U. Schühle, J. B. Pallix, and C. H. Becker, *J. Am. Chem. Soc.* **110,** 2323 (1988).

16. R. N. Zare, J. H. Hahn, and R. Zenobi, *Bull. Chem. Soc. Jpn.* **61,** 87 (1988).

17. F. Engelke, J. H. Hahn, W. Henke, and R. N. Zare, *Anal. Chem.* **59,** 909 (1987).

18. J. H. Hahn, R. Zenobi, and R. N. Zare, *J. Am. Chem. Soc.* **109,** 2842 (1987).

19. G. S. Hurst, M. H. Nayfeh, and J. P. Young, *Appl. Phys. Lett.* **30,** 229 (1977).

20. G. I. Bekov, V. S. Letokhov, and V. N. Radaev, *J. Opt. Soc. Am.* **B2,** 1554 (1985).

21. U. Krönert, J. Bonn, H.-J. Kluge, W. Ruster, K. Wallmeroth, P. Peuser, and N. Trautmann, *App. Phys.* **B38,** 65 (1985).

22. G. I. Bekov and V. S. Letokhov, *Appl. Phys.* **B30,** 161 (1983).

23. G. S. Hurst and M. G. Payne, *Principles and Applications of Resonance Ionisation Spectroscopy* (Adam Hilger, Bristol, 1988).

24. G. S. Hurst and M. G. Payne, *Spectrochim. Acta* **43B,** 715 (1988).

25. C. E. Moore, *1948 Atomic Energy Levels,* NBS Circular 467 (U.S. Government, Washington, DC, 1948).

26. N. Thonnard, J. E. Parks, R. D. Willis, L. J. Moore, and H. F. Arlinghaus *Surf. Int. Anal.* **14,** 751 (1989).

27. E. B. Saloman, *Spectrochim. Acta* **45B,** 37 (1990).

28. E. B. Saloman, *Spectrochim. Acta* **46B,** 319 (1991).

29. M. Towrie, S. L. T. Drysdale, R. Jenning, A. P. Land, K. W. D. Ledingham, P. T. McCombes, R. P. Singhal, M. H. C. Smyth, and C. J. McLean, *Int. J. Mass Spectrom. Ion Proc.* **26,** 309 (1990).

30. P. T. McCombes, I. S. Borthwick, R. Jennings, K. W. D. Ledingham, and R. P. Singhal, in *Optogalvanic Spectroscopy* (Institute of Physics, Bristol, 1991).

31. J. Grotemeyer and E. W. Schlag, *Acc. Chem. Res.* **22,** 399 (1989).

32. A. Marshall, A. Clark, R. Jennings, K. W. D. Ledingham, C. J. McLean, and R. P. Singhal, *RIS 90, IoP Conf. Series* **114,** 173 (1991).

33. A. Marshall, private communication.

34. V. S. Antonov, V. S. Letokhov, and A. N. Shibanov, *Sov. Phys. Usp.* **27,** 81 (1984).

35. S. W. Downey and R. S. Hozak, *J. Vac. Sci. Technol.* **A8,** 791 (1990).

36. S. W. Downey, A. B. Emerson, R. F. Kopf, and J. M. Kuo *Surf. Int. Anal. 15,* 781 (1990).

37. H. F. Arlinghaus, M .T. Sparr, and N. Thonnard, *J. Vac. Sci. Technol.* **A8,** 2318 (1990).

38. R. Tembreull, C. H. Sin, P. Li, H. M. Pang, and D. M. Lubman, *Anal. Chem.* **57,** 1186 (1985).

39. R. Tembreull and D. M. Lubman, *Anal. Chem.* **56,** 1962 (1984).

40. B. A. Bushaw and G. K. Gerke, *RIS 88, IoP Conf. Series* **94,** 277 (1989).

41. B. A. Bushaw and J. T. Munley, *RIS 90, IoP Conf. Series* **114,** 387 (1991).

42. D. M. Lubman and R. N. Zare, *Anal. Chem.* **54,** 2117 (1982).

43. H. F. Arlinghaus, N. Thonnard, M. T. Sparr, R. A. Sachleben, F. W. Larimer, R. S. Foote, R. P. Woychik, G. M. Brown, F. V. Sloop, and K. B. Jacobson, *Anal. Chem.* **63,** 402 (1991).

44. C. J. McLean, J. H. Marsh, A. P. Land, A. Clark, R. Jennings, K. W. D. Ledingham, P. T. McCombes, A. Marshall, R. P. Singhal, and M. Towrie, *Int. J. Mass Spectrom. Ion Proc.* **96,** R1 (1990).

45. F. R. Verdun, G. Krier, and J. F. Muller, *Anal. Chem.* **59,** 1383 (1987).

46. H.-M. Pang and E. S. Yeung, *Anal. Chem.* **61,** 2546 (1989).

47. L. Wang, I. S. Borthwick, R. Jennings, P. T. McCombes, K. W. D. Ledingham, R. P. Singhal, and C. J. McLean, *Appl. Phys.* B 53, 34 (1991).

48. I. S. Borthwick, K. W. D. Ledingham, and R. P. Singhal, *Spectrochimica Acta* B, 1992, in press.

49. L. Wang, K. W. D. Ledingham, C. J. McLean, and R. P. Singhal, *Appl. Phys.* B *54,* 71 (1992).

50. K. W. D. Ledingham and R. P. Singhal, *J. Anal. Atom. Spectrom.* **6,** 73 (1991).

10
Ultrafast Spectroscopic Methods

P. A. Anfinrud, C. K. Johnson, R. Sension, and R. M. Hochstrasser

10.1. General Introduction

Ultrafast spectroscopic methods are now capable of exposing the intimate details of condensed-phase reaction dynamics. Recent years have seen a large number of applications of pulsed laser techniques in chemistry[1] and biology.[2] Some indication of the current vitality and growth of this field of research may be gleaned from *Ultrafast Phenomena,* which has appeared every 2 years since 1978.[3] At present, experiments may be carried out on time scales that are so small that the nuclei of molecules and liquids may be considered effectively fixed during the experiment.

This article describes a variety of methods for carrying out transient optical spectroscopy, fluorescence spectroscopy, Raman scattering, and infrared spectroscopy on subpicosecond time scales. In each case a standard approach is described in some detail and the uses of such apparatus are illustrated by examples from recent work. No attempt has been made to provide a review of all the important work in this field. It is hoped that the references will provide those interested with an entry to the fuller literature of ultrafast phenomena.

10.2. Ultrafast Optical Absorption Spectroscopy

10.2.1. Introduction

Electronic absorption techniques are among the most widely applicable methods used to investigate transient species and fast processes. The ubiquitous near-UV, visible, or near-IR electronic absorption spectra of most transient polyatomic species, combined with their relatively large signals, make the technique effective in studying many problems. Millisecond to nanosecond transient absorption techniques have been applied to the study of a wide variety of photochemical problems involving the identification of intermediate species and their lifetimes. With the development of picosecond laser technology it became possible to apply time-resolved absorption techniques to the study of shorter-lived intermediates and to the investigation of more sophisticated physical problems involving, for example, rotational diffusion and solvent relaxation.

Over the last decade, femtosecond laser technology has developed to permit the routine measurement of even faster processes. Current laser technology provides usable pulses as short as 6 fs at selected wavelengths[4-6] and widely tunable, routinely usable pulses of less than about 90 fs.[7-14] The advances in femtosecond laser technology have provided the opportunity to apply time-resolved absorption spectroscopy to the investigation of many primary photochemical and photophysical processes that occur on time scales that were not previously accessible. Among the questions that are now being directly addressed are the following:

1. *Energy disposal in condensed-phase photochemical reactions.* In general, the excitation process that initiates a photochemical reaction results in the deposition of a significant amount of excess energy that must be lost to the solvent, either directly during the process of reaction or more slowly through the vibrational cooling of the product molecule(s).

2. *Partitioning between primary reaction pathways.* In photochemical reactions where more than one product is formed it is now possible to study the motion on the excited-state surface at short enough times to begin to investigate the questions of partitioning on the excited-state surface and the effects of the external environment on this partitioning.

3. *Fast energy transfer, electron transfer, and proton transfer reactions in "two-level" systems.* With very short laser pulses it is possible to prepare coherent superposition states of molecules and to use transient absorption and anisotropy measurements to observe the loss of coherence in the initially prepared states.

4. *Relative alignments of the product and reactant molecules in photochemical reactions.* If a photochemical reaction occurs quickly with respect to the overall rotational diffusion of the reactant and product molecules, po-

larized absorption spectroscopy may be used to determine average align-ments between reactant and product molecules. These measurements place significant constraints on the proposed reaction coordinates and aid in the investigation of environmental effects on chemical reactions.

5. *Electronic and vibrational dephasing in condensed phase environments.* With pulses of 6–50 fs, investigations of such dephasing processes are now possible.

10.2.2. Principles of Transient Absorption Spectroscopy

In a pump-probe transient absorption experiment the experimentally measured quantity is the change in absorbance at a given probe wavelength as a function of time following an excitation pulse. If the initial sample is transparent at the probe wavelength, the signal is due to absorption by, or stimulated emission from, the species (one or more) pro-duced by the pump pulse. If the initial sample absorbs the probe pulse, the bleaching of the ground state absorption (due to removal of ground state population) also contributes to the observed change in absorbance.

The time-dependent transient absorption signal is proportional to the time-dependent number of absorbing molecules and their molar extinction coefficients. A gain, or negative absorbance, is produced when the probe pulse stimulates emission from an excited electronic state population. The gain signal produced by an emitting species is proportional to the number of excited-state molecules, the overall radiative rate (k_r) for the transition, and the emission probability at the probe wavelength. It is not dependent on the integrated f fluorescence quantum yield, which may be very small due to nonradiative decay processes.

Fast transient absorption measurements of isotropic samples are gener-ally dependent on the relative polarizations of the pump and probe pulses. The data are normally obtained for parallel, perpendicular, or "magic-an-gle" (MA = 54.74°) orientations of the pump and probe polarizations. The parallel and perpendicular measurements depend on both the population kinetics and the time-dependent relative orientations of the pumped and probed transition dipoles, while the magic-angle measurement depends only on the population kinetics. The population and orientation informa-tion is separated by obtaining the following quantities:

$$I_T(t) = I_{MA}(t) = (I_{\parallel}(t) + 2I_{\perp}(t))/3 \tag{10.1}$$

$$r(t) = \frac{I_{\parallel}(t) - I_{\perp}(t)}{I_{\parallel}(t) + 2I_{\perp}(t)} = \frac{2}{5} <P_2(\cos\theta(t))> \tag{10.2}$$

where $r(t)$ is called the anisotropy, P_2 is the second Legendre polynominal $P_2(x) = (3x^2 - 1)/2$, $\theta(t)$ is the angle between the pumped and probed transition dipoles, and the angular brackets indicate an ensemble average over the distribution of angles between the pumped and probed transition dipoles. For a signal composed of contributions from several species:

$$I_T(t) = A_1(t) + A_2(t) + A_3(t) + \cdots \tag{10.3}$$

$$r(t) = \frac{A_1(t)}{I_T(t)} r_1(t) + \frac{A_2(t)}{I_T(T)} r_2(t) + \frac{A_3(t)}{I_T(T)} r_3(t) \cdots \tag{10.4}$$

where $A_n(t)$ is the signal as a function of time due to species n and is proportional to the signed amplitude (i.e., gain or absorption) of the signal from species n multiplied by its population kinetics; $r_n(t)$ is the anisotropy that would be obtained if only the signals from species n were being detected. If all the species contributing to the total signal can be considered to have approximately the same time constant (τ_r) for overall rotational diffusion, equation (10.4) may be simplified somewhat:

$$r(t) = \left(\frac{A_1(t)}{I_T(t)} r_1(0) + \frac{A_2(t)}{I_T(t)} r_2(0) + \frac{A_2(t)}{I_T(t)} r_3(0) \cdots \right) e_r^{-t/\tau} \tag{10.5}$$

If the data are simultaneously fitted to equations (10.3) and (10.5), the "initial" anisotropies of the various species may be extracted. This is particularly useful for measurements of photochemical reactions, where the anisotropy of the photoproduct may be determined. Here $r_p(0) = 0.4 < P_2$ (cos θ_{RP})>, where θ_{RP} is the angle between the pumped transition dipole of the reactant and the probed transition dipole of the product. If the directions of the transition dipoles in the molecule-fixed frames are known or can be estimated with reasonable certainty for both the reactant and product, the anisotropy measurement provides a significant constraint regarding the average reaction path. This measurement, in combination with theoretical calculations, can be used to distinguish between alternative mechanisms and to test proposed reaction coordinates.

10.2.3. Techniques and Instrumentation

This section discusses the techniques and instrumentation used for transient absorption spectroscopy, with emphasis placed on the discussion of instrumentation for the routine measurement of time-resolved absorption spectra on timescales of ~100 fs.

A flexible instrument available for subpicosecond time-resolved absorption spectroscopy is based on an argon ion laser-pumped colliding-pulse mode-locked (CPM) dye laser, as shown in Figure 10.1.[7] The CPM laser provides stable, short pulses (50–70 fs) at approximately 624 nm with a 100-MHz repetition rate and subnanojoule pulse energies. The output beam of this laser is amplified in a four-stage dye amplifier pumped by a 20-Hz Q-switched Nd:YAG laser.[8] The first three stages are pumped transversely and the fourth stage is pumped longitudinally, with the Nd:YAG

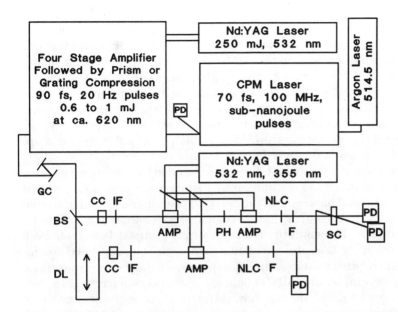

Figure 10.1 Schematic diagram of tunable subpicosecond transient absorption spectrometer. PD = photodiode, GC = grating compressor, BS = beam splitter, DL = delay time, CC = continuum cell, IF = interference filter, AMP = amplifier cell, PH = pinhole (500–800 μm), NLC = nonlinear optical crystal (KDP or BBO), and SC = sample cell. Mirrors and lenses have been omitted from the diagram. A mirror is implied when the beam changes direction and lenses are necessary to focus and recollimate the beam for continuum generation, amplification, and frequency doubling as well as to focus the beams into the sample cell.

and the amplified CPM pulses counterpropagating. The amplified pulse is recompressed using either a prism pair or a grating pair. This laser system will routinely provide <100-fs pulses with pulse energies of 600 μJ to 1 mJ at 20 Hz around 624 nm. The pulse-to-pulse stability is approximately 13%. Most important, once aligned the CPM and the amplifier will usually run reliably for several (3–10) hours without readjustment. With these noise conditions it takes about 15 minutes of unnormalized averaging to yield an absorbance change accuracy of 10^{-3}.

At the expense of pulse energy, somewhat shorter pulses may be obtained by compressing the amplified CPM pulse.[15–16] The pulse compression is achieved by focusing the amplified pulse through a 500-μm aperture to obtain a Gaussian beam profile, producing a linear chirp by passing the beam through a short piece of some nonlinear optical material such as quartz, using a second aperture and lens to recollimate the beam and a prism pair or a grating pair to recombine the various wavelength components in time. This procedure provides 20–50-fs pulses of ≥100 μJ at 20 Hz around 624 nm.[15]

Tunable pump and probe beams are obtained as shown in Figure 10.1. The amplified CPM pulses are split into two beams with a beamsplitter (usually 50:50 but sometimes 70:30 or 30:70), and one of the beams is reflected through a computer-controlled variable delay line. Tunable probe pulses are obtained by focusing the 624-nm probe beam into a 1-cm quartz cell of H_2O or D_2O to generate a white light continuum. A bandpass interference filter is placed after the continuum to select the desired probe wavelength. Standard filters with 10–12 nm full width at half maximum (FWHM) bandpass are sufficient for most transient absorption experiments. The direct continuum provides usable probe pulses between approximately 1.3 μm in the near-IR and 380 nm in the near-UV.

Probe pulses deeper into the UV may be obtained by amplifying narrow-wavelength regions of the continuum in a single-stage amplifier pumped by the excess second harmonic of the Nd:YAG laser used to pump the main amplifier, or by the second or third harmonic of a secondary Nd:YAG laser, as discussed later. The amplifier is a transversely pumped 1-cm quartz cell with an appropriate laser dye circulating through it. The amplified continuum is then frequency-doubled in a thin potassium dihydrogen phosphate (KDP) or β-barium borate (BBO) nonlinear optical crystal (crystals are typically 0.1–1 mm thick). Alternatively, difference-frequency generation in lithium iodate, lithium niobate, silver gallium sulfide, or silver gallium selenide crystals may be used to obtain IR probe pulses between 12 μm and 1.3 μm.[17] IR pulses are generated by difference mixing 624-nm radiation with a slice of amplified continuum of the appropriate wavelength.

Tunable pump pulses are obtained by amplifying narrow-wavelength slices of white light continuum, as discussed earlier. To generate sufficient pulse energy, one, two, or three amplification stages are used (one stage is generally sufficient with visible pump wavelengths), separated by pinhole apertures to minimize amplified spontaneous emission (ASE). The amplification stages are pumped by the second or third harmonic of a second Nd:YAG laser that is operated as a slave laser having its flashlamps triggered by the oscillator-synchronized output of the master laser. Both laser Q switches are triggered by the same trigger pulse synchronized to the CPM laser. Proper timing between the master and slave laser is obtained by varying the length of the cable transmitting the Q-switch trigger pulse to the slave laser. UV pump wavelengths are obtained by frequency-doubling the amplified continuum in KDP or BBO. In this manner it is possible to obtain pump pulses with energies of at least 2 μJ and often more, between approximately 900 nm and 200 nm. Polarizers are placed in both the pump and probe arms to clean up the final polarizations. A single-order quartz half-wave plate, a multiple-order mica half-wave plate or a Soleil–Babinet compensator is used in either the pump or the probe arm to control the relative polarizations of the two beams. Data may be collected with the polarizations of the pump and probe pulses parallel, perpendicular, or at the magic angle (54.74°).

Several detection schemes have been used with this laser system at the University of Pennsylvania. For single-wavelength probe experiments three amplified photodiodes are used to monitor the pump beam after the sample, and the probe beam both before and after the sample. The photodiode signals are integrated, processed, and stored in a computer. Variable time delay is obtained by using a computer-controlled mechanical translation stage on either the pump or the probe arm of the experiment. Pulse-to-pulse normalization is performed to compensate for pulse-to-pulse fluctuations and 15–50 laser shots are averaged at each time delay. Software discrimination routines eliminate laser shots that fall outside a predetermined intensity range, and recollect data for averaging if the accuracy of the absorbance data falls below an acceptable limit. Several scans of the delay time are performed to average out long-term fluctuations in the laser intensity.

Broadband absorption spectra may be obtained as a function of time delay by using multichannel detection.[9-14] Large sections of white-light continuum, produced by focusing the amplified CPM pulse into H_2O, are selected by broadband cutoff filters. A Schott RG665 filter provides usable continuum over the range of 830–635 nm and a Corning 5–56 filter provides usable continuum over the range of 610–380 nm. These filters are necessary to attenuate the intense 624-nm CPM fundamental, which otherwise causes significant stray light problems in the spectrograph and multichannel detector. To probe regions deeper into the UV, the broadband continuum may be frequency-doubled in a very thin (0.1–0.3 mm) nonlinear optical crystal such as KDP. The use of a thin crystal permits simultaneous doubling of a fairly wide range of the continuum. Reference and signal beams are collected by using quartz fibers and analyzed using a spectrograph and an intensified dual-diode array detector.

Additionally, fluorescence detection may be used to detect absorption due to a single species in a multicomponent system. In this experimental arrangement two pulses of nearly equal energy are used. The initial pulse prepares the excited electronic state of the reactant molecule, and the second pulse excites the transient product molecule to a fluorescent state. The product fluorescence is then detected by a photomultiplier with bandpass filters to eliminate stray light and fluorescence from other species.

The major advantage of the instrument described earlier is its flexibility. Instrument response functions of <1 ps FWHM may be obtained for pump wavelengths ranging from 900 to 200 nm (pulse energy ≥ 2 μJ) and probe wavelengths ranging from 12 μm to 200 nm. Shorter instrument response functions (~50–200 fs) may be obtained over a somewhat more limited frequency range by pulse compression techniques, as described earlier. The major limitations of this instrument are its low repetition rate and the frequency chirp introduced in the continuum generation process. This later limitation is not significant for narrowband probe experiments but is a significant limitation in the time resolution available for broadband spectral measurements. The time spread produced during continuum generation in 1 cm of H_2O is approximately 1 ps/50 nm.

A similar pump-probe subpicosecond transient absorption spectrometer based on a XeCl* excimer laser pumped amplifier has been described in detail in the literature.[18,19] The spectrometer described by Ernsting and

Kaschke operates at a repetition rate of 10 Hz and has most of the same limitations as the spectrometer described earlier. Theoretically, however, it should be possible to increase the repetition rate of the XeCl* amplified system to ~100 Hz.[20]

Higher-repetition-rate CPM-based transient absorption spectrometers may be constructed by using a variety of other amplification schemes, involving several types of pump lasers. In particular, copper vapor laser–based (CVL)[21] and excimer laser–based amplifiers are fairly widely used.[20] In general, there is a trade-off between repetition rate and pulse energy. A high-repetition-rate CPM-based transient absorption spectrometer may be constructed using an 8-kHz CVL to amplify the CPM pulses.[21] Such a system has the advantage of an 8-kHz repetition rate but the disadvantage of low pump energies (~1–5 μJ at 624 nm). This system is ideal for pulse compression techniques used to obtain instrument response functions on very short time scales. Ultrashort pulses are obtained[4] by coupling a portion of the amplified CPM pulse into a polarization-preserving quartz optical fiber (to broaden the pulse spectrally) and recompressing the pulses in a combined grating and prism sequence. Pulses as short as 6 fs have been generated and used in transient absorption experiments.[5-6]

10.2.4. Applications

Ultrafast laser technology has been used to perform time-resolved absorption studies of a large number of systems. A number of examples will be presented briefly to demonstrate the usefulness of the various techniques. The reader is referred to the original literature for more detailed discussion of the interpretation and analysis of the data presented.

The tunable subpicosecond time-resolved absorption spectrometer described earlier has been used to study a number of systems. In particular, an extensive study of the photoisomerization reactions of *cis*-stilbene has been performed.[11-13,22] *Cis*-stilbene excited into its lowest singlet excited electronic state reacts to form *trans*-stilbene and dihydrophenanthrene (DHP) with quantum yields of 0.35 and 0.10, respectively.[23,24] At room temperature, the ground electronic state product molecules are produced within 1–2 ps of the excitation of the *cis* molecule, even in relatively high-friction environments.[22,25]

Pump-probe measurements of the kinetics of product formation, shown in Figure 10.2, demonstrate the disappearance of the excited electronic state *cis* molecules within 1–2 ps and the appearance of the product molecule absorption signals on approximately the same time scale. The data shown in Figure 10.3 were obtained using 312-nm and 250-nm pump wavelengths. The data obtained with 312-nm excitation exhibits at most a very short delay (≤300 fs) between the decay of the excited state *cis** and the rise of the product absorption.

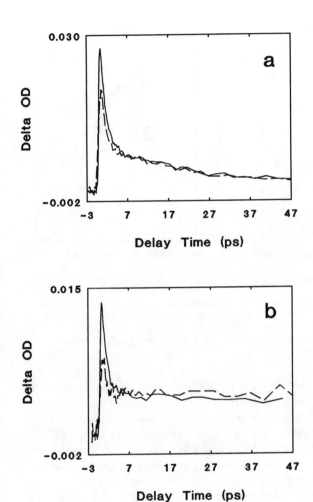

Figure 10.2 Magic angle time-resolved absorption of *cis*-stilbene in hexadecane. The data in (a) were obtained using a 312-nm pump (solid line) or a 250-nm pump (dashed line) and a 330-nm probe wavelength. The primary photoproduct observed at 330 nm is *trans*-stilbene. The data in (b) were obtained using a 312-nm pump and a 480-nm probe wavelength (solid line) or a 250-nm pump and a 460-nm probe wavelength (dashed line). The primary photoproduct observed at 480 and 460 nm is DHP.

The combination of kinetic and spectral measurements represents a powerful tool in the identification of transient or product species. Spectral measurements performed in the regions of the *trans* and DHP product absorptions, shown in Figure 10.3, demonstrate that the persistent absorption signals observed in Figure 10.2 are indeed due to the expected product molecules. An anisotropy measurement of the product absorption at 312 nm is shown in Figure 10.4. This anisotropy measurement, along with similar measurements at other wavelengths, demonstrates that the product

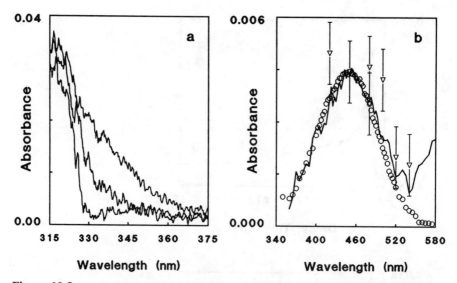

Figure 10.3 (a) Absorption spectrum of hot *trans*-stilbene obtained at 6 ps, 15 ps, and 50 ps. The pump wavelength is 312 nm. The 50-ps spectrum is indistinguishable from the spectrum of ground-state *trans*-stilbene obtained using the same apparatus: (b) DHP absorption at ~5 ps (data points with error bars), 200 ps (solid line), and 50 ms (circles). The pump wavelength is 312 nm.

absorption observed at early times is attributable to a distribution of vibrationally hot *trans* molecules rather than to molecules in a distinct intermediate state such as the postulated perpendicular phantom state.[13] In addition, the anisotropy measurements provide information on the relative alignments of the reactant and product molecules and provide a significant constraint to be considered when discussing possible reaction coordinates for the *cis* to *trans* reaction in condensed-phase environments. The *trans* product anisotropy has been measured to be 0.20 ± 0.04, which is larger than the value expected if a simple torsional motion around the ethylenic double bond is the only reorientation. It is necessary to consider more complicated nuclear motions and the overall effect of the solvent environment to account for the higher than expected degree of alignment between *cis* reactant and *trans* product molecules.[13,26]

Transient absorption spectra and pump-probe kinetics measurements of the excited states of C^{60} are shown in Figures 10.5 and 10.6. Figure 10.5 shows the absorption spectrum obtained in neat toluene at 500 ps. The peak that grows in at 740 nm is attributed to absorption from the T_1 state of C_{60}.[14] Pump-probe measurements of the rise of the triplet absorption probed at 740 nm and pumped at 312 nm, 520 nm, and 624 nm are shown in the inset in Figure 10.5. The charge transfer complexation of C_{60} with N,N-dimethylaniline (DMA) quenches the formation of the triplet state when DMA is present. A three-dimensional plot of the transient absorption spectrum of C_{60} in DMA versus delay time is shown in Figure 10.6. These

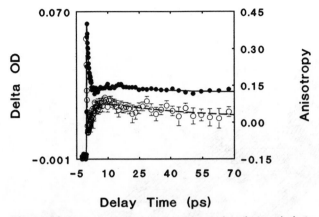

Delay Time (ps)

Figure 10.4 Magic-angle transient absorption (heavy circles) and anisotropy (open circles) obtained from a sample of *cis*-stilbene pumped at 312 nm and probed at 312 nm. The solid lines represent a fit of the data to a model consisting of an instrument-limited coherent spike with an anisotropy of 0.40, a signal due to *cis* ground-state bleaching and *cis** absorption, which decays with a time constant of 1.3 ps, and a signal due to the product absorptions, which appears on a time scale of 1.3 ps and persists. The overall rotational decay of the anisotropy is on a time scale of 80 ps.

data are interpreted as representing a 1–2-ps time scale for the formation of the ion pair $DMA^+:C_{60}^-$ and a ~20-ps time scale for the geminate charge recombination process that results in recovery of the original ground state absorption spectrum.[12]

The same transient absorption spectrometer has also been used to study the reaction center of the photosynthetic bacterium *Rhodobacter spheroides*.

Figure 10.5 15 ps and 500 ps absorption spectra of C_{60} in toluene obtained with a 520-nm pump. The inset contains transient absorption data obtained using a 740-nm probe pulse and a 312-nm (triangles), 520-nm (circles) or 624-nm (squares) pump pulse. The data are fitted to a model consisting of a 650-ps decay of the singlet state and a 650-ps rise of the triplet state.

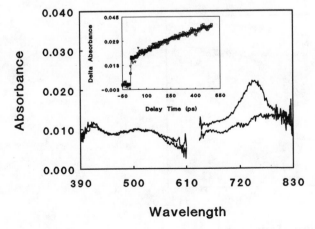

Wavelength

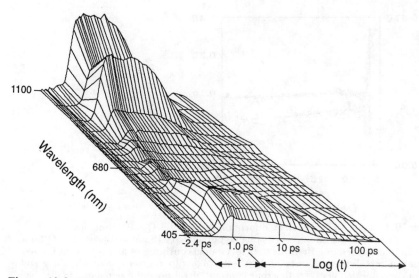

Figure 10.6 Time-resolved absorption spectra of C_{60} in *N,N*-dimethylaniline pumped at 624 nm (from Ref. 12). By 200 ps the entire spectrum has decayed to baseline.

Samples of the reaction center were pumped with 4-μJ pulses at 860 nm and probed at 920 nm and 940 nm. This corresponds to pumping the special-pair absorption directly and monitoring the decay of the excited state by measuring the gain signal induced by the probe beam. The typical measurements shown in Figure 10.7 were fitted to a functional form involving an instrument-limited rise followed by a decay of the gain signal with a

Figure 10.7 Transient gain signal from reaction centers from *Rhodobacter spheroides*. The pump wavelength is 860 nm and the probe wavelengths are 920 nm (circles) and 940 nm (diamonds). The data are fitted to a functional form: $A_1 e^{-t/\tau} + A_2$, where A_1 is the amplitude of the special-pair gain signal, $\tau = 3.1$ ps is the lifetime of the electronically excited special pair, and A_2 is a small amount of persistent bleach at 920 nm and absorption at 940 nm.

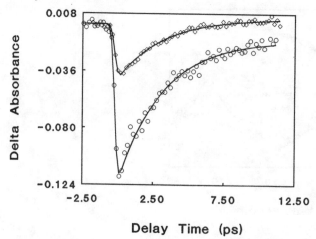

characteristic time of 3.1 ps. These measurements are consistent with those reported by other workers.[27,28]

Transient absorption measurements at higher time resolution may be obtained for selected pump and probe wavelengths using the high-repetition-rate CVL amplified CPM-based system described earlier (See Section 10.6.3 for more details.) Figure 10.8 shows a 305-nm pump, 610-nm probe measurement of the *cis*-stilbene excited-state transient absorption in isopropanol obtained with a 60-fs instrument function.[29] These data fit a functional form with a *cis** decay of 680 fs.

The cutting edge in ultrafast electronic transient absorption spectroscopy is demonstrated by the transient absorption spectra of bacteriorhodopsin (BR) obtained with a 60-fs pump pulse at 618 nm and a 6-fs probe pulse, also centered around 620 nm.[6] The laser system used was a CVL-amplified CPM laser with the 6-fs pulses produced by using the fiber compression technique discussed earlier. A spectrometer and diode array detector were used to analyze the absorption of the various frequency components of the 6-fs probe pulse after the sample. These results show the rapid torsional dynamics on the excited-state surface on a 100-fs time scale followed after 200 fs by the appearance of the ground electronic state of the photoproduct.[30] Another superfast example concerns the transient absorption spectra of Nile Blue obtained using 6–9 fs pump and probe pulses centered at 620 nm.[5,31]

10.2.5. Summary of Future Goals

The prospects for ultrafast transient absorption spectroscopy are quite bright. The development of tunable, high-energy subpicosecond transient absorption spectrometers such as the one described earlier permit the application of ultrafast transient absorption spectroscopy to a wide variety of problems that have not been addressed previously. In particular, the availability of deep UV (200–300-nm) pump pulses should permit the study of a large number of condensed-phase small-molecule reactions. The availability of tunable pump and probe pulses in the near-UV, visible, and near-IR will permit the study of a number of systems of biological importance with greater detail than has hitherto been possible.

The further development of ultrafast laser technology will permit the utilization of extremely ultrashort pulses (<10 fs) in the study of a larger number of systems. For this technique to realize its full potential, however, it is necessary to develop more tunable laser sources. The investigations to date have been confined to a region of approximately 3000 cm^{-1} around the fundamental of a CPM laser (600–630 nm). One of the most interesting possibilities for future progress in ultrafast laser technology involves the use of titanium:sapphire (TiS) for the generation and amplification of <100-fs laser pulses. This will eliminate the need for many of the expensive and messy laser dyes used currently in CPM lasers, synchronously

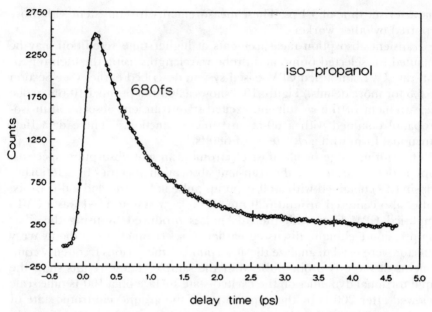

Figure 10.8 Transient absorption of *cis*-stilbene in isopropanol. (From Ref. 29.)

mode-locked dye lasers, and various amplification schemes. In addition, it may be possible to use much of the tuning range of TiS in the visible and near-IR to extend the currently accessible spectral range for <10-fs experiments.

10.3. Ultrafast Fluorescence Studies

10.3.1. Introduction

Fluorescence techniques are widely applicable as probes of transient species and of ultrafast processes. The three key fluorescence parameters are the decay profile, the spectrum at a given instant, and the anisotropy at each time and emission wavelength. The technical challenge is to obtain high signal-to-noise (S/N) measurements of each of these properties at the best possible time resolution. High S/N is needed because the decay functions need to be determined rather than assumed, and the samples may generate multicomponent emissions that need to be sorted out.

Ultrafast processes that may be tracked by fluorescence measurements are ubiquitous. Molecules in liquids undergo sticklike reorientational motion by diffusion on the time scale $\tau_R \sim V\eta/k_B T$. (V is the hydrodynamic volume and η is the viscosity.) This corresponds to times of around 10 ps for molecules having the size of benzene derivatives in solvents having viscosities comparable with hexane. Many molecules, especially symmetric ones like benzene, undergo even faster motion, such as sliplike rotation or inertial rotation.[32,33] The inertial motion arises because at any instant a molecule in a liquid has on average an angular velocity of $(k_B T/I)^{1/2}$, where I is its moment of inertia. Such motions go on uninterrupted for ~ 100 fs, so that with picosecond time resolution only the diffusional relaxation that corresponds to many of these smaller inertial steps is observed. Energy transfer can be extremely rapid, and this can lead to fluorescence or anisotropy decay.

Solvation processes in liquids are among the fastest processes to have been observed with fluorescence techniques.[34,35] Here the response of the solvent to electronic excitation of the solute molecule causes a time-dependent Stokes shift of the fluorescence, which takes only about 0.5 ps for a solvent such as water.[36] Of course intramolecular relaxation processes such as internal conversion in optically excited molecules can also be extremely rapid, in the femtosecond regime.

The optimum methods used to time-resolve fluorescence processes depend on the required time resolution. For processes occurring on time scales longer than 20 ps, time-correlated single-photon counting is the method of choice. For better time resolution, optical gating methods or streak cameras must be used. Streak cameras have improved significantly over the past few years and now have subpicosecond time resolution, though not routinely. However, to utilize a streak camera it is necessary to have an appropriately short laser pulse, in which case the optical gating method may be rather straightforward to set up and with care will yield significantly better time resolution. This gating technique, first introduced by Mahr and Hirsch[37] and used so effectively by Topp et al.[38] with picosecond pulses, has been employed at 50-fs time resolution to study the UV fluorescence of semiconductors.[39] A more detailed description of this method is given later.

10.3.2. Principles of Measurement of Time-Resolved Emission

The constant k_r characterizing spontaneous emission from an excited state is known as the *radiative rate coefficient* and is the sum of the radiative rate constants corresponding to all possible radiative decay pathways. If *nonradiative* processes are occurring, characterized by a rate coefficient k_{nr}, then the quantity $[k_{nr} + k_r]^{-1}$ is the *emission lifetime* $\tau_f = 1/k_f$. An ensemble of molecules will radiate such that the instantaneous intensity of emitted radiation (photons $m^{-2}\,s^{-1}$) detected at time t after excitation is $I(t) = k_r n_e(t)/4\pi r^2$, where the detector or collecting lens is located at distance r from the very small excitation volume containing $n_e(t)$ excited molecules. The decay function for $n_e(t)$ is:

$$n_e(t) = n_e(0) \exp[-k_f t] \tag{10.6}$$

and the total yield of photons is k_r/k_f. This yield influences the signal strength in a time-integrated experiment, but it does not affect the signal at times short compared with k_{nr}^{-1} in a time-gated experiment that measures $I(0) \propto k_r n_e(0)$. Thus, the only molecular parameter that influences the time-gated signal is the *radiative rate*. The extension of this discussion to mixtures is straightforward.

In the condensed phase the electronic dephasing is very fast, $T_2 \sim 20$–500 fs, so that the memory of the phase of the excitation beam is rapidly lost. The emission is therefore mainly "dephased emission." The spontaneous emission radiated while coherence is preserved is resonance Raman scattering, but this is much weaker than the dephased emission and has a lifetime determined by T_2.[40,41] The ratio of the Raman to the dephased emission intensity is approximately $2k_r T_2$. As the time resolution of emission experiments gets closer to the range of T_2 values common for moderately sized molecules in ordinary solutions, that is about 30 fs, it will become feasible to measure these Raman decays at ambient temperatures.

The anisotropy of the spontaneous emission is a powerful source of information not only on the orientational dynamics of transition dipoles but also on the composition of radiating mixtures. Reorientation of transition dipoles can occur by virtue of overall molecular motion, energy transport, electron transport, or any other process that causes the initially formed transition dipole to differ from the transition dipole detected by spontaneous emission. The anisotropy measurement also provides independent information on the kinetics governing the interchange and formation of the various species whose emission is detected.

The fluorescence anisotropy, $r(t)$, takes the following form in cases where the radiative rate constants are time-independent:

$$r(t) = \sum_i \alpha_i(\lambda_d, \lambda_e; t) r_i (t) \tag{10.7}$$

where $\alpha_i(\lambda_d, \lambda_e; t)$ is the fraction of the observed signal at detection and excitation wavelengths λ_d and λ_e, at time t, and $r_i(t)$ is the anisotropy

$$r_i(t) = 0.4 <P_2[\mathbf{u}(0) \cdot \mathbf{u}_i(t)]> \tag{10.8}$$

Here $\mathbf{u}(0)$ and $\mathbf{u}(t)$ are unit vectors in the coordinate frame of the initially excited molecule corresponding to the initial transition dipole direction $\mu(0)$, and the radiating transition dipole direction on the ith species, $\mu_i(t)$, at time t.

10.3.3. Upconversion Technique for Optical Gating

The principle of upconversion for optical gating involves combining the fluorescent radiation at wavelength λ_e and intensity I_e with a short laser pulse at wavelength λ_L and intensity I_L in a nonlinear crystal. When both fields are present in the crystal at the same time, nonlinear mixing takes place and radiation at the sum or difference frequencies, having wavelengths $\lambda_L \lambda_e / (\lambda_e \pm \lambda_L)$ are generated. One of these signals is detected, and its intensity, which is proportional to the product $I_L I_e$, is measured as a function of the variable delay of the laser pulse arrival at the nonlinear crystal. The signal is actually

$$S(\tau) = c \int_{\infty}^{-\infty} dt \, I_L(t - \tau) I_e(t) \tag{10.9}$$

where c is a constant, depending on the nonlinear susceptibility of the unconverting material. The fraction of the (weak) fluorescence signal that is upconverted is independent of its intensity, and it is not difficult to upconvert almost the total amount present in the crystal when sufficiently intense laser pulses are used. However, such a situation is undesirable, and it is best to work at less than 30% conversion to avoid saturation phenomena. The integral in equation (10.9) expresses the convolution of the laser pulse with the rise and decay function $I_e(t)$, with the laser pulse center arriving at time $t = \tau$, with $t = O$ defined at the overlap of pump and probe pulses in the crystal.

Experiments of this nature with femtosecond time resolution have been carried out in a number of ways. There are two important experimental limits, depending on whether detection of the upconverted slice of fluorescence is significantly shot noise-limited. When using a moderately high energy pulse to excite and upconvert the fluorescence, the effect of photon statistics on the S/N of the experiment is negligible and the accuracy is limited by the laser pulse fluctuations. On the other hand, with the relatively low energies usually obtained in higher repetition rate experiments, the shot noise also must be overcome. At extremely high repetition rates, such as 10–100 MHz, the method of photon counting naturally becomes superior. The method described in detail later to time-resolve UV fluorescence uses an amplified dye laser running at 20 Hz. There are approximately 100–1,000 upconverted photons generated in each laser shot from 10-μJ pump pulses at 302 nm of width ~150 fs. This experiment is much easier and has better S/N if visible fluorescence is detected, simply because of the availability of conventional filters to isolate a larger fraction of the sum-frequency signal. When upconversion with 604 nm is used with UV fluorescence at, say, 302 nm, the detected signal is at 201 nm. The time resolution is limited by the thickness of the crystals generating the UV light. The following is a readily tunable system incorporating a dye laser constructed by Pereira[33,42] based on the laser described by Dawson et al.[43]

The laser design is composed of a double-jet dye laser (Rhodamine 610 in ethylene glycol as the gain medium and a 3:2 DQOCI/DODCI mixture as the saturable absorber solution) with two prisms for intracavity dispersion correction. The pump laser is a continuously pumped, acousto-optically mode-locked Nd:YAG laser, frequency-doubled in KTP. The two prisms used for intracavity dispersion compensation disperse the light transversely, allowing tunability just by the insertion of a slit in the optical path, which controls the frequency and bandwidth of operation. The dye laser routinely produces 40 mW of average power at 604 nm with a pulse width of 95 fs.

The dye laser pulses were amplified at 20 Hz by a Q-switched Nd:YAG pumped three-stage amplifier with Sulforhodamine 640 in methanol/water as the amplifier dye, as shown in Figure 10.9. Pulse broadening in the amplifier was reduced by carefully adjusting the concentration in the three stages of the dye amplifier. The 604-nm amplified pulses had energies of ~400 μJ and an approximately Gaussian autocorrelation width of 200 fs. The amplified beam was sent through a beamsplitter, and half of it was focused with a long-focal-length lens and passed through a 0.5-mm KDP crystal to generate the second harmonic needed for excitation. The cross correlation of the amplified dye fundamental and the second harmonic, determined by sum-frequency generation in BBO, was ~250 fs in width and Gaussian in shape.

The 302-nm light (10–20 μJ/pulse) was focused into the sample with a 10-cm f1 lens after passing through a polarizer and a half-wave plate to allow excitation with either vertically or horizontally polarized light. Fluo-

Figure 10.9 Diagram of UV fluorescence up-conversion gate experiment.

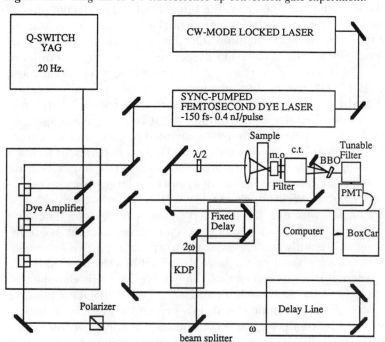

m.o.=reflective microscope objective
c.t.= Cassegrian Telescope

rescence was collected collinearly with the pump by a reflective microscope objective (74× magnification, focal length 2.6 mm, aperture 0.65) collimated for filtering purposes, then focused again by a Cassegrian telescope into a BBO crystal 200 μm thick. In the region where the fluorescence was collimated, a solution filter rejected the residual pump beam. The purpose of collinear fluorescence collection was to minimize the transit time spread in the fluorescence originating from different parts of the illuminated volume. The fluorescence has to be collimated to interact with the filter solution; otherwise, different rays will travel different paths, resulting in loss of time resolution. Collinearity also makes it easier to align the experiment. Reflective objects were chosen because lenses are chromatic and their use results in having different components of the fluorescence focusing at slightly different points, also resulting in loss of time resolution. The thin BBO crystal was the shortest available at the time of the experiment and was chosen to minimize the pulse broadening because of dispersion while the fluorescence and the gating pulse interact inside this nonlinear material. Recently, much thinner crystals of BBO have been used.[39]

The remaining fundamental light at 604 nm traversed a variable delay line and was weakly focused and overlapped with the fluorescence in the mixing BBO crystal after reflection from a mirror positioned very close to the fluorescence path. The fluorescence and the gating beam were slightly noncollinear to allow isolation of the unconverted signal by spatial filtering. The signal generated at the sum frequency was passed through a small monochromator and detected with a solar blind photomultiplier and a gated integrator. Since the intensity generated at the sum frequency is proportional to the intensity of the fluorescence, the fluorescence decay can be mapped by measuring the upconverted intensity as a function of the optical delay traversed by the 604-nm gating pulse. For each pulse the pump beam was sampled and the ratio between upconverted fluorescence and pump beam intensity was recorded. The sensitivity is easily good enough to isolate a strong signal from the C–H Raman line of hydrocarbon solvents, which could be removed by angle tuning the crystal to either side of the line.

The overall time resolution of the experiment is determined by the temporal widths of the excitation and gating pulses, and the interaction between the fluorescence and the gating pulse in the BBO crystal. The gating pulse and the fluorescence travel at different velocities in the crystal due to dispersion, so the upconverted signal results from the moving overlap of the fluorescence and gating pulse along the length of interaction. This is why a short crystal is required. Upconversion experiments have limited spectral capabilities due to phase-matching conditions. The instrument function (the signal that would be observed from a sample with a delta-function response) was obtained from the upconversion of the Raman line from isopentane, and its FWHM measures 300 fs.

When the fluorescence spectrum is upconverted and detected with a given orientation of the gating crystal, the signal may change with time as the fluorescence spectrum evolves. This effect depends on the bandwidth of the upconverted radiation, and it disappears when the bandpass of the detector is sufficiently large. The evolution of the spectral characteristics of the fluorescence is frequently caused by the time-dependent Stokes shift.

10.3.4. Applications of Ultrafast Fluorescence

Recent studies of ultrafast fluorescence have used both UV and visible excitation. It is currently possible to obtain high-quality fluorescence signals with <100-fs lifetimes that can be interpreted without recourse to deconvolution. A significantly greater challenge involves obtaining anisotropy decays with a high S/N ratio. Three examples concerned with anisotropy will be discussed: the overall rotations of aniline,[31,44] tryptophan dynamics,[45] and ultrafast energy transfer in bifluorene.[46] Each of these examples involves UV excitation and the upconversion of UV emission, and they are representative of the most challenging experiments yet carried out. In each of these examples the radiative lifetimes are relatively long (~20 ps), so that the signal strengths are not as large as those for dyes absorbing in the visible range.

Figure 10.10 shows the anisotropy decay for aniline in isopentane[33] excited into the 0–0 band region at 330 nm. In this case the fluorescence decay is on the nanosecond time scale, but the anisotropy decays in less than 1 ps, so the best achievable time resolution was needed for this study. Also shown in Figure 10.10 is the instrument response function obtained by detecting the Raman spectrum of the solvent C—H stretching mode. The decay function does not fit predictions from hydrodynamics because the motion of aniline in this solvent is not diffusive. It was suggested that the motion has some inertial character and that aniline rotates through as much as 20° before it suffers a collision that effectively randomizes the angular momentum.

Figure 10.11 shows the decay of $I_\parallel$ and $I_\perp$ for tryptophan in water after excitation at 300 nm.[45] In this case the initial rapid drop in anisotropy sig-

Figure 10.10 Time-resolved fluorescence of aniline in isopentane: (a) $I_{\perp(t)}$; (b) $I_{\parallel(t)}$; (c) instrument function from Raman C—H stretch of isopentane.

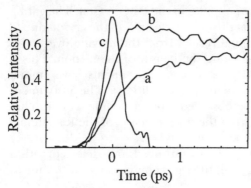

nals the relaxation of tryptophan from the initially excited L_b state to the lower-energy L_a state. The magnitude of the drop in anisotropy provides information on the angle between the L_a and L_b transition dipoles, and the decay time relates to internal conversion. The authors used a high-repetition-rate UV femtosecond pulse source.[45] Many applications of this type of experiment are possible to condensed phase dynamics.

The final example concerns energy transfer between identical chromophores.[46] Bifluorene consists of two fluorenes connected through a single C—C bond (9,9'-). A pulse at 320 nm excites one-half of the molecule. Later, the excitation jumps to the other half, then back to the original ring, and so on. In the overdamped limit the jumps of the excitation are random in time. Since the two halves of the molecule do not have the same orientation in the laboratory frame, the anisotropy decays. In this case the exponential decay time of the anisotropy (400 fs) shown in Figure 10.12 is $\tau_j/2$, where τ_j is the mean time between excitation jumps.[46]

10.3.5. Summary

Fluorescence measurements over ultrashort time scales provides many outstanding research opportunities. As the time resolution becomes shorter, the fluorescence measurement yields more informative functional forms and a measure of the changing structure and energy distribution of the probed molecules and of the surrounding solvent. Since the final state of the fluorescence process is the ground state potential surface,

Figure 10.11 Comparison of the measured parallel and perpendicular fluorescence decays from the 335-nm emission of tryptophan excited with 300-nm light to the simulated curves generated by the single-angle model that includes sample heterogeneity with $\theta = 90°$. (After Ref. 45 with permission.)

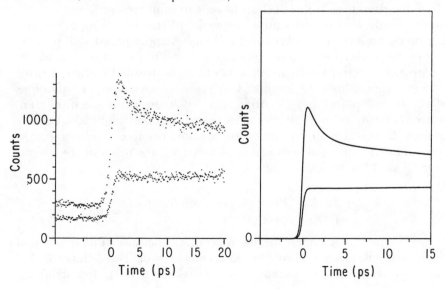

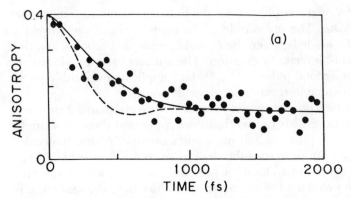

Figure 10.12 Fluorescence anisotropy of bifluorene in isopentane. The dashed and solid lines are for models in which the solvent dephasing of the electronic transition of bifluorene is 70 fs and 100 fs, respectively. (From Ref. 46.)

such rapidly evolving spectral and temporal changes have an excellent chance of being related to theoretical descriptions of condensed phase dynamics.

The prognosis for UV-induced ultrafast fluorescence seems excellent. Therefore, the previously unexplored time scale lying between that deduced from Raman scattering and that corresponding to substantial motion along a reaction coordinate seems accessible with continued effort.

10.4. Raman Spectroscopy

10.4.1. Introduction

Many structural and dynamical properties of molecules are closely correlated with their vibrational frequencies. These properties are defined on time scales as short as tens to hundreds of femtoseconds. Consequently, to study structural properties of states evolving over ultrafast times, time-resolved vibrational techniques are required with picosecond or subpicosecond time resolution. Furthermore, vibrational bands offer greater spectral selectivity than is often obtainable by absorption or fluorescence methods. In addition, with resonant excitation, the vibrations of the chromophore and electronic state of interest are specifically enhanced, so that, for example, the active site of a protein might be selectively probed. In this section, we treat ultrafast time-resolved Raman spectroscopy (TRRS) with readily available ultrafast light sources in the visible, near-IR, and UV regions of the spectrum.

10.4.2. Principles of Time-Resolved Raman Spectroscopy

Raman scattering is generated because of modulation of the polarizability of a molecule by vibrational motion. The differential Raman cross section (in m^2 molecule^{-1} sr^{-1}) is related to the polarizability $\alpha_{\alpha\beta}$

of the molecule by[47]

$$\frac{d\sigma_{\alpha\beta}}{d\Omega} = \left(\frac{2\pi}{\lambda}\right)^4 |\alpha\beta|^2 \rho_{gg} \tag{10.10}$$

where ρ_{gg} is the diagonal density matrix element for the lower level, the polarizability is given by

$$\alpha_{\alpha\beta} = \sum_i \frac{<g|\mu_\alpha|i><i|\mu_\beta|i> + <g|\mu_\beta|i><i|\mu_\alpha|f>}{\hbar(\omega_{ig} - \omega_L)} \tag{10.11}$$

and $d\sigma/d\Omega$ is the differential Raman cross section, typically of the order of 10^{-33} m^{-2} molecule^{-1} sr^{-1} for strong nonresonant Raman scatterers. The cross section can be enhanced by several orders of magnitude by resonant excitation. The time resolution of TRRS in the condensed phase is determined by laser pulsewidths and, in contrast to fluorescence decay measurements, does not rely on the response time of the detector or optical gate. The reason lies in the separation of the time scales that typically characterize the Raman and fluorescent components of spontaneous emission in condensed phases.[40,48,49] The time scale of nonresonant Raman scattering given by $\Delta E/h$ is typically of the order of 10^{-16} s. The time scale characteristic of resonant Raman scattering is determined by the time scale of electronic dephasing, which (in the rapid fluctuation limit) is much faster than the vibrational lifetime.

The development of picosecond and subpicosecond time-resolved Raman techniques has been linked closely not only with the availability of suitable ultrafast laser pulses, but also with the development of sensitive multichannel detectors. A series of papers appeared in the 1960s and 1970s reporting work with television camera tubes (vidicon and image intensifiers).[50,51] Spontaneous Raman spectra of equilibrium states obtained with picosecond laser pulses were reported in the late 1970s.[52–55] Time-resolved Raman studies of species generated by a pump pulse were also reported on longer time scales.[52] The potential of time-resolved Raman methods began to receive widespread attention with the first International Conference on Time-Resolved Vibrational Spectroscopy.[56]

The first picosecond studies of intermediates used single beams of mode-locked laser pulses, which played the roles of pump and probe simultaneously. With low intensity, the contribution of the ground state equilibrium species is predominant. With high intensity, contributions from transient states generated within the pulsewidth by optical pumping increase, and by careful spectral subtraction these contributions were identified. Such spectra are characteristic of a fixed mean time delay after creation of an excited state. Experiments of this type were obtained for transient intermediate states of hemoglobin,[57–61] bacteriorhodopsin,[62,63] and rhodopsin.[64] These studies produced important new understandings of structural features and conformational dynamics associated with ultrafast events and led to a widespread realization of the potential of the technique for the study of fast events.

Picosecond Raman experiments with distinct pump and probe pulses were first reported by Gustafson et al. in 1983.[65] Cavity-dumped pulses from a synchronously pumped dye laser were amplified at 760 kHz in a double-pass dye amplifier. The second harmonic of the amplified pulses was used to excite *trans*-stilbene, while the fundamental at 593 nm was used to generate Raman scattering from the excited state S_1 with 25-ps resolution. The Raman scattering was detected with an intensified Reticon photodiode array. Another approach, based on the harmonics of an active/passive hybrid mode-locked Nd:YAG laser, was introduced by Hochstrasser et al.[66] With 355-nm pump pulses and 532-nm probe pulses, vibrational bands attributed to the 15-ps lifetime S_1 state of benzophenone were observed with a cooled vidicon SIT multichannel detector.

10.4.3. Techniques and Instrumentation:
Time-Resolved Spontaneous Raman Spectroscopy

The signal detected in a time-resolved Raman experiment is

$$S = \frac{d\sigma}{d\Omega} \Delta Q N_t I_p \Phi_e R \qquad (10.12)$$

where $\Delta\Omega$ is the solid angle of collected emission, N_t is the density of transient states, I_p is the probe pulse intensity, Φ_e is the product of the efficiencies of the spectrograph and the detector, and R is the laser repetition rate. Several strategies can be followed to optimize the signal-to-noise ratio obtained in an experiment. The most significant choice is that of the laser system. Here the trade-off between low pulse energies at high repetition rate and higher pulse energies at low repetition rate must be considered carefully. Some transient states are sufficiently photolabile that probe pulse energies must be limited to avoid generation of new states by the probe pulse. (This possibility is in fact the basis of single-pulse transient Raman methods.) A maximum usable probe pulse energy can be estimated by the photolability parameter,[67] given by $F = \Phi_{ph}\sigma(\lambda)I(\lambda)$, where Φ_{ph} is the photochemical quantum yield, $\sigma(\lambda)$ is the absorption cross section, and $I(\lambda)$ is the laser peak intensity. The photolability parameter should be low enough that $F\tau_p \leq 0.1$ (where τ_p is the laser pulse width) to avoid artifacts from optical pumping by the probe pulse. Furthermore, high probe pulse intensities may cause sample heating, or nonlinear frequency or linewidth dependences.[68] The probe intensity must also be maintained below the threshold for stimulated Raman scattering. These advantages of low probe intensities argue for the use of high-repetition-rate systems. Although the correspondingly low pump intensities generate lower values of N_t, it is often possible to compensate by tighter focusing of the pump and probe beams. The range of repetition rates and average pump/probe powers employed in time-resolved Raman experiments is summarized in Table 10.1.

Table 10.1 Instrumentation Used in Ultrafast Time-Resolved Raman Systems

Laser System	Wavelengths	Pulse Width	Pulse Energy	Repetition Rate	Detector	Ref.
Dual cavity-dumped synchronously pumped dye lasers; triple-pass amplifier	210–900 nm	3 ps	200 nJ (590 nm)	10^6 Hz	CCD	82
Dual cavity-dumped synchronously pumped dye lasers	5565–610 nm	6 ps	10 nJ (590 nm)	10^6 Hz	Reticon	123,124
Synchronously pumped dye laser; triple-stage amplifier pumped	283.5–567.0 nm	3 ps	80 μJ (283.5 nm)	50 Hz	Reticon	95
Mode-locked Nd: YAG; regenerative amplifier; pulse compressor	532 nm, 355 nm, 266 nm	10 ps	300 μJ (532 nm)	10^3 Hz	Photomultiplier	103–106
Mode-locked Q-switched Nd: YAG	532 nm, 355 nm	100 ps	1 μJ (542 nm)	800 Hz (pulse trains)	Reticon	85–87
Active-passive mode-locked Nd: YAG	532 nm, 355 nm	25 ps	1 mJ (532 nm)	20 Hz	Intensified vidicon	66

The detectors used in representative time-resolved Raman experiments are also summarized in Table 10.1. The importance of the sensitivity and noise level of the detector is illustrated by the link between the development of time-resolved Raman methods and the availability of multichannel detectors. Initially, vidicon tubes were used. The sensitivity of these tubes can be enhanced by intensifiers (SIT and ISIT tubes), and the dark noise can be reduced by cooling. Increased sensitivity became available with Reticon photodiode arrays, which also could be coupled with an intensifier. These intensifiers can be cooled and gated with gate widths as short as 5 ns for reduction of background noise levels. A disadvantage of these detectors in some applications is that they consist of a linear rather than a two-dimensional array of channels, with a pixel height of 2.5 mm. Hence, the effective slit height available is limited to 2.5 mm, or less if the spectrometer is astigmatic.

More recently, charge-coupled devices (CCDs) have become the multichannel detector of choice for many workers. Although these devices cannot be gated, they can be cooled to temperatures where dark noise levels are extremely low. CCDs are two-dimensional detectors and offer wide flexibility in horizontal or vertical binning of channels. In addition, they are sensitive to longer wavelengths (~1,000 nm) than Reticon detectors. The relative merits of CCD and Reticon detectors depend on the application. Another option, which has not been widely used, is photon-counting image acquisition with a position-sensitive detector. This detector is sensitive to very low light levels. However, this device is not a multichannel detector, and the probability of coincidence of photons within the detector response time must be considered.

Either single- or triple-stage spectrometers can be used with multichannel detectors. The latter allow sharp rejection of Rayleigh scattering by an input double-monochromator stage with canceling dispersion. However, a price is paid in overall throughput. Often, a single spectrograph with a good optical filter for Rayleigh rejection is sufficient. Some workers have preferred to use a scanning double monochromator with single-channel photomultiplier detection. Both 90° and 180° scattering have been employed in TRRS experiments. By use of both cylindrical and spherical lenses, rectangular sample illumination can be achieved in a backscattering geometry and imaged onto the entrance slit of the spectrometer.

10.4.4. New Techniques

New approaches to time-resolved spontaneous Raman scattering can be contemplated, based on recent developments in instrumentation and methodology. For example, the gain in sensitivity by surface-enhanced Raman scattering (SERS)[69] should allow studies of ultrafast processes on metal surfaces such as electrodes or colloidal aggregates. With the availability of ultrashort laser pulses in the UV, time-resolved resonance

Raman techniques can be extended to small molecules in solution. Examples of picosecond time-resolved UV resonance Raman experiments are discussed in the next section. Many more ultrafast studies with UV Raman excitation can be anticipated.

Another attractive recent development in spontaneous Raman methodology is the use of near-IR nonresonant excitation,[70,71] which circumvents limitations imposed by sample fluorescence or sample photolability in resonance Raman experiments. It has become apparent that Fourier-transform (FT) instrumentation possesses substantial advantages for Raman spectroscopy relative to conventional dispersive instruments in this spectral region.[72] These are attractive advantages for time-resolved Raman experiments as well. The availability of mode-locked pulses at 1,064 nm suggests the possibility of nonresonant FT–Raman experiments. The extent of overlap of the pulses from the two branches of the interferometer determines the spectral resolution possible. Since the laser pulse profile is approximately Gaussian, the effect is to weight the interferogram with a Gaussian function, in effect imposing Gaussian apodization, which contributes an instrumental width of $\sim 1/L$, where L is the maximum optical path difference in the interferometer.[73] For 100-ps pulses the achievable resolution is in principle <1 cm^{-1}. The availability of step-scan interferometers will allow FT–Raman experiments to be implemented with pulsed laser excitation.

10.4.5. Coherent Raman Methods

Time-resolved coherent Raman methods can be classified into two groups:

1. Pump-probe experiments where the pump pulse generates a transient state, which is probed by coherent Raman spectroscopy;
2. Time-resolved experiments where one of the time delays between the interactions of the system with the three input fields in a four-wave mixing process is varied.

The latter group includes time-resolved coherent anti-Stokes or Stokes–Raman spectroscopic (CARS or CSRS) measurements of vibrational dephasing times,[74-76] and related techniques. The former type of experiment will be considered here. These methods in principle constitute six-wave mixing experiments, which can, however, be considered as a distinct pump-pulse and a four-wave-mixing probing process.

Coherent Raman methods such as CARS or CSRS generate a signal with a well-defined direction and frequency. Consequently, these methods can be used to discriminate against fluorescence. The direction and frequency of the probing beams can be chosen to select resonances with excited-state or ground-state vibrational modes as well as resonance with excited electronic states. The flexibility comes at the cost of increased complexity. In addition to the pump pulse, beams of at least two different frequencies are required for the probe. In a typical CARS experiment, the ω_1 and ω_3 fields in $\chi^{(3)}$ $(\omega_{as}; \omega_1, \omega_2, \omega_3)$ are generated by the same pulse, and ω_2 is chosen such that $\omega_1 - \omega_2 = \omega$, a vibrational frequency. To record the CARS spectrum at a given time delay, $\omega_1 - \omega_2$ can be tuned through the vibrational region of interest. An alternative approach is to use a broadband ω_2 beam and to record the anti-Stokes (in CARS) or Stokes (in CSRS) beam on a multichannel detector. This method was introduced by Goldberg.[77] A broadband dye laser probe pulse has also been used.[78,79]

An experimental difficulty in coherent Raman measurements in condensed phases is often caused by the non-resonant background susceptibility χ_{nr}. Since the signal generated is determined by $|\chi_v^{(3)} + \chi_{nr}^{(3)}|^2$, interferences are sometimes observed that make interpretation of spectra difficult. The nonlinear background can be suppressed by selecting appropriate polarizations of input beams and the signal beam.[80,81]

10.4.6. Applications

To illustrate the kind of research that can be done with ultrafast time-resolved Raman methods, applications in three areas are considered: (1) excited states, (2) photochemical processes, and (3) ultrafast processes in proteins. In each area, TRRS has had a substantial impact, and the examples cited demonstrate the capability of this technique to contribute to investigations of a wide variety of short-lived states and fast dynamical processes.

10.4.6.1. Excited States

The first pump-probe TRRS experiments on the picosecond time scale concerned the generation of resonance Raman spectra of the S_1 state of *trans*-stilbene in hexane by Gustafson et al.[65,82] Spectra at several time delays were generated by probe pulses at 593 nm, in resonance with an $S_n \leftarrow S_1$ transition. Several S_1 Raman bands were observed, anti-Stokes scattering from S_1 was reported,[83] the excited-state Raman spectrum of diphenylbutadiene was presented,[84] and mode-specific, solvent-dependent relaxation was observed via the excited-state Raman spectrum.[82]

Another application of TRRS has been to the lowest excited singlet and triplet states of carotene and carotenoids. In a series of papers, Koyama and Atkinson et al. have investigated the $2^1A_g^-$ excited state of β-carotene and related compounds.[85–90] The C=C stretching mode observed around $1,777$ cm^{-1} was used to identify the lowest excited singlet state as $2^1A_g^-$ in the carotenoids in photosynthetic systems.[91,92]

Excited states have also been probed by time-resolved CARS. These methods were first applied by Payne and Hochstrasser.[78] Time-resolved resonance CARS has since been employed to study the T_1 state of Ni-octaethylporphyrin.[93,94]

10.4.6.2. Photochemistry

TRRS can provide structural information as well as kinetic data on transient photochemical intermediates. A recent example is the observation of the photochemical ring opening in 1,3-cyclohexadiene by TRRS excited with 3-ps pulses at 283.5 nm.[95] The appearance of a vibrational spectrum assigned to *cis*-hexatriene was observed in 8 ps. Similar observations have been reported for electrocyclic rearrangements in 1,3,5-cyclooctatriene and α-phellandrene, where similar time constants of ∼10 ps were observed.[96]

Dissociation and recombination have been investigated in a series of experiments on I_2. In these experiments, I_2 was dissociated with a 532-nm pulse, and probed in the UV.[97–99] Vibrational relaxation was observed in the X state of I_2 over a period of 100 ps,[97] while transient shifts in CC_4 solvent bands were assigned to heating of the solvent cage within the laser pulsewidth.[98,99]

Organometallic complexes have also been the subject of picosecond TRRS studies. This work extends the earlier nanosecond TRRS experiments on [Ru(bipy)$_3$]$^{2+}$ into the picosecond time regime.[100–102] In the picosecond investigations, Raman scattering was excited at 355 nm, either in single-pulse experiments or after excitation with 532-nm pump pulses.[103–106] Transient Raman bands of the ligands bipyridine, 1,10-phenanthroline, and others reduced by metal-to-ligand electron transfer were observed in several Ru(II) complexes.

10.4.6.3. Protein Dynamics

One of the applications of TRRS that has attracted the most activity and interest in recent years has been in studies of photoactivated transient states in proteins. Picosecond TRRS experiments have been particularly relevant to understanding dynamics in the photoactivated proton pumping protein bacteriorhodopsin, and in studies of ligand dissociation and rebinding in heme proteins. The first applications of picosecond TRRS were in fact in these areas.[57–63]

The goal of time-resolved studies of the heme proteins has been an understanding of the factors controlling ligand binding and dissociation, and the mechanism of the cooperative binding of ligands such as O_2. Time-resolved Raman work on these proteins has focused on the heme core marker bands, and on the Fe–histidine stretching mode.[107,108] In pioneering picosecond TRRS studies of hemoglobin, a species resembling deliganded hemoglobin (Hb) appeared following photolysis of carboxyhemoglobin (HbCO), but several core-size marker modes were shifted to lower frequency relative to their frequencies in equilibrium Hb.[57–61] These bands were found to have relaxed to their equilibrium frequencies not before 20 ns, but within 300 ns.[109–110] In the monomeric heme protein myoglobin (Mb), similar frequency shifts in the core-size marker modes were reported with 25-ps single-pulse TRRS experiments.

The first application of picosecond pump-probe TRRS methods to biological molecules showed that the time scale of relaxation of this metastable state is significantly faster in photolyzed MbCO than in HbCO.[112,113] These results are shown in Figure 10.13. In Raman spectra recorded at 400 ps after photolysis by the 355-nm pump pulse, the core marker modes have already relaxed to their equilibrium positions. A similar result (not shown) was also obtained for a 1-ns delay. In contrast, the Raman spectrum of photolyzed HbCO at 8 ns confirms that the analogous relaxation has not yet taken place. The marker modes detected in these experiments have been correlated with the heme core size, with lower frequencies reflecting an expanded core.[114] The expanded core size following photolysis suggests that full relaxation of the iron out of the heme plane is restricted for at least 30 ps, but for less than 400 ps in Mb, and for at least 20 ns in Hb. This result is consistent with a recent molecular dynamics simulation of MbCO dissociation that predicts that, although the initial motion of the Fe out of the heme plane is very fast (<1 ps), relaxation is not complete by the end of the 100-ps trajectory.[115]

The Fe–histidine stretching mode at ~ 220 cm^{-1} in Mb and Hb has also been the subject of several picosecond Raman studies.[116,117] The frequency of this band was shown to be sensitive to tertiary and quaternary structural changes in photolyzed Hb on nanosecond and microsecond time scales and was suggested to correlate with the tilt of the proximal histidine with respect to the heme plane.[118] In contrast to its evolution in Hb, this mode appears at its equilibrium frequency in metastable Mb 30 ps after photolysis.[118] Thus, like the core marker modes, the Fe–histidine stretching mode responds quite differently in Mb from what it is in Hb.

A concern has been raised about the possible effect of heme heating by the pump pulse, since the photolysis pulse deposits some 11,000 cm^{-1} (at 532 nm) of excess energy into the heme. Molecular dynamics simulations predict cooling of the heme in tens of picoseconds after photolysis.[119] Subpicosecond TRRS experiments have detected a shift in the oxidation-state marker band (v_4) to lower frequency in 200 fs, and subsequent changes on 10-ps and 30-ps time scales.[120,121] The evolution of this band was attributed in part to heme cooling. Vibrationally hot heme has been observed recently in time-resolved Stokes and anti-Stokes spectra of deoxyhemoglobin following excitation at 532 nm.[122] Vibrational cooling is complete in <50 ps, in reasonable agreement with molecular dynamics results.

The early application of picosecond Raman methods to bacteriorhodopsin (BR) by El-Sayed et al. was a key in implicating ultrafast isomerization as the primary ultrafast event.[62,63] (For a review of ultrafast events in BR, see Ref. 108.) Subsequently, time-resolved pump-probe experiments by At-

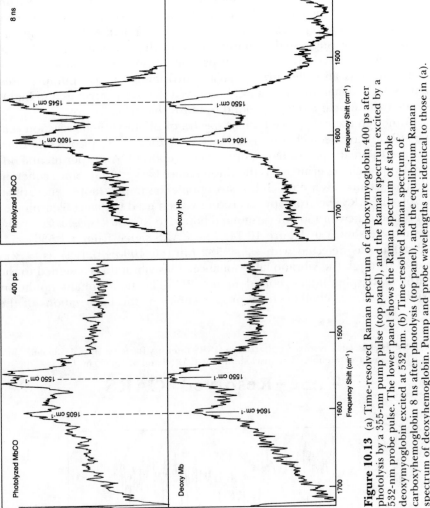

Figure 10.13 (a) Time-resolved Raman spectrum of carboxymyoglobin 400 ps after photolysis by a 355-nm pump pulse (top panel), and the Raman spectrum excited by a 532-nm probe pulse. The lower panel shows the Raman spectrum of stable deoxymyoglobin excited at 532 nm. (b) Time-resolved Raman spectrum of carboxyhemoglobin 8 ns after photolysis (top panel), and the equilibrium Raman spectrum of deoxyhemoglobin. Pump and probe wavelengths are identical to those in (a).

kinson et al.[123,124] with 5-ps resolution have generated the Raman spectrum of the K610 intermediate of BR. Recently, subpicosecond transient Raman experiments were reported, confirming isomerization in <1 ps.[125] The fingerprint region suggested a conformation of the polyene chain different from that observed after about 10 ps. Time-resolved anti-Stokes Raman scattering was also measured.[126] With sufficiently intense pump pulses, anti-Stokes Raman intensities were observed that were attributed to highly excited vibrational levels of ground state BR that relax in 7 ps.

Although the resonance Raman spectra of photosynthetic reaction centers (RCs) have been characterized by CW methods,[127,128] picosecond time-resolved Raman spectra have not yet been reported. An experimental difficulty in such experiments is the fluorescence background. Since coherent Raman techniques are much less susceptible to sample fluorescence, time-resolved CARS spectroscopy has recently been used to detect bleaching of a ground state vibration of carotenoid bound to RCs of *R. spheroides*.[129] The results are shown in Figure 10.14. The CARS signal from a 1,536 cm^{-1} vibration was detected with ω_1 = 532 nm (80–100-ps pulses) as ω_2 was tuned through the vibrational resonance. This vibration is assigned to the carotenoid spheroidene bound to the RC. The nonresonant nonlinear background was eliminated by proper choice of the polarizations of the

Figure 10.14 Time-resolved polarization CARS intensity for $\omega_1 - \omega_2 = 1,536$ cm^{-1} in a reaction center carotenoid. The inset shows the CARS band at 1,536 cm^{-1}.

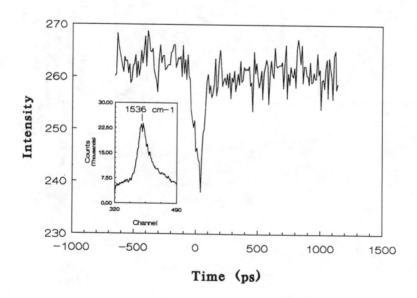

CARS beams: ω_1 and ω_2 were set at $\pm$ 30°, and the anti-Stokes signal was detected through an analyzer with polarization set to 90°. The sample was excited with 532-nm pulses. Fast recovery of the electronic ground state due to energy transfer following excitation is demonstrated by bleaching and recovery with the pulsewidth.

10.4.7. Concluding Remarks

Ultrafast TRRS has come of age during the last decade and is now able to address problems in a wide range of chemical applications, including the structures of excited states and photochemical intermediates, vibrational relaxation processes, and fast events in proteins. Expansion into the UV and near-IR spectral regions will increase its range of applicability, and time resolution can be expected to approach the limit at which vibrational frequencies can be measured. In practical terms, however, the viability of new applications of TRRS continues to depend on advances in detectors and ultrafast light sources.

10.5. Time-Resolved Spectroscopy

10.5.1. Introduction

Because of recent technological advances, time-resolved mid-IR spectra may be recovered with subpicosecond time resolution. The extension of femtosecond spectroscopy to the IR portion of the electromagnetic spectrum represents a significant breakthrough: The IR spectrum reveals vibrational features that can often be assigned to unique molecular structures. Therefore, time-resolved IR spectroscopy can reveal the identity of short-lived photo-generated intermediates as well as their dynamics. This rapidly developing field of research is spawning incisive investigations into photophysical, photochemical, and photobiological processes.

This section will focus on picosecond and subpicosecond time-resolved IR spectroscopy because most photo-initiated molecular transformations occur on this time scale. Microsecond time-resolved techniques typically address only diffusion-controlled processes. Time-resolved IR spectroscopy, in the context of this section, refers to the recovery of transient IR spectra following ultraviolet or visible excitation.

Femtosecond IR methods have achieved a time resolution approaching 200 fs (the instrument response function FWHM), spectral coverage as broad as 2.3–5.5 μm, and noise levels of $\sim 10^{-5}$ (in optical density units).

The most important technological challenges remaining in this field involve improvements in spectral coverage and sensitivity. Improvements in time resolution will require increased spectrometer sensitivity.

10.5.2. Principles of Time-Resolved IR Absorption

Mid-IR (2.5–15.4 μm) absorption spectra reveal rather well-resolved vibrational features that are indicative of specific functional groups (even in the condensed phase). Indeed, a portion of the mid-IR spectrum (8.0–15.4 μm) is often referred to as the "fingerprint" region. Femtosecond time-resolved IR spectra may therefore reveal both the identity and dynamics of short-lived intermediates.

Conventional IR spectra reveal rovibrational features whose center frequencies and lineshapes are influenced by their bonding, environment, and motional dynamics. The spectrum observed therefore reveals static as well as dynamical information. Indeed, in the Heisenberg picture, a vibrational spectrum is simply the Fourier transform of the transition dipole time correlation function[130]:

$$I(\omega) = \frac{1}{2\pi} \int_{-\infty}^{\infty} e^{-i\omega t} <\mathbf{u}(0) \cdot \mathbf{u}(t)> \, dt \tag{10.13}$$

where $\mathbf{u}$ is a vector along the direction of the dipole moment and ω corresponds to the center frequency for the transition. While the limits for integration extend from $t = -\infty$ to ∞, the integrand is nonzero only over a finite range of times about $t = 0$ due to damping of the dipole correlation function. The more rapid the damping, the broader the spectral linewidth. Transform-limited spectra [i.e., spectra that obey equation (10.13)] can be obtained when the illumination period is longer than the damping time. Illumination periods shorter than the damping time cause the spectrum to broaden. For example, a molecule that quickly traverses the illuminated area experiences a short illumination period, resulting in a "transit time" broadened spectrum.

What are the consequences of time-resolved IR measurements when the transitions probed are destroyed, created, and/or evolve on ultrafast time scales? The differential transient absorption spectrum (pumped minus unpumped absorbance) reveals both negative- and positive-going features,[131] where a negative feature *(bleach)* corresponds to a transition destroyed at time $t = 0$. The spectrum of the bleach reproduces the conventional absorption spectrum of the ground-state species. Newly formed species, however, exhibit broadened spectra, analogous to transit-time broadening. Their spectral widths decrease and their maxima grow with time (the integrated intensity is constant during the spectral evolution) for a period of the order of their dephasing time. Consequently, the spectral features for new absorptions are broad (and small) at short times, adversely affecting their detectability. For example, the spectrum of an intermediate existing for 100 fs must be at least 50 cm^{-1} broad but a species living 1 ps may reveal a spectral linewidth as narrow as 5 cm^{-1}, on the order of the narrowest vibrational features observed in condensed media at ambient temperatures. Since the detectability of a feature is related to its peak absorbance, improved time resolution is practical only to the extent that sensitivity can be increased.

Table 10.2 summarizes the frequencies, linewidths, and oscillator strengths for several electronic and vibrational transitions. The linewidths correspond to the full-width at half-maximum (FWHM), ε_{max} refers to the molar absorption coefficient at the peak of the transition, and the integrated absorbance is simply $\int \varepsilon(\bar{\nu}) \, d\bar{\nu}$. The linewidths for vibrational transitions are typically much narrower than those for electronic transitions. For example, the CO linewidths in Table 10.2 are about two orders of magnitude narrower than the Soret $\pi \to \pi^*$ linewidth. Narrow spectroscopic features allow subtle environmental factors, such as the those that partition CO oscillators into A and B states, to be resolved: the A states arise from differing bond orientations with the heme iron.[132,133] Moreover, narrower linewidths enlarge ε_{max}. This is important because the magnitude of ε_{max}, not the integrated absorption, dictates whether or not a spectroscopic transition can be observed above the experimental noise level. (Note that ε_{max} for CO is only about 200 times smaller than that for the Soret band in spite of the $\sim 10^5$ difference in their integrated absorption).

The comparisons made between electronic and vibrational transitions in Table 10.2 involve the C—O stretch, one of the strongest vibrational transitions known. The CO transition moment is further enhanced (~ 20 times) upon bonding to the heme iron in HbCO and MbCO: The covalent bond with iron decreases the bond strength of C—O and increases its dipole moment. Since experimental sensitivity is dependent, in part, on the magnitude of ε_{max}, the earliest time-resolved IR investigations probed mainly CO stretching vibrations in metal carbonyls.

10.5.3. Time-Resolved IR Methodology: Pump-Probe Spectroscopy

Photodetectors that respond to electromagnetic radiation in the mid-IR region do not possess sufficient bandwidth to follow picosecond or subpicosecond intensity modulation. Consequently, ultrafast IR spectroscopy must employ the pump-probe technique. Briefly, a pump pulse is focused into the sample to initiate a photoreaction. A variably delayed probe pulse is routed through the photoexcited sample volume and

Table 10.2 Summary of Porphyrin[134,135] and CO Transitions for Carboxyhemoglobin (HbCO), Carboxymyoglobin (MbCO), and Solid CO

Species	Feature	Transition Wavenumber (cm⁻¹)	Linewidth (FWHM, cm⁻¹)	ε_{max} (m² mol⁻¹)	Integrated Absorbance (m² mol⁻¹ cm⁻¹)
HbCO	Soret band ($\pi \rightarrow \pi^*$)	23,900	~1 × 10⁻³	19,200	2 × 10⁷ᵃ
	Q_0 band	17,600		1,390	
	Q_v band ($\pi \rightarrow \pi^*$)	18,600		1,390	
	CO stretch[136]	1,951	8	370	3,400
MbCO	CO stretch[110]				3,040
	A_0	1,969	8		
	A_1	1,944.8	8		
	A_3^b	1,927.0	17		
Mb·CO	CO stretch[137]				140
	B_0	2,146	12		
	B_1	2,130.5	5		
	B_2	2,119	4.5		
Solid CO	CO stretch (30 K)[138]	2,138	2.8	84ᵃ	250

Note: Mb·CO refers to photodissociated MbCO with the ligand located within the heme pocket. The integrated CO absorption given for MbCO and Mb·CO is distributed over the *A* and *B* states, respectively.

ᵃCalculated from other two parameters; Gaussian lineshape assumed.

ᵇThe 1,927.0 transition has been relabeled as A_3 to be consistent with recent usage.

monitored with a photodetector. The photodetector signal is usually normalized with respect to the incident probe pulse energy to obtain T^p and T^u, the normalized probe pulse transmission with and without photolysis (pumped and unpumped). These signals are used to calculate

$$\Delta A = \log \left(\frac{T^u}{T^p} \right) = \sum_i \sigma_i(\bar{\nu}) \, [N_i^p(t) - N_i^u] \tag{10.14}$$

where ΔA is the differential absorbance (pumped minus unpumped), σ_i is the cross section associated with species i at wavenumber ñ, and $N_i(t)$ is the number of species i divided by the cross-sectional area of the illuminated volume at time t. The quantity ΔA is recorded as a function of probe wavelength and relative delay to yield time-resolved spectra. The time resolution of the pump-probe method is determined by the cross correlation of the pump and probe pulses, not the photodetector bandwidth: The photodetectors need not be fast because the information required to determine the sample transmission is contained in the integrated probe pulse energy, not the temporal envelope of the pulse.

The pump-probe method has been employed extensively in time-resolved ultraviolet and visible spectroscopy. Successful application of this method to the IR region, however, is more difficult. Significantly more pump pulse energy could be needed in transient IR experiments because IR radiation cannot be focused as tightly as UV or visible light. The diffraction-limited spot size for a focusing lens is proportional to wavelength. Since the number of photons needed to bleach the sample is of the order of the number of molecules within the illuminated volume, a transient IR experiment requires more pump pulse energy than is needed for UV-visible pump-probe experiments carried out at the diffraction limit.

The fact that IR transitions are generally much weaker than electronic transitions introduces an additional complication. The experimental sensitivity is greatest when the relative cross section for pumped and probed transitions is of the order of 0. 5. Consequently, a sample that is pumped and probed collinearly may require excitation near the edge of the electronic transition. Alternatively, the sample cell can be pumped transversely. In this case, the sample concentration is adjusted so that the penetration depth of the pump pulse approximates to the probe beam waist, and the cell path length is set to achieve an optical density of the order of 0.5 at the probe wavelength. The pump pulse is then focused with a cylindrical lens along the side of the cell such that the excited volume envelopes the probe beam waist. To preserve ultrafast time resolution, however, transverse pumping must employ a traveling wave cell design,[139] so that the pump photons arrive along the length of the sample cell at a rate corresponding to the group velocity of the probe pulse.

Two basic experimental approaches, differing primarily in the method of IR detection, have been used to recover picosecond and subpicosecond IR spectra. They are the *pulsed IR probe method* and the *IR upconversion method*.

10.5.3.1. Pulsed IR Probe Method   An ultrashort pump pulse is used to excite the sample and a variably delayed IR pulse is routed through the illuminated volume. If the IR pulse is broadband, it must be dispersed in a monochromator before impinging on a cooled IR photodetector; otherwise the transmitted probe pulse is detected without additional filtering. Time-resolved spectra are constructed, point by point, by recording the pump induced change in sample absorbance as a function of the probe wavelength and delay time. This method can achieve subpicosecond time resolution only when the pump and IR probe pulses are derived from the same optical pulse. (Light pulses generated in independent lasers invariably exhibit timing jitter on the picosecond time scale.)

Various methods for generating ultrashort mid-IR pulses have been developed. Glownia et al. generated 160-fs broadband near-IR pulses by Raman shifting an intense 308-nm pulse in Ba vapor.[140] The pulses were not tunable, but covered the entire spectral range from 2.2 to 2.7 μm. The detection scheme employed by this group did not use IR detectors. Instead, the broadband IR was upconverted into the visible and detected with an optical multichannel analyzer (OMA). Their experimental apparatus is quite elaborate and includes the following: a colliding-pulse mode-locked (CPM) dye laser; a femtosecond dye amplifier; a frequency doubler for generating femtosecond 308-nm pulses; a 308-nm excimer (XeCl) amplifier; a Ba vapor downconverter for frequency shifting 308 nm to the IR; an Rb upconverter for shifting the IR back into the visible by mixing with the output from an excimer (XeCl) pumped dye laser; a high-pressure H_2 Raman cell for frequency shifting 308-nm radiation to 248.5 nm and a 248.5-nm excimer (KrF) amplifier.

Moore and Schmidt reported the first broadly tunable source for femtosecond IR pulses.[141] A femtosecond pulse from a CPM laser was amplified to 25 μJ in a three-stage amplifier that was pumped with an excimer (XeCl) laser. A portion of the amplified pulse was focused into a 1-cm cell containing 200-proof ethanol to generate a femtosecond continuum. The continuum was subsequently difference-frequency mixed with the other portion of the amplified pulse in a $LiNbO_3$ crystal. The downconversion process provided nanojoule energy levels and tunability over a range extending from 1.7 to 4.0 μm. The mid-IR pulse was less than 200 fs in duration and possessed a bandwidth of 80–100 cm^{-1}.

Jedju and Rothberg, using a similar approach, downconverted a water-cell-generated continuum into the mid-IR by difference-frequency mixing with a femtosecond pulse in $LiIO_3$.[142,143] The femtosecond pulses were generated in a single-jet dye laser that was pumped with compressed and doubled output from a CW mode-locked Nd:YAG laser. The dye laser pulses (~350 fs) were amplified to 0.5–1.0 mJ in a three-stage amplifier that was pumped with the frequency-doubled output from a Q-switched Nd:YAG laser. The scheme produced ~350 fs ~500 cm^{-1} broad IR pulses that were tunable from 2.3 to 5.5 μm and contained ~5 nJ of energy (with amplification of the continuum).

Vanherzeele reported on difference-frequency mixing and parametric generation–amplification in $KTiOPO_4$.[144] He generated 2–5 ps pulses with energies in the hundreds of microjoules that were tunable in the 0.6–4.0-μm range. The necessary wavelengths for mixing were generated with a CW mode-locked Nd:YLF laser, a synchronously pumped picosecond dye laser, and a seeded Nd:YLF regenerative amplifier.

Elsaesser et al. generated 8-ps IR pulses with a spectral bandwidth of ~6.5 cm over a range extending from 3.9 to 9.4 μm by difference-frequency mixing in $AgGaS_2$. The pump and signal pulses used in the downconversion process were derived from the amplified output of a pulse-selected passively mode-locked Nd:YAG laser and a broadly tunable (1,200–1,460 nm) traveling-wave IR dye laser, respectively. Because of parametric amplification in the nonlinear crystal, the generated mid-IR pulse is intense (pump energy is converted with efficiencies of a few percent) and highly collimated.

Spears et al. developed an efficient two-step downconversion scheme for generating tunable narrowband mid-IR pulses.[146] The first step produced a near-IR pulse by difference-frequency mixing the amplified output from a short-cavity dye laser with an intense IR pulse in $LiLO_3$. The near-IR pulse was subsequently downconverted to the mid-IR by difference-frequency mixing with a second intense IR pulse in $AgGaS_2$. A pulse-selected passively mode-locked Nd:YAG laser provided the pulse energies needed to pump the short-cavity dye laser and the two nonlinear mixing processes. The temporal duration and spectral bandwidth of the tunable IR pulses were 23 ps and ~5 cm^{-1}, respectively. A spectral range from 5.2 to 6.4 μm was achieved by tuning the Rhodamine 6G short-cavity dye laser from 579 to 591 nm while simultaneously rotating the two nonlinear crystals to maintain optimum phase-matching. Tunability over a range extending from 4 to 11 μm is, in principle, possible by changing the dyes in the short-cavity dye laser oscillator–amplifier.

We recently reported a vibrational T_l time measurement made with an apparatus which generates independently tunable pump and probe pulses at 500 Hz.[147] The idlers for each of two $LiLO_3$ optical parametric oscillators (OPO) were provided by two dye lasers pumped by a mode-locked, Q-switched Nd:YAG laser, which also seeded a regenerative amplifier. Half of the frequency-doubled output of the regenerative amplifier was used to amplify the output from one dye laser while the other half sequentially pumped the two OPOs. The infrared output of the first OPA was a 250-nJ pulse at 1,975 cm^{-1} with 25-ps duration, which was used to excite the sample vibrationally. The second OPO, which was not optimized for output power, provided an estimated 1-nJ probe pulse at 1,963 cm^{-1}. The tuning range of this apparatus was not explored, but infrared output tunable over the useful range of $LiLO_3$ should be obtainable with available dyes. The characteristics of all the preceding schemes are summarized in Table 10.3.

10.5.3.2. IR Upconversion Method The second approach to recovering ultrafast time-resolved spectra employs time-gated upconversion of a tunable CW (or quasi-CW) probe. An upconversion method was used to measure subpicosecond time-resolved *electronic* spectra using an argon laser and a high-repetition-rate dye laser.[148] This principle was recently extended into the fundamental IR region, and time-resolved mid-IR spectra have been recovered with picosecond[149] and femtosecond[150] time resolution.

In the upconversion approach (see Figure 10.15) the sample is pumped with an ultrashort pulse and probed with narrowband radiation from a tunable IR source. The transmitted IR beam is combined with a variably delayed replica of the pump pulse in a nonlinear crystal that is phase-matched for sum-frequency generation. The portion of the IR radiation that arrives in coincidence with the ultrashort pulse is "upconverted" into the visible region of the electromagnetic spectrum and can be monitored with a conventional photomultiplier tube (PMT). Nonlinear IR upconversion can be quite efficient and yields three useful results. The intensity of the IR probe beam can be measured over a time interval as short as the upconverting pulse, the upconverted probe can be detected with visible photodetectors instead of less sensitive liquid N_2-cooled IR detectors, and the spectral resolution attained is limited only by the spectral bandwidth of the probe source. Time-resolved IR spectra are constructed by recording the differential (pumped minus unpumped) upconverted intensity as a function of probe wavelength and relative delay between the pump and upconverting pulses.

The time resolution of this technique is determined by the cross correlation of the pump and upconverted pulses. When the upconversion method employs a single light pulse for both pumping the sample and gating the IR detection, the cross correlation may be as narrow as the ultrashort pulse autocorrelation. Several criteria must be met to achieve this limit. First, the upconversion efficiency must be maintained within the linear conversion regime: High-efficiency sum-frequency generation yields an upconverted pulse that is broader than the upconverting pulse and therefore compromises the time resolution. Second, the path length through the nonlinear medium must be short enough to minimize broadening of the upconverted pulse due to group velocity dispersion (GVD): GVD causes the IR radiation and upconverting pulse to propagate at different rates in the nonlinear medium and broadens the duration of the upconverted pulse. Third, the nonlinear interaction must be properly phase-matched

Table 10.3 Characteristics of Ultrashort Mid-IR Pulses Generated by Nonlinear Optical Methods

Nonlinear Medium	Pulse Duration (ps)	Spectral Bandwidth (cm^{-1})	Tuning Range (μm)	Pulse Energy (μJ)	Repetition Rate (Hz)	Ref.
Ba vapor	0.16	3,700	2.2–2.7	4,000		160
$LiNbO_3$	0.2	80–100	1.7–4.0	0.002	100	141
$LiIO_3$	0.35	500	2.0–5.5	0.005	10	142,143
$KTiOPO_4$	2–5	Nearly transform-limited	0.6–4.0	100	10	144
$AgGaS_2$	8	6.5	3.9–9.4	50		145
$LiIO_3$ + $AgGaS_2$ (two stages)	23	5	5.2–6.4	35	10	146
$LiIO_3$	25	6	2.0–5.5	0.25	500	147

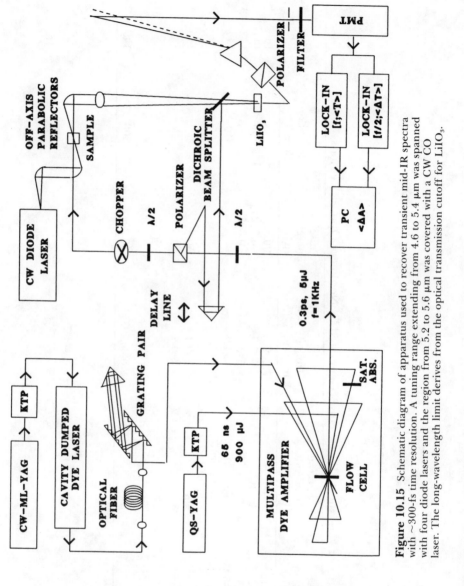

Figure 10.15 Schematic diagram of apparatus used to recover transient mid-IR spectra with ~300-fs time resolution. A tuning range extending from 4.6 to 5.4 μm was spanned with four diode lasers and the region from 5.2 to 5.6 μm was covered with a CW CO laser. The long-wavelength limit derives from the optical transmission cutoff for LiIO₃.

over the entire spectral bandwidth of the ultrashort pulse: an acceptance bandwidth that is narrower than the upconverting pulse bandwidth yields a broadened upconverted pulse. When these conditions are met, the up-converted pulse envelope is a replica of the pump pulse envelope and the time resolution is limited by the pump pulse autocorrelation.

The upconversion approach can be accomplished with a variety of IR sources. One should adhere to several criteria, however, to ensure effective implementation of the method. First, the source should generate an IR beam that can be focused to a beam waist commensurate with the amount of pump pulse energy available. Second, the intensity of the source should be sufficiently high to reduce shot-noise to a level that is less than or com-parable with other noise sources (the shot-noise being inversely propor-tional to the square root of the number of upconverted photons detected). However, the average probe flux must not be too great, as excessive heating of the sample may occur when the probe is tuned to a sample absorption. Heating problems may be alleviated by using a pulsed (quasi-CW) source or by reducing the duty cycle of the CW source with acousto-optic modu-lators coated for IR wavelengths. Third, the spectral bandwidth of the probe laser should be nearly transform-limited (i.e., the spectral bandwidth of the probe should be equivalent to the Fourier transform of its temporal width). When the probe is not transform-limited, its intensity envelope is modulated because of interference among longitudinal modes. This mod-ulation appears as pulse-to-pulse fluctuations in the upconverted intensity and limits the ability to record the differential sample transmission accu-rately.

Since the time resolution for upconversion experiments is not dictated by the probe duration, the IR probe may be CW or quasi-CW. Conse-quently, IR upconversion approaches have been based upon a number of IR sources, including CW[149,150] and pulsed[151] diode lasers, CO lasers,[152] and IR pulses generated via nonlinear downconversion.[153] Other potential can-didates include F-center lasers or free-electron lasers (FEL). When the mid-IR source is pulsed, the pulse duration must be broader than the timing jitter between the IR source and the upconverting pulse.

The upconversion technique pioneered by Hochstrasser et al.[149] was based upon a CW diode laser probe beam. Cryogenically cooled lead-salt diode lasers generate narrow-bandwidth IR and can be "compositionally tuned" over a broad spectral range extending from 2.7 to over 30 μm.[154] A particular composition may be temperature- and/or current-tuned over a range of some tens to hundreds of cm^{-1}. The shot-noise associated with efficient subpicosecond upconversion of single-mode CW Pb-salt lasers (emitting from 1 to 1,000 μW) is large enough to limit the spectrometer sensitivity. Higher peak power and hence lower shot noise is attained by pulsing the diode laser. However, pulsed operation is usually accompanied by multimode diode laser emission, which reduces the spectral resolution. Moreover, interference among randomly phased longitudinal modes introduces an additional source of noise. The noise from these random sources has been reduced to an acceptable level ($<10^{-4}$ in optical density units) through extensive signal averaging.[151]

Hochstrasser et al. also incorporated a grating-tuned CW CO laser into their femtosecond IR spectrometer.[152] The CO laser generated tens to hundreds of milliwatts of single-mode IR power and therefore provided improved sensitivity relative to diode laser sources. The tunability, however, was quite limited. For example, waveguide CO lasers operating near ambient temperature are line tunable from ~5 to 6.4 μm (the rovibrational lines are spaced by ~4 cm^{-1}).

Difference-frequency mixing schemes provide perhaps the most versatile source of tunable IR radiation, and several methods have already been discussed in Section 10.5.3.1. However, IR pulses appropriate for use with the upconversion method should be spectrally narrow, their duration should be broad relative to the timing jitter between the IR and upconverting pulses, and pulse-to-pulse energy fluctuations should be minimized. Rothberg et al. developed a difference-frequency mixing scheme for generating ~70-ps mid-IR probe pulses that were tunable from ~4.4 to 5.5 μm, exhibited a linewidth of less than 0.4 cm^{-1}, were well synchronized with tunable picosecond upconverting pulses, and featured amplitude fluctuations of ± 19%.[153] The tuning range could be extended to wavelengths shorter than 4.4 μm by changing the dye in the tunable source used for downconversion. The long-wavelength limit of 5.5 μm, however, was a consequence of the transmission characteristics for the nonlinear medium used ($LiIO_3$).

F-center lasers offer continuous tunability over the broad homogeneous linewidth of the color center and can achieve single-mode power in the tens to hundreds of milliwatts. Their spectral range, however, is limited to the short-wavelength portion of the mid-IR spectrum. For example, RbCl:Li is tunable from ~2.6 to 3.3 μm[155]; other color centers provide tunable output at shorter wavelengths.

Another potential candidate, the free electron laser (FEL), may be capable of generating 10-ps IR pulses over a spectral range extending from 3 to 50 μm with as much as 100 μJ of energy per pulse (3–8 μm) at a repetition frequency of 36.6 MHz (preliminary specifications).[156] The spectral coverage achieved with femtosecond upconversion of a FEL source would be restricted, however, because of limitations in the phase-matching range for nonlinear media.

10.5.4. Time-Resolved IR Results

The first femtosecond near-IR investigation did not involve the fundamental region, but measured a rate of internal conversion between two electronic states.[157] Since that time a number of other picosecond and femtosecond IR approaches in the fundamental region have been exploited to reveal the orientation of vibrational dipoles, the dynamics of CO motion in heme proteins, the photodissociation and geminate rebinding dynamics of transition metal carbonyls, and carrier relaxation in semiconductors. These studies have only begun to tap the full potential of ultrafast time-resolved IR methodologies.

Femtosecond near-IR spectroscopy was used to determine the internal conversion rate of photoexcited 1,4-diazabicyclo[2.2.2]octane (DABCO) vapor.[157] The sample was pumped into the $\tilde{B}$-state with a 250-fs UV pulse (248.5 nm) and probed with a 160-fs broadband IR continuum (2.2–2.7 μm). The transmitted IR continuum was upconverted into the visible via stimulated electronic Raman scattering in Rb vapor and detected with an OMA. An infrared absorption at ~2.5 μm appeared within 500 fs and was interpreted as a $\tilde{B} \leftarrow \tilde{A}$ transition between vibrationally excited levels. This implied that $\tilde{B} \rightarrow \tilde{A}$-state internal conversion occurred in less than 500 fs.

The iron-carbonyl geometries in MbCO, HbCO, and protoporphyrin-CO (PCO) have been determined by means of picosecond time-resolved IR spectroscopy.[132,149,158] Polarized visible and IR beams were used to photodissociate the heme—CO bond and probe the bound CO absorbance, respectively. The ratio of parallel and perpendicular polarized IR absorbance (relative to the pump polarization) following laser photolysis is related to the angle between the vibrational dipole and the heme normal. The polarized IR absorbance was recorded with picosecond time resolution via upconversion of a CW IR probe beam from a tunable diode laser: the calculated bond geometries are summarized in Table 10.4. Time-resolved IR absorption measurements were necessary because the anisotropy created by polarized excitation must be measured before it is destroyed by rotational diffusion: The rotational diffusion time for PCO in an aqueous solution containing 70% methanol at ambient temperature is ~300 ps.[158] In Table 10.4 α refers to the inverse cosine of the ensemble-averaged RMS angle between the vibrational dipole and the heme normal. For a dipole direction randomly distributed within a cone that is oriented along the direction of the heme normal, the semiangle of this cone is given by β. In recent work of 200-fs time resolution the heme was shown to have a liberation amplitude of less than 8°, indicating crystal-like rigidity in the solution phase protein.

Table 10.4 Iron–Carbonyl Bond Geometry of Carboxyhemes

Species	HbCO	MbCO		PCO		
		A_1	A_3	90% Glycerol	60% Ethylene glycol	70% Methanol
$\nu(CO)/cm^{-1}$	1951	1,944	1,933	1962	1962	1966
α/degrees	18[132],17 ± 2[159]	20	35	25	21	16
β/degrees	24	29	51	36	30	23

The dynamics of ligand motion following photolysis of HbCO were investigated with ~300-fs time resolution by direct detection of 300-fs IR probe pulses[160,161] and by upconversion of a quasi-CW IR probe.[150–152] The femtosecond IR probe method reported a bleach at 1,952 cm^{-1}, a metastable intermediate at 2,009 cm^{-1} that disappeared with a 2-ps time constant, and a "free" CO absorption at 2,132 cm^{-1} that was observed after 5 ps and persisted for at least 500 ps. The upconversion approach offered nearly an order of magnitude better sensitivity because the data were accumulated at a 1-kHz rate rather than at 10 Hz, the repetition frequency of the femtosecond IR probe method. A kinetic study showed that the 1,951 cm^{-1} bleaching recovery was slow: less than 4% of dissociated CO rebound in 1 ns.

The metastable feature found with the IR probe pulse method was not observed with the more sensitive upconversion method. Rather, a broad short-lived absorption (see Figure 10.16) appeared and disappeared with the excitation pulse. This signal is now known to correspond to two-photon absorption[162] of the IR pulse induced by the presence of the pump. It is observed over a wide spectral region from 1,820 to 2,100 cm^{-1}. The spectrum of the "free" CO appeared within 250 fs, was ~25 cm^{-1} broad, and approximately bracketed the *B* states observed in matrix-isolation studies

Figure 10.16 IR optical two-photon absorption of deoxyhemoglobin carried out with a variable delay between a 300-fs optical pulse (527 nm) and a gated IR CO laser pulse at 1,855 cm.

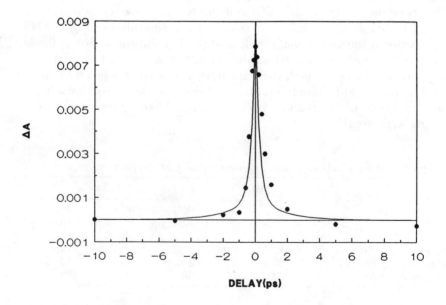

1855cm^{-1}

DELAY(ps)

of photodissociated HbCO.[163] However, the spectral resolution attained with pulsed diode lasers (~ 12 cm^{-1}) was insufficient to resolve individual *B* states. The integrated intensity of the "free" CO diminished by $\sim 15 \pm 10\%$ at 1 ns, suggesting that a small portion of dissociated CO escapes from the heme pocket on this time scale.

The photochemistry of [CpFe(CO)$_2$]$_2$[Cp = η^5-C$_5$H$_5$] in cyclohexane has been investigated with picosecond[164] and femtosecond[165,166] time-resolved IR spectroscopy. An ultrashort visible pulse photolyzed the sample and transient IR spectra in the carbonyl stretching region were recovered using gated upconversion of a tunable CW IR diode laser (see Figure 10.17) or a CO laser probe. The picosecond investigation revealed a spectrum at long times (> 100 ps) that resembled that observed with microsecond[167] time-resolved IR spectroscopy. Spectra accumulated at earlier times, however, revealed broadened features that were indicative of new transient intermediates. These features were resolved in the femtosecond time-resolved studies, and some of the primary processes in the photodissociation of the dinuclear transition metal carbonyl were identified. Absorption of a visible photon (580 nm) produced an excited dimer that persisted for ~ 1 ps. The excited molecule dissociated with high quantum efficiency by breaking one or more bonds, including the doubly bridged CO bonds and the Fe—Fe bond. Approximately 50% of the dissociated molecules recombined geminately to form "hot" ground-state molecules that cooled by transferring vibrational energy to the solvent with a time constant of ~ 30 ps. The other photochemical pathway involved homolytic cleavage to form hot "naked" radicals that became solvated within 5–10 ps. The solvated radicals cooled,

Figure 10.17 Transient IR spectra of Fe$_2$(CO$_4$)Cp$_2$ follow optical excitation using a diode laser source. (After Ref. 166.)

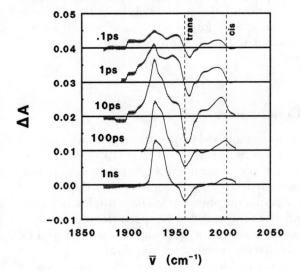

$\bar{v}$ (cm^{-1})

diffused apart from one another, and ultimately dimerized on a bimolecular (microsecond) time scale. These experiments dramatize the importance of ultrafast IR methods in determining the primary processes of photoreactions in solution.

Spears et al. investigated the photochemistry of $Cr(CO)_6$ in nonpolar cyclohexane with picosecond IR spectroscopy.[168] The sample was photolyzed with ultraviolet (266-nm) pulses and probed with tunable infrared (5.2–6.4 μm) pulses. The pulse duration in either case was ~20 ps, and the bandwidth of the probe was ~5 cm^{-1}. Photolysis at UV wavelengths in condensed phases resulted in the dissociation of a single carbonyl and presumably generated a "naked" $Cr(CO)_5$ molecule. They recorded the transient IR absorbance at 1,986 cm^{-1}, 1,980 cm^{-1}, 1,920 cm^{-1}, and 1,961 cm^{-1}, which reportedly correspond to the IR absorption maxima for $Cr(CO)_6$, square pyramidal (SP) $Cr(CO)_5$, trigonal bipyramid (TBP) $Cr(CO)_5$, and $[Cr(CO)_5] \cdot C_6H_6$ (solvated fragment), respectively. Their data suggested that the yield of TBP is approximately twice that of SP, and TBP becomes solvated 2.3 times faster than SP. Additional studies in polar tetrahydrofuran (THF)[169] suggested that the SP reacted with the O site of THF 37 times faster than the H site, while the TBP complex reacted with the O site only 2.7 times faster than the H site. Of the two "naked" complexes, TBP $Cr(CO)_5$ was more reactive to THF.

A picosecond IR study of *trans*-polyacetylene provided evidence for the existence of photogenerated charged solitons.[170] The sample was pumped with a polarized visible (575-nm) pulse and probed with an ultrashort mid-IR (2.5–5.5-μm) pulse. That work was extended into the femtosecond regime with oriented *trans*-polyacetylene,[171] so that the absorption dynamics following interchain and intrachain excitation could be independently examined. Incident polarization parallel to the aligned chain yielded intrachain charged soliton pairs that recombined geminately in < 2 ps. Perpendicularly polarized light created, in addition to intrachain charged soliton pairs, interchain electron–hole pairs. The electron–hole pairs evidently led to the formation of polarons.

10.5. Conclusions

Ultrafast time-resolved IR spectrometers have enabled incisive investigations into the internal conversion of DABCO, the molecular geometry of carboxyhemes, the dynamics of ligand motion in photodissociated Hb–CO, the photochemistry of transition metal carbonyls, and the evolution of charge carriers in semiconductors. These studies, however, only begin to tap the potential of time-resolved IR spectroscopy.

The experimental approaches devised for carrying out time-resolved IR studies are quite varied, each having advantages and disadvantages. The 160-fs broadband IR probe generated by Sorokin et al. offers unsurpassed time resolution.[140] However, the spectral region is confined to 2.2.–2.7 μm, limiting its experimental utility. The IR probe pulse generated by Jedju and Rothberg[142,143] is broadly tunable, but the sensitivity of their apparatus was limited due to low repetition-rate operation (10 Hz). The CW diode laser upconversion approach pioneered by Hochstrasser et al. offers superior sensitivity due to high-repetition-rate data accumulation (~1 kHz).[149] Moreover, the time resolution is excellent, being limited only by the ultrashort pulse duration. High sensitivity operation could, however, only be attained with unusually high-power diode lasers and was usually achieved at the expense of spectral resolution. The downconversion—upconversion approach developed by Rothberg et al. and by Iannone et al. remedies many of the problems associated with diode and CO laser probes and appears quite promising.[153,147] The apparatus developed by Spears et al. provides broad tunability but somewhat limited sensitivity due to low repetition rate data collection (10 Hz).[146]

Further improvements in experimental methodology are needed to realize the full potential of femtosecond IR spectroscopy. Broader spectral coverage, improved spectral resolution, and greater sensitivity are required. The upconversion method could benefit greatly from a broadly tunable, stable, suitably intense, transform-limited IR source such as the FEL. On the other hand, the primary handicap for pulsed IR probe methods is the low rate of data accumulation. The recent advent of high-repetition-rate regenerative amplifiers should resolve this issue.[147,172] Such improvements extend the capabilities of time-resolved IR spectroscopy to include IR transitions that are much weaker than CO. This breakthrough will open the door to wide-ranging ultrafast investigations in chemistry, physics, and biology.

10.6. Some Nonlinear Spectroscopic Methods

The types of spectroscopy discussed so far have involved mainly direct measurements of linear absorption, gain, or spontaneous decay. Nonlinear methods are also useful for determining time-dependent population changes and anisotropies. In fact, when ultrashort pulses are used in pump–probe experiments, many of the techniques developed for nonlinear spectroscopy can be extremely helpful.

10.6.1. Principles of Nonlinear Spectroscopy

The nonlinear interactions between various electromagnetic fields and a molecular system give rise to new fields that radiate with predictable directions, frequencies, and phases. Such radiation is termed *coherently generated light,* and its properties can be predicted from considering the interaction of classical electromagnetic fields with quantum levels of the medium. The coherent, classical-like radiation originates in the macroscopic polarization $P(t)$ in the presence of which Maxwell's equations demand that there exist an electromagnetic field called the generated field. The part of $P(t)$ that is linear in the applied field gives rise to ordinary absorption. The first generally occurring nonlinear polarization appears in the third-order of the expansion of $P(t)$ in powers of the incident fields. The sequence of events in such nonlinear spectroscopy involves first interacting a quantized level system with three fields in some particular time order to generate a macroscopic dipole $P(t)$. This polarization is generally damped, so that it diminishes in time in a manner determined by the system response. Response functions for a wide variety of level systems are routinely available from a variety of sources, and rapid diagrammatic methods have been developed to recover the formulas needed for most physical problems that have so far been investigated.[80,173] It should be realized that the third-order polarization is in fact the basis of not only coherent light generation processes but also of all pump–probe-type gain and loss experiments. This is readily supported by the following argument.[173] The gain or loss in a medium is proportional to $<\dot{P}\cdot E>$, which for a third-order polarization is proportional to E^4 or I^2, which coincides with attenuation due to light scattering or fluorescence. The signal current (intensity) of the coherently generated light is proportional to P^2 or I^3, which corresponds to CARS, the Kerr effect or other coherent four-wave process.

In Section 10.4.3, coherent anti-Stokes Raman scattering was discussed in relation to vibrational Raman spectra. In that case the three fields interacting generally correspond to the ordering ω_1, ω_2, ω_1, and the generated wave at $\omega_3 = 2\omega_1 - \omega_2$ is directed along $2\mathbf{k}_1 - \mathbf{k}_2$, where $\mathbf{k}_1$ and $\mathbf{k}_2$ are the wave vectors of the ω_1 and ω_2 fields. When ultrashort light pulses are used, these frequencies may all fall within the spectrum of one pulse, and the effect or irradiation is then to initiate coherent oscillations in the material system at the vibrational frequency $\omega_v = \omega_1 - \omega_2$. These oscillations have been observed by many workers in a variety of different experimental configurations.[174-177] The same physical phenomena as observed through such experiments in transparent media can be deduced from Raman and Rayleigh lineshape analysis. When the pumping field is on resonance with an electronic transition, there are new phenomena to be observed, since the electronic spectral widths of conventional spectra (due to electronic dephasing and inhomogeneous broadening) obscure the vibrational spectral and dynamical information. This is where higher-order responses become critically valuable. One widely applicable experimental approach involves the optical Kerr effect.

10.6.2. Transient Measurements Using the Optical Kerr Effect

In the presence of an electric field an isotopic medium can become anisotropic. This means that optical constants of the material such as the index of refraction or the absorption become different for different choices of light direction or polarization. This so-called Kerr effect occurs with static or alternating fields, and the name *optical Kerr effect* implies the effects on the sample of an electrical field oscillating at optical frequencies. In molecular samples the measurement of the anisotropy induced by such an alternating field leads to novel spectroscopic information, and when laser pulses are used in such an experiment, information regarding molecular dynamics is obtained on the time scale of the pulses.

A useful method to study optical Kerr effects is by means of a polarization configuration in which the sample is placed between crossed polarizers, as in Figure 10.18. The spectroscopic probe source cannot pass through the crossed polarizers unless the sample is caused to be anisotropic. This can be accomplished by irradiation of the sample with a pump field that is usually required to be quite intense. The pump field, polarized in the direction $\hat{x} + \hat{y}$, introduces anisotropy in the absorption coefficient $\Delta\alpha = (\alpha_\parallel - \alpha_\perp)$, and a birefringence $\Delta k = (k_\parallel - k_\perp)$, where $k_i = \omega n_i/c_0$. The outgoing polarizer can be very slightly misalgned by θ toward the positive $\hat{x}$ direction, so that to order first in small quantities, the outgoing field of frequency ω incident on the detector is

$$2E_s = \varepsilon_p(t - \tau)e^{-i\omega t}(i\Delta kl + \Delta\alpha l/2 + 2\theta) \qquad (10.15)$$

where $\varepsilon_p(t - \tau)$ is the slowly varying envelope of the probe field centered at $t = \tau$. When short pulses are used to pump and then probe the sample, both Δk and $\Delta\alpha$ are dependent on the delay time. The signal current depends on $|E_s|^2$ and therefore has a part linear in $\Delta\alpha$ (i.e., $\theta l\Delta\alpha$) called the heterodyne signal. The usual homodyne part ($\Delta K^2 + \Delta\alpha^2$) can be arranged to be smaller by judicious choice of θ.

The quantities $\Delta\alpha$ and Δk depend on the pump intensity $I = Jc\varepsilon_0|E|^2$. Thus, the field E_s depends on the product of the three fields ε_p, ε, and ε^*,

Figure 10.18 Schematic diagram of an optical Kerr effect experiment.

Optical Kerr Experiment

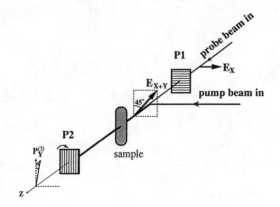

indicating that the signal is calculable from the third-order nonlinear polarization. The generated field takes the form

$$E_s = a[i\chi' - \chi'']\varepsilon_p \qquad (10.16)$$

where χ' and χ'' are the real and imaginary parts of the third-order nonlinear susceptibility $\chi^{(3)}$. The factors Δk and $\Delta \alpha$ can therefore be associated
with χ' and χ'', respectively. The appropriate forms of χ' and χ'' can be
obtained from an expansion of the density matrix in the usual manner for
the particular dynamical situation prevailing.[176,177]

An alternative way to carry out this type of experiment if measurements
of Δk are required is to exactly cross the polarizers ($\theta = 0°$) and insert a
wave plate after the sample that introduces some additional birefringence
λ: Now, in the absence of dichroism, the generated field is proportional to
$[i\Delta kL + i\lambda L]$ and a heterodyne signal proportional to λ and Δk is obtained
when the field interacts with the detector photocathode. A $\lambda/4$ plate inter-
changes the role of birefringence and dichroism at the detector.

Birefringence arises whenever the index of refraction is changed by the
pump or when the response is entirely nonresonant. The $\Delta \alpha$ terms arise
when there is a resonance of some type, such as a Raman resonance in the
stimulated Raman effect, or a linear absorption.

10.6.3. Applications of the Kerr Effect Configuration

The technique was widely used in picosecond experi-
ments and more recently with ultrafast light pulses to study Rayleigh and
Raman responses[178–180] and vibrational coherences in resonant excitation

Figure 10.19 Optical Kerr response of bifluorene excited at 310 nm and probed at
620 nm.

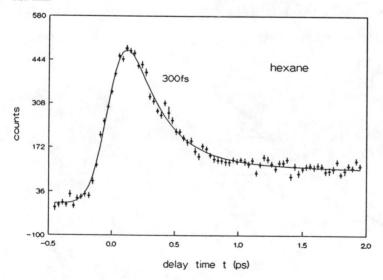

experiments.[181] One example in which $\Delta\alpha(t)$ is measured is shown in Figure 10.19.[182] These data correspond to the same energy transfer experiment as described using fluorescence detection in Section 10.3.4. In this case the sample is pumped with a pulse at 305 nm and probed at 610 nm via excited-state absorption. The time resolution is ~70 fs. The rapid decay of $\Delta\alpha(t)$ to a plateau is quite evident in these spectra which have significantly better S/N than the corresponding fluorescence anisotropy data on the sub-picosecond time scale. (See Figure 10.13.) The nonnegligible solvent response is subtracted in these experiments by plotting the difference between heterodyned signals at θ and $-\theta$. For an induced birefringence in the solvent of Δk_s the signals at $\pm\theta$ are proportional to $|i\Delta k_s + \Delta\alpha \pm \theta|^2$, so that their difference is $2\theta\Delta\alpha(t)$. The stilbene result in Figure 10.8 was also recorded in this way using the induced dichroism.[182]

Acknowledgment
This research was supported by NSF and NIH.

REFERENCES

1. G. R. Fleming, *Chemical Applications of Ultrafast Spectroscopy* (Oxford University Press, Oxford, 1986).

2. W. Kaiser, ed. *Ultrashort Laser Pulses and Applications* (Springer, Berlin, 1986).

3. C. B. Harris, E. P. Ippen, G. A. Mourou, and A. H. Zewail, eds., *Ultrafast Phenomena*, Vol. 7 (Springer, Berlin, 1990).

4. R. L. Fork, C. H. Brito-Cruz, P. C. Becker, and C. V. Shank, *Opt. Lett.* **12,** 483 (1987).

5. H. L. Fragnito, J.-Y. Bigot, P. C. Becker, and C. V. Shank, *Chem. Phys. Lett.* **160,** 101 (1989).

6. R. A. Mathies, C. H. Brito-Cruz, W. T. Pollard, and C. V. Shank, *Science* **240,** 777 (1988).

7. R. L. Fork, B. I. Greene, and C. V. Shank, *Appl. Phys. Lett.* **38,** 671 (1981).

8. R. L. Fork, C. V. Shank, and R. T. Yen, *Appl. Phys. Lett.* **41,** 224 (1982).

9. M. A. Kahlow, W. Jarzeba, T. DuBruil, and P. F. Barbara, *Rev. Sci. Instrum.* **59,** 1098 (1988).

10. W. T. Lotshaw, D. McMorrow, T. Dickson, and G. A. Kenny-Wallace, *Opt. Lett.* **14,** 1195 (1989).

11. S. T. Repinec, R. J. Sension, A. Z. Szarka, and R. M. Hochstrasser, *J. Phys. Chem.*, in press.

12. R. J. Sension, A. Z. Szarka, G. R. Smith, and R. M. Hochstrasser, *Chem. Phys. Lett.*, in press.

13. R. J. Sension, S. T. Repinec, A. Z. Szarka, and R. M. Hochstrasser, manuscript in preparation.

14. R. J. Sension, C. M. Phillips, A. Z. Szarka, W. J. Romanow, A. R. McGhie, J. P. McCauley, A. B. Smith, and R. M. Hochstrasser, *J. Phys. Chem.* **95,** 6075 (1991).

15. C. Rolland and P. B. Corkum, *J. Opt. Soc. Am. B* **5,** 641 (1988).

16. S. L. Chin, C. Rolland, P. B. Corkum, and P. Kelly, *Phys. Rev. Lett.* **61,** 153 (1988).

17. A. Laubereau, in Ref. 2, p. 118.

18. N. P. Ernsting and M. Kaschke, *Rev. Sci. Instrum.* **62,** 600 (1991).

19. N. P. Ernsting and T. Arthen-Engeland, *J. Phys. Chem.* **95,** 5502 (1991).

20. C. Rolland and P. B. Corkum, *Opt. Commun.* **59,** 64 (1986).

21. W. B. Knox, M. C. Downer, R. L. Fork, and C. V. Shank, *Opt. Lett.* **9,** 552 (1984).

22. S. A. Abrash, S. T. Repinec, and R. M. Hochstrasser, *J. Chem. Phys.* **93,** 1041 (1990).

23. K. A. Muszkat and E. Fischer, *J. Chem. Soc.* (B), 662 (1967).

24. T. Wismonski-Knittel, G. Fischer, and E. Fischer, *J. Chem. Soc. Perkin* II, 1930 (1974).

25. D. C. Todd, J. M. Jean, S. J. Rosenthal, A. J. Ruggerio, D. Yang, and G. R. Fleming, *J. Chem. Phys.* **93,** 8658 (1990).

26. R. J. Sension, S. T. Repinec, and R. M. Hochstrasser, *J. Phys. Chem.* **95,** 2946 (1991).

27. J.-L. Martin, J. Breton, A. J. Hoff, A. Migus, and A. Antonetti, *Proc. Natl. Acad. Sci. USA* **83,** 957 (1986).

28. W. Zinth, J. Dobler, and W. Kaiser in *Ultrafast Phenomena,* Vol. V, G. R. Fleming and A. E. Siegman, eds. (Springer, Berlin, 1986), p. 379.

29. F. Zhu and R. M. Hochstrasser, unpublished results.

30. W. T. Pollard, C. H. Brito-Cruz, C. V. Shank, and R. A. Mathies, *J. Chem. Phys.* **90,** 199 (1989).

31. W. T. Pollard, H. L. Fragnito, J.-Y. Bigot, C. V. Shank, and R. A. Mathies, *Chem. Phys. Lett.* **168,** 239 (1990).

32. B. J. Berne and R. Pecora, *Dynamic Light Scattering* (Wiley, New York, 1976).

33. M. Pereira, P. E. Share, M. J. Sarisky, and R. M. Hochstrasser, *J. Chem. Phys.* **94,** 2513 (1991).

34. M. Maroncelli, J. MacInnes, and G. R. Fleming, *Science* **243,** 1674 (1989).

35. P. F. Barbara and W. Jarzeba, *Adv. Photochem.* **15,** 1 (1990).

36. W. Jarzeba, G. C. Walker, A. E. Johnson, and P. F. Barbara, *J. Phys. Chem.* **92,** 7039 (1988).

37. M. Mahr and M. D. Hirsch, *Opt. Commun.* **13,** 96 (1975).

38. L. A. Halliday and M. R. Topp, *Chem. Phys. Lett.* **46,** 8 (1977).

39. D. Edelstein, E. S. Wachman, L. K. Cheng, W. R. Bosenberg, and C. L. Tang, *Appl. Phys. Lett.* **52,** 2211 (1989).

40. F. A. Novak and R. M. Hochstrasser, *Chem. Phys. Lett.* **53,** 3 (1978).

41. R. M. Hochstrasser, F. A. Novak, and C. A. Nyi, *Israel J. Chem.* **16,** 250 (1977).

42. M. Pereira, Ph.D. dissertation, University of Pennsylvania, (1990).

43. M. D. Dawson, T. F. Borges, D. W. Garvey, and A. Smirl, *Opt. Lett.* **11,** 721 (1986).

44. A. B. Myers, M. A. Pereira, P. L. Holt, and R. M. Hochstrasser, *J. Chem. Phys.* **86,** 5146 (1987).

45. J. E. Hansen, S. J. Rosenthal, and G. R. Fleming, *J. Phys. Chem.*, in press.

46. Y. R. Kim, P. Share, M. Pereira, M. Sarisky, and R. M. Hochstrasser, *J. Chem. Phys.* **91**, 7557 (1989).

47. M. D. Levenson, *Introduction to Nonlinear Laser Spectroscopy* (Academic, New York, 1982), p. 130.

48. R. Kubo, *Pure Appl. Chem.* **57**, 201 (1985), and references therein.

49. F. A. Novak, J. M. Friedman, and R. M. Hochstrasser, in *Laser and Coherence Spectroscopy*, J. I. Steinfeld, ed. (Plenum, New York, 1978), p. 451.

50. M. Bridoux and M. Delhaye, in *Advances in Infrared and Raman Spectroscopy*, Vol. 2, R. J. H. Clark and R. E. Hester, ed. (Heyden, London, 1976), p. 140, and references therein.

51. W. H. Woodruff and S. Farquharson, *Anal. Chem.* **50**, 1389 (1978).

52. M. Bridoux, A. Deffontaine, and C. Reiss, *Compt. Rend. Acad. Sci.* **282**, 771 (1976).

53. R. F. Dallinger, W. H. Woodruff, and M. A. J. Rodgers, *Appl. Spectrosc.* **33**, 522 (1979).

54. P. Valat and H. Tourbez, *J. Raman Spectrosc.* **8**, 139 (1979).

55. T. H. Bushaw, F. E. Lytle, and R. S. Tobias, *Appl. Spectrosc.* **32**, 585 (1978).

56. G. H. Atkinson, ed., *Time-Resolved Vibrational Spectroscopy* (Academic, New York, 1983), and references therein.

57. J. Terner, T. G. Spiro, M. Nagumo, M. F. Nicol, and M. A. El-Sayed, *J. Am. Chem. Soc.* **102**, 3238 (1980).

58. M. Coppey, H. Tourbez, P. Valat, and B. Alpert, *Nature* **284**, 568 (1980).

59. J. Terner, J. D. Strong, T. G. Spiro, M. Nagumo, M. Nicol, and M. A. El-Sayed, *Proc. Natl. Acad. Sci. USA* **78**, 1313 (1981).

60. M. Nagumo, M. Nicol, and M. A. El-Sayed, *J. Phys. Chem.* **85**, 2435 (1981).

61. J. Terner, D. F. Coss, C. Paddock, R. B. Miles, and T. G. Spiro, *J. Phys. Chem.* **85**, 859 (1982).

62. C. L. Hsieh, M. Nagumo, and M. A. El-Sayed, *J. Phys. Chem.* **85**, 2714 (1981).

63. C. L. Hsieh, M. A. El-Sayed, M. Nicol M. Nagumo, and J. H. Lee, *Photochemistry* **38**, 831 (1983).

64. G. H. Hayward, W. Carlsen, A. Siegman, and L. Stryer, *Science* **211**, 942 (1981).

65. T. L. Gustafson, D. M. Roberts, and D. A. Chernoff, *J. Chem. Phys.* **79**, 1559 (1983).

66. C. K. Johnson, G. A. Dalickas, S. A. Payne, and R. M. Hochstrasser, *Pure Appl. Chem.* **57**, 195 (1985).

67. M. Stockburger, W. Klusmann, H. Gattermann, G. Massig, and R. Peters, *Biochemistry* **18**, 4886 (1979).

68. B. Dick and R. M. Hochstrasser, *J. Chem. Phys.* **81**, 2897 (1984).

69. T. M. Cotton, in *Spectroscopy of Surfaces*, R. J. H. Clark and R. E. Hester, eds. (Wiley, Chichester, 1988), p. 91, and references therein.

70. B. Chase, *Anal. Chem.* **59**, 881A (1987).

71. V. M. Hallmark, C. G. Zimba, J. D. Swalen, and J. F. Rabolt, *Spectroscopy* **2**, 40 (1987).

72. T. Hirschfeld and B. Chase, *Appl. Spectrosc.* **40,** 133 (1986).

73. P. R. Griffiths and J. A. de Haseth, *Fourier Transform Infrared Spectrometry* (Wiley, New York, 1986), p. 19.

74. F. Ho, W. S. Tsay, J. Trout, and R. M. Hochstrasser, *Chem. Phys. Lett.* **83,** 5 (1981).

75. D. D. Dlott, *Ann. Rev. Phys. Chem.* **37,** 157 (1986).

76. T. C. Chang and D. D. Dlott, *J. Chem. Phys.* **90,** 3590 (1989).

77. L. S. Goldberg, in *Picosecond Phenomena*, Vol. 3, K. B. Eisenthal, R. M. Hochstrasser, W. Kaiser, and A. Laubereau, eds. (Springer, Berlin, 1982), p. 94.

78. S. A. Payne and R. M. Hochstrasser, *Opt. Lett.* **11,** 285 (1986).

79. S. A. Payne and R. M. Hochstrasser, *J. Phys. Chem.* **90,** 2068 (1986).

80. Y. R. Shen, *The Principles of Nonlinear Optics* (Wiley, New York, 1984).

81. V. F. Kamalov, N. I. Koroteev, and B. N. Toleutev, in *Time Resolved Spectroscopy*, R. J. H. Clark and R. E. Hester, eds. (Wiley, Chichester, 1989), p. 255.

82. T. L. Gustafson, I. Iwata, W. L. Weaver, L. A. Huston, and R. L. Benson, in *Laser Diagnostics of Biological Molecules: Linear and Nonlinear Methods*, S. A. Akhmanov, ed. *Proc. SPIE* **1403,** (1990).

83. T. L. Gustafson, D. M. Roberts, and D. A. Chernoff, *J. Chem. Phys.* **81,** 3438 (1984).

84. T. L. Gustafson, J. F. Palmer, and D. M. Roberts, *Chem. Phys. Lett.* **127,** 505 (1986).

85. H. Hashimoto and Y. Koyama, *Chem. Phys. Lett.* **154,** 321 (1989).

86. H. Hashimoto and Y. Koyama, *Chem. Phys. Lett.* **162,** 532 (1989).

87. H. Hashimoto and Y. Koyama, *Chem. Phys. Lett.* **163,** 251 (1989).

88. T. Noguchi, H. Hayashi, M. Tasumi, and G. H. Atkinson, *Chem. Phys. Lett.* **175,** 163 (1990).

89. T. Noguchi, S. Kolaczkowski, C. Arbour, S. Aramaki, G. H. Atkinson, H. Hayashi, and M. Tasumi, *Photochem. Photobiol.* **50,** 603 (1989).

90. H. Hayashi, S. V. Kolaczkowski, T. Nofuchi, D. Blanchard, and G. H. Atkinson, *J. Am. Chem. Soc.* **112,** 4664 (1990).

91. M. Kuki, H. Hashimoto, and Y. Koyama, *Chem. Phys. Lett.* **165,** 417 (1990).

92. H. Hashimoto and Y. Koyama, *Biochim. Biophys. Acta* **1017,** 181 (1990).

93. V. F. Kamalov, V. V. Kvach, N. I. Koroteev, B. N. Toleutaev, A. Y. Chikiskev, and A. P. Shkurinov, *Opt. Spectrosc. (USSR)* **64,** 460 (1988).

94. A. Y. Chikishev, V. F. Kamaloc, N. I. Koroteev, V. V. Kvach, A. P. Shkurinov, and B. N. Toleutaev, *Chem. Phys. Lett.* **144,** 90 (1988).

95. P. J. Reid, S. J. Doig, and R. A. Mathies, *Chem. Phys. Lett.* **156,** 163 (1989).

96. P. J. Reid, S. J. Doig, and R. A. Mathies, *J. Phys. Chem.* **94,** 8396 (1990).

97. X. Xu, R. Lingle, Jr., S. C. Yu, U. J. Chang, and J. B. Hopkins, *J. Chem. Phys.* **92,** 2106 (1990).

98. R. Lingle, Jr., X. Xu, S. C. Yu, Y. J. Chang, and J. B. Hopkins, *J. Chem. Phys.* **92,** 4628 (1990).

99. R. Lingle, Jr., X. Xu, S. C. Yu, H. Zhu, and J. B. Hopkins, *J. Chem. Phys.* **93,** 5667 (1990).

100. P. G. Bradley, N. Kress, B. A. Hornberger, R. F. Dallinger, and W. H. Woodruff, *J. Am. Chem. Soc.* **103**, 7441 (1981).

101. M. Forster and R. E. Hester, *Chem. Phys. Lett.* **81**, 42 (1981).

102. J. V. Caspar, T. D. Westmoreland, G. H. Allen, P. G. Bradley, T. J. Meyer, and W. H. Woodruff, *J. Am. Chem. Soc.* **106**, 3492 (1984).

103. L. K. Orman, Y. J. Chang, D. R. Anderson, T. Yabe, X. Xu, S. C. Yu, and J. B. Hopkins, *J. Chem. Phys.* **90**, 1469 (1989).

104. T. Yabe, D. R. Anderson, L. K. Orman, Y. J. Chang, and J. B. Hopkins, *J. Phys. Chem.* **93**, 2302 (1989).

105. Y. J. Chang, X. Xu, T. Yabe, S. C. Yu, D. R. Anderson, L. K. Orman, and J. B. Hopkins, *J. Phys. Chem.* **94**, 729 (1990).

106. T. Yabe, L. K. Orman, D. R. Anderson, and J. B. Hopkins, *J. Phys. Chem.* **94**, 7128 (1990).

107. T. G. Spiro, ed., *Biological Applications Raman Spectroscopy*, Vol. 3 (Wiley, New York, 1988), and references therein.

108. R. M. Hochstrasser and C. K. Johnson, in *Ultrashort Laser Pulses and Applications*, W. Kaiser, ed. (Springer, Berlin, 1988), p. 357.

109. S. Dasgupta and T. G. Spiro, *Biochemistry* **25**, 5941 (1986).

110. P. Stein, J. Terner, and T. G. Spiro, *J. Phys. Chem.* **86**, 168 (1982).

111. S. Dasgupta, T. G. Spiro, C. K. Johnson, G. A. Dalickas, and R. M. Hochstrasser, *Biochemistry* **24**, 5295 (1985).

112. C. K. Johnson, G. A. Dalickas, R. M. Hochstrasser, S. Dasgupta, and T. G. Spiro, unpublished results (1985).

113. C. K. Johnson and R. M. Hochstrasser, in *Advances in Laser Science*, Vol. 1, W. C. Stwalley and M. Lapp, eds. (American Institute of Physics, New York, 1986), p. 686.

114. S. Choi, T. G. Spiro, K. C. Langry, K. M. Smith, D. L. Budd, and G. N. LaMar, *J. Am. Chem. Soc.* **104**, 4345 (1982).

115. J. W. Petrich, J. C. Lambry, K. Kuczera, M. Karplus, C. Poyart, and J. L. Martin, *Biochemistry* **30**, 3975 (1991).

116. E. W. Findsen, T. W. Scott, M. R. Chance, and J. M. Friedman, *J. Am. Chem. Soc.* **107**, 3355 (1985).

117. E. W. Findsen, J. M. Friedman, M. R. Ondrias, and S. R. Simon, *Science* **229**, 661 (1985).

118. J. M. Friedman, *Science* **228**, 1273 (1985).

119. E. R. Henry, W. A. Eaton, and R. M. Hochstrasser, *Proc. Natl. Acad. Sci. USA* **83**, 8982 (1986).

120. J. W. Petrich, J. L. Martin, D. Houde, C. Poyart, and A. Orszag, *Biochemistry* **26**, 7914 (1987).

121. J. W. Petrich and J. L. Martin, *Chem. Phys.* **131**, 31 (1989).

122. R. Lingle, Jr., X. Xu, H. Zhu, S. C. Yu, and J. B. Hopkins, *J. Am. Chem. Soc.* **113**, 3992 (1991).

123. G. H. Atkinson, D. Blanchard, T. L. Brack, D. Blanchard, and G. Rumbles, *Chem. Phys.* **131**, 1 (1989).

124. T. L. Brack and G. H. Atkinson, *J. Mol. Struct.* **214**, 289 (1989).

125. R. van den Berg, D. J. Jang, H. C. Bitting, and M. A. El-Sayed, *Biophys. J.* **58,** 135 (1990).

126. T. L. Brack and G. H. Atkinson, *J. Phys. Chem.* **95,** 2351 (1991).

127. B. Robert, E. Nabedryk, and M. Lutz, in *Time Resolved Spectroscopy,* R. J. H. Clark and R. E. Hester, eds. (Wiley, Chichester, 1989), p. 301.

128. M. Lutz and B. Robert, in *Biological Applications of Raman Spectroscopy,* Vol. 3, T. G. Spiro, ed. (Wiley, New York, 1987).

129. J. Qian, C. Wan, and C. K. Johnson, unpublished results.

130. R. G. Gordon, *Adv. Magn. Reson.* **3,** 1 (1968).

131. P. A. Anfinrud, C. Han, T. Lian, and R. M. Hochstrasser, *J. Phys. Chem.* **94,** 1180 (1990).

132. J. N. Moore, P. A. Hansen, and R. M. Hochstrasser, *Proc. Natl. Acad. Sci. USA* **85,** 5062 (1988).

133. P. Ormos, D. Braunstein, H. Frauenfelder, M. K. Hong, S.-L. Lin, T. B. Sauke, and R. D. Young, *Proc. Natl. Acad. Sci. USA,* **85,** 8492 (1988).

134. W. A. Eaton and J. Hofrichter, *Meth. Enzymol.* **76,** 175 (1981).

135. W. A. Eaton, L. K. Hanson, P. J. Stephens, J. C. Sutherland, and J. B. R. Dunn, *J. Am. Chem. Soc.* **100,** 4991 (1978).

136. J. C. Maxwell and W. S. Caughey, *Meth. Enzymol.* **54,** 302 (1978).

137. J. O. Alben, D. Beece, S. F. Bowne, W. Doster, L. Eisenstein, H. Frauenfelder, D. Good, J. D. McDonald, M. C. Marden, P. P. Moh, L. Reinisch, A. H. Reynolds, E. Shyamsunder, and K. T. Yue, *Proc. Natl. Acad. Sci. USA* **79,** 3744 (1982).

138. G. J. Jiang, W. B. Person, and K. G. Brown, *J. Chem. Phys.* **62,** 1201 (1975).

139. J. Hebling and J. Kuhl, *Opt. Lett.* **14,** 278 (1989).

140. J. H. Glownia, J. Misewich, and P. P. Sorokin, *Opt. Lett.* **12,** 19 (1987).

141. D. S. Moore and S. C. Schmidt, *Opt. Lett.* **12,** 480 (1987).

142. T. M. Jedju and L. Rothberg, *Appl. Opt.* **27,** 615 (1988).

143. T. M. Jedju, L. Rothberg, and A. Labrie, *Opt. Lett.* **13,** 961 (1988).

144. H. Vanherzeele, *Appl. Opt.* **29,** 2246 (1990).

145. T. Elsaesser, H. Lobentanzer, and A. Seilmeier, *Opt. Commun.* **52,** 355 (1985).

146. K. G. Spears, X. Zhu, X. Yang, and L. Wang, *Opt. Commun.* **66,** 167 (1988).

147. M. Iannone, B. R. Cowen, R. Diller, S. Maiti, and R. M. Hochstrasser, *Appl. Opt.,* in press.

148. J. M. Wiesenfeld and E. P. Ippen, *Chem. Phys. Lett.* **67,** 213 (1979).

149. J. N. Moore, P. A. Hansen, and R. M. Hochstrasser, *Chem. Phys. Lett.* **138,** 110 (1987).

150. P. Anfinrud, P. A. Hansen, J. N. Moore, R. M. Hochstrasser, in *Ultrafast Phenomena,* Vol. 6, T. Yajima, K. Yoshihara, C. B. Harris, and S. Shionoya, eds. (Springer, Berlin, 1988), p. 442.

151. P. A. Anfinrud, C. Han, and R. M. Hochstrasser, *Proc. Natl. Acad. Sci. USA* **86,** 8387 (1989).

152. R. M. Hochstrasser, P. A. Anfinrud, R. Diller, C. Han, M. Iannone, T. Lian, and B. Locke, in *Ultrafast Phenomena,* Vol. 7, C. B. Harris, E. P. Ippen, G. A. Mourou, and A. H. Zewail, eds. (Springer, Berlin, 1990), p. 429.

153. T. M. Jedju, M. W. Roberson, and L. Rothberg, *Appl. Opt.*, submitted (1991).

154. R. S. Eng, J. F. Butler, and K. J. Linden, *Opt. Engl.* **19**, 945 (1980).

155. L. F. Mollenhauer and D. H. Olsen, *J. Appl. Phys.* **46**, 3109 (1975).

156. *Combustion Dynamics Facility: Scientific Program Summary*, PUB-5284, Sandia National Laboratories and Lawrence Berkeley Laboratory, June 1990)

157. J. H. Glownia, J. Misewich, and P. P. Sorokin, *Chem. Phys. Lett.* **139**, 491 (1987).

158. P. A. Hansen, J. N. Moore, and R. M. Hochstrasser, *Chem. Phys.* **131**, 49 (1989).

159. R. B. Locke, T. Lian, and R. M. Hochstrasser, *Chem. Phys.*, in press.

160. T. M. Jedju, L. Rothberg, and A. Labrie, *Opt. Lett.* **13**, 961 (1988).

161. L. Rothberg, T. M. Jedju, and R. H. Austin, *Biophys. J.* **57**, 369 (1990).

162. T. Lian, R. B. Locke, and R. M. Hochstrasser, in press.

163. F. G. Fiamingo and J. O. Alben, *Biochemistry* **24**, 7964 (1985).

164. J. N. Moore, P. A. Hansen, and R. M. Hochstrasser, *J. Am. Chem. Soc.* **111**, 4563 (1989).

165. P. A. Anfinrud, C. Han, T. Lian, and R. M. Hochstrasser, in *Ultrafast Phenomena*, Vol. 7, C. B. Harris, E. P. Ippen, G. A. Mourou, and A. H. Zewail, eds. (Springer, Berlin, 1990), p. 489.

166. P. A. Anfinrud, C. Han, T. Lian, and R. M. Hochstrasser, *J. Phys. Chem.*, **95**, 574 (1991).

167. B. D. Moore, M. B. Simpson, M. Poliakoff, and J. J. Turner, *J. Chem. Soc. Chem. Commun.* 972 (1984).

168. L. Wang, X. Zhu, and K. G. Spears, *J. Am. Chem. Soc.* **110**, 8695 (1988).

169. L. Wang, X. Zhu, and K. G. Spears, *J. Phys. Chem.* **93**, 2 (1989).

170. L. Rothberg, T. M. Jedju, S. Etemad, and G. L. Baker, *Phys. Rev. Lett.* **57**, 3229 (1986).

171. L. Rothberg, T. M. Jedju, P. D. Townsend, S. Etemad, and G. L. Baker, *Phys. Rev. Lett.* **65**, 100 (1990).

172. I. N. Duling III, T. Norris, T. Sizer II, P. Bado, and G. A. Mourou, *J. Opt. Soc. Am. B* **2**, 616 (1985).

173. B. Dick, R. M. Hochstrasser, and H. P. Trommsdorff, in *Nonlinear Optical Properties of Organic Molecules and Crystals*, Vol. 2, D. S. Chemla and J. Zyss, eds. (Academic, New York, 1986), p. 159.

174. A. Laubereau and W. Kaiser, *Rev. Mod. Phys.* **50**, 607 (1978).

175. T. J. Trout, S. Velsko, and R. M. Hochstrassser, *J. Chem. Phys.* **79**, 2114 (1983).

176. S. D. Silvestri, J. G. Fujimoto, E. P. Ippen, E. B. Gamble, L. R. Williams, and K. A. Nelson, *Chem. Phys. Lett.* **116**, 146 (1985).

177. S. Ruhman, A. G. Joly, and K. A. Nelson, *IEEE J. Quantum Electron.* QE**24**, 443 (1988).

178. B. I. Greene and R. C. Farrow, *J. Chem. Phys.* **77**, 4779 (1982).

179. J. M. Halbout and C. K. Tang, *Appl. Phys. Lett.* **40**, 765 (1982).

180. D. McMorrow, W. T. Lotshow, and G. A. Kenney-Wallace, *IEEE J. Quantum Electron* QE**24**, 443 (1988).

181. J. Chesnoy and A. Mokhami, *Phys. Rev. A.* **38**, 3566 (1988).

182. F. Zhu and R. M. Hochstrasser, unpublished results.

APPENDIX
Acronyms in Laser Spectroscopy

AAS	Atomic absorption spectroscopy
AFS	Atomic fluorescence spectroscopy
AO	Acousto-optic
ASE	Amplified spontaneous emission
BBO	Beta barium borate
BOXCARS	Box (geometry) coherent anti-Stokes Raman scattering
BW	Bandwidth
CARS	Coherent anti-Stokes Raman scattering
CCD	Charge-coupled device
CD	Circular dichroism
CDR	Circular differential Raman (spectroscopy)
CPM	Colliding-pulse mode-locked (laser)
CSARS	Coherent subharmonic anti-Stokes Raman scattering
CSRS	Coherent Stokes Raman scattering
CVL	Copper vapor laser
CW	Continuous-wave
DCS	Data collection system
DE	Display electronics
DFDL	Distributed-feedback dye laser
DFT	Discrete Fourier transform
DIAL	Differential absorption lidar
DL	Diffraction limited
DLS	Dynamic light scattering
DOAS	Differential optical absorption spectroscopy
DRIMS	Double-resonance ionization mass spectrometry
EL	Electroluminescent
EO	Electro-optic
FDS	Fluorescence dip spectrometry
FEL	Free-electron laser

461

FFT	Fast Fourier transform
FIR	Far infrared
FPI	Fabry–Perot interferometer
FSR	Free spectral range
FTIR	Fourier-transform infrared
FWHM	Full width at half-maximum
GVD	Group velocity dispersion
HFS	Hyperfine structure
HORSES	High-order Raman spectral excitation studies
IC	Internal conversion
ICP	Inductively coupled plasma
IR	Infrared
IRED	Infrared emitting diode
IRIS	Infrared interferometric spectrometer
IRMPD	Infrared multiphoton dissociation
ISC	Intersystem crossing
ISIT	Intensified silicon intensifier target
IVR	Intramolecular vibrational (energy) randomisation
KDP	Potassium dihydrogen phosphate
KTP	Potassium titanyl phosphate
LADAR	Laser detection and ranging
LAMMA	Laser microprobe mass analysis
LAMMS	Laser microprobe mass spectrometry
LAMS	Laser mass spectrometer
LAS	Laser absorption spectrometer
LEAFS	Laser-excited atomic fluorescence spectroscopy
LEF	Laser-excited fluorescence
LIA	Lock-in amplifier
LIBS	Laser-induced breakdown spectroscopy
LIDAR	Light detection and ranging
LIF	Laser-induced fluorescence
LIMA	Laser-induced mass analysis
LIMS	Laser ionization mass spectrometry
LIPF	Laser-induced predissociation fluorescence
LITD	Laser-induced thermal desorption
LMR	Laser magnetic resonance
LOG	Laser optogalvanic (spectroscopy)
LPS	Laser photoacoustic (or photoionisation) spectroscopy
MEGL	Modified exponential gap law
MIM	Metal–insulator–metal (diode)
MPI	Multiphoton ionisation
MPRI	Multiphoton resonance ionization
NDT	Nondestructive testing
NEP	Noise equivalent power
NIR	Near infrared

NLO	Nonlinear optics
OA	Optical activity
OGS	Optogalvanic spectroscopy
OMA	Optical multichannel analyser
OODR	Optical–optical double resonance
OPD	Optical path difference
OPO	Optical parametric oscillator
ORD	Optical rotatory dispersion
PARS	Photoacoustic Raman spectroscopy
PAS	Photoacoustic spectroscopy
PC	Photocathode
PCS	Photon correlation spectroscopy
PDS	Photodischarge spectroscopy
PHOPHEX	Photofragment excitation
PM	Polarization (or phase) modulation
PMT	Photomultiplier tube
PRF	Pulse repetition frequency
PSD	Phase-sensitive detector
QLS	Quasi-elastic light scattering
RAS	Remote active spectrometer
REMPI	Resonance-enhanced multiphoton ionisation
RIKES	Raman-induced Kerr effect spectroscopy
RIMS	Resonance ionization mass spectrometry
RIS	Resonance ionization spectroscopy
RLA	Resonant laser ablation
ROA	Raman optical activity
RRE	Resonance Raman effect
RRS	Resonance Raman scattering
SALI	Surface analysis by laser ionization
SEP	Stimulated emission pumping
SERS	Surface-enhanced Raman scattering
SFG	Sum-frequency generation
SHG	Second harmonic generation
SIRIS	Sputter-initiated resonant ionization spectroscopy
SIRS	Spectroscopy by inverse Raman scattering
SIT	Silicon intensifier target
SLAM	Scanning laser acoustical microscope
SNR	Signal-to-noise ratio
SPM	Self-phase modulation
SRL	Stimulated Raman loss
SRS	Stimulated Raman scattering
SVL	Single vibronic level (fluorescence)
TDL	Tunable diode laser
TDS	Time domain spectroscopy
TEA	Transversely excited atmospheric (pressure)

TEM	Transverse electromagnetic mode
THG	Third harmonic generation
TIR	Total internal reflection
TOF	Time of flight
TPA	Two-photon absorption
TPD	Two-photon dissociation; temperature-programmed desorption
TPF	Two-photon fluorescence
TRRS	Time-resolved Raman spectroscopy
UHV	Ultrahigh vacuum
USLS	Ultrafast supercontinuum laser source
UV	Ultraviolet
VUV	Vacuum ultraviolet
XUV	Extended ultraviolet
YAG	Yttrium aluminum garnet
YLF	Yttrium lanthanum fluoride

Index